Pitman Research Notes in Mathematics Series

3 0116 00480 9842

Submission of proposals for consideration

Suggestions for publication, in the form of outlines and representative samples, are invited by the Editorial Board for assessment. Intending authors should approach one of the main editors or another member of the Editorial Board, citing the relevant AMS subject classifications. Alternatively, outlines may be sent directly to the publisher's offices. Refereeing is by members of the board and other mathematical authorities in the topic concerned, throughout the world.

Preparation of accepted manuscripts

On acceptance of a proposal, the publisher will supply full instructions for the preparation of manuscripts in a form suitable for direct photo-lithographic reproduction. Specially printed grid sheets can be provided and a contribution is offered by the publisher towards the cost of typing. Word processor output, subject to the publisher's approval, is also acceptable.

Illustrations should be prepared by the authors, ready for direct reproduction without further improvement. The use of hand-drawn symbols should be avoided wherever possible, in order to maintain maximum clarity of the text.

The publisher will be pleased to give any guidance necessary during the preparation of a typescript, and will be happy to answer any queries.

Important note

In order to avoid later retyping, intending authors are strongly urged not to begin final preparation of a typescript before receiving the publisher's guidelines. In this way it is hoped to preserve the uniform appearance of the series.

Addison Wesley Longman Ltd
Edinburgh Gate
Harlow, Essex, CM20 2JE
UK
(Telephone (0) 1279 623623)

Titles in this series. A full list is available from the publisher on request.

151 A stochastic maximum principle for optimal control of diffusions
U G Haussmann

152 Semigroups, theory and applications. Volume II
H Brezis, M G Crandall and F Kappel

153 A general theory of integration in function spaces
P Muldowney

154 Oakland Conference on partial differential equations and applied mathematics
L R Bragg and J W Dettman

155 Contributions to nonlinear partial differential equations. Volume II
J I Díaz and P L Lions

156 Semigroups of linear operators: an introduction
A C McBride

157 Ordinary and partial differential equations
B D Sleeman and R J Jarvis

158 Hyperbolic equations
F Colombini and M K V Murthy

159 Linear topologies on a ring: an overview
J S Golan

160 Dynamical systems and bifurcation theory
M I Camacho, M J Pacifico and F Takens

161 Branched coverings and algebraic functions
M Namba

162 Perturbation bounds for matrix eigenvalues
R Bhatia

163 Defect minimization in operator equations: theory and applications
R Reemtsen

164 Multidimensional Brownian excursions and potential theory
K Burdzy

165 Viscosity solutions and optimal control
R J Elliott

166 Nonlinear partial differential equations and their applications: Collège de France Seminar. Volume VIII
H Brezis and J L Lions

167 Theory and applications of inverse problems
H Haario

168 Energy stability and convection
G P Galdi and B Straughan

169 Additive groups of rings. Volume II
S Feigelstock

170 Numerical analysis 1987
D F Griffiths and G A Watson

171 Surveys of some recent results in operator theory. Volume I
J B Conway and B B Morrel

172 Amenable Banach algebras
J-P Pier

173 Pseudo-orbits of contact forms
A Bahri

174 Poisson algebras and Poisson manifolds
K H Bhaskara and K Viswanath

175 Maximum principles and eigenvalue problems in partial differential equations
P W Schaefer

176 Mathematical analysis of nonlinear, dynamic processes
K U Grusa

177 Cordes' two-parameter spectral representation theory
D F McGhee and R H Picard

178 Equivariant K-theory for proper actions
N C Phillips

179 Elliptic operators, topology and asymptotic methods
J Roe

180 Nonlinear evolution equations
J K Engelbrecht, V E Fridman and E N Pelinovski

181 Nonlinear partial differential equations and their applications: Collège de France Seminar. Volume IX
H Brezis and J L Lions

182 Critical points at infinity in some variational problems
A Bahri

183 Recent developments in hyperbolic equations
L Cattabriga, F Colombini, M K V Murthy and S Spagnolo

184 Optimization and identification of systems governed by evolution equations on Banach space
N U Ahmed

185 Free boundary problems: theory and applications. Volume I
K H Hoffmann and J Sprekels

186 Free boundary problems: theory and applications. Volume II
K H Hoffmann and J Sprekels

187 An introduction to intersection homology theory
F Kirwan

188 Derivatives, nuclei and dimensions on the frame of torsion theories
J S Golan and H Simmons

189 Theory of reproducing kernels and its applications
S Saitoh

190 Volterra integrodifferential equations in Banach spaces and applications
G Da Prato and M Iannelli

191 Nest algebras
K R Davidson

192 Surveys of some recent results in operator theory. Volume II
J B Conway and B B Morrel

193 Nonlinear variational problems. Volume II
A Marino and M K V Murthy

194 Stochastic processes with multidimensional parameter
M E Dozzi

195 Prestressed bodies
D Iesan

196 Hilbert space approach to some classical transforms
R H Picard

197 Stochastic calculus in application
J R Norris

198 Radical theory
B J Gardner

199 The C^*-algebras of a class of solvable Lie groups
X Wang

200 Stochastic analysis, path integration and dynamics
K D Elworthy and J C Zambrini

201 Riemannian geometry and holonomy groups
   S Salamon
202 Strong asymptotics for extremal errors and
   polynomials associated with Erdös type weights
   D S Lubinsky
203 Optimal control of diffusion processes
   V S Borkar
204 Rings, modules and radicals
   B J Gardner
205 Two-parameter eigenvalue problems in ordinary
   differential equations
   M Faierman
206 Distributions and analytic functions
   R D Carmichael and D Mitrovic
207 Semicontinuity, relaxation and integral
   representation in the calculus of variations
   G Buttazzo
208 Recent advances in nonlinear elliptic and
   parabolic problems
   P Bénilan, M Chipot, L Evans and M Pierre
209 Model completions, ring representations and the
   topology of the Pierce sheaf
   A Carson
210 Retarded dynamical systems
   G Stepan
211 Function spaces, differential operators and
   nonlinear analysis
   L Paivarinta
212 Analytic function theory of one complex variable
   C C Yang, Y Komatu and K Niino
213 Elements of stability of visco-elastic fluids
   J Dunwoody
214 Jordan decomposition of generalized vector
   measures
   K D Schmidt
215 A mathematical analysis of bending of plates with
   transverse shear deformation
   C Constanda
216 Ordinary and partial differential equations.
   Volume II
   B D Sleeman and R J Jarvis
217 Hilbert modules over function algebras
   R G Douglas and V I Paulsen
218 Graph colourings
   R Wilson and R Nelson
219 Hardy-type inequalities
   A Kufner and B Opic
220 Nonlinear partial differential equations and their
   applications: Collège de France Seminar.
   Volume X
   H Brezis and J L Lions
221 Workshop on dynamical systems
   E Shiels and Z Coelho
222 Geometry and analysis in nonlinear dynamics
   H W Broer and F Takens
223 Fluid dynamical aspects of combustion theory
   M Onofri and A Tesei
224 Approximation of Hilbert space operators.
   Volume I. 2nd edition
   D Herrero
225 Operator theory: proceedings of the 1988
   GPOTS–Wabash conference
   J B Conway and B B Morrel

226 Local cohomology and localization
   **J L Bueso Montero, B Torrecillas Jover and
   A Verschoren**
227 Nonlinear waves and dissipative effects
   D Fusco and A Jeffrey
228 Numerical analysis 1989
   D F Griffiths and G A Watson
229 Recent developments in structured continua.
   Volume II
   D De Kee and P Kaloni
230 Boolean methods in interpolation and
   approximation
   F J Delvos and W Schempp
231 Further advances in twistor theory. Volume I
   L J Mason and L P Hughston
232 Further advances in twistor theory. Volume II
   L J Mason, L P Hughston and P Z Kobak
233 Geometry in the neighborhood of invariant
   manifolds of maps and flows and linearization
   U Kirchgraber and K Palmer
234 Quantales and their applications
   K I Rosenthal
235 Integral equations and inverse problems
   V Petkov and R Lazarov
236 Pseudo-differential operators
   S R Simanca
237 A functional analytic approach to statistical
   experiments
   I M Bomze
238 Quantum mechanics, algebras and distributions
   D Dubin and M Hennings
239 Hamilton flows and evolution semigroups
   J Gzyl
240 Topics in controlled Markov chains
   V S Borkar
241 Invariant manifold theory for hydrodynamic
   transition
   S Sritharan
242 Lectures on the spectrum of $L^2(\Gamma \backslash G)$
   F L Williams
243 Progress in variational methods in Hamiltonian
   systems and elliptic equations
   M Girardi, M Matzeu and F Pacella
244 Optimization and nonlinear analysis
   A Ioffe, M Marcus and S Reich
245 Inverse problems and imaging
   G F Roach
246 Semigroup theory with applications to systems
   and control
   N U Ahmed
247 Periodic-parabolic boundary value problems and
   positivity
   P Hess
248 Distributions and pseudo-differential operators
   S Zaidman
249 Progress in partial differential equations: the Metz
   surveys
   M Chipot and J Saint Jean Paulin
250 Differential equations and control theory
   V Barbu

251 Stability of stochastic differential equations with respect to semimartingales
X Mao
252 Fixed point theory and applications
J Baillon and M Théra
253 Nonlinear hyperbolic equations and field theory
M K V Murthy and S Spagnolo
254 Ordinary and partial differential equations. Volume III
B D Sleeman and R J Jarvis
255 Harmonic maps into homogeneous spaces
M Black
256 Boundary value and initial value problems in complex analysis: studies in complex analysis and its applications to PDEs 1
R Kühnau and W Tutschke
257 Geometric function theory and applications of complex analysis in mechanics: studies in complex analysis and its applications to PDEs 2
R Kühnau and W Tutschke
258 The development of statistics: recent contributions from China
X R Chen, K T Fang and C C Yang
259 Multiplication of distributions and applications to partial differential equations
M Oberguggenberger
260 Numerical analysis 1991
D F Griffiths and G A Watson
261 Schur's algorithm and several applications
M Bakonyi and T Constantinescu
262 Partial differential equations with complex analysis
H Begehr and A Jeffrey
263 Partial differential equations with real analysis
H Begehr and A Jeffrey
264 Solvability and bifurcations of nonlinear equations
P Drábek
265 Orientational averaging in mechanics of solids
A Lagzdins, V Tamuzs, G Teters and A Kregers
266 Progress in partial differential equations: elliptic and parabolic problems
C Bandle, J Bemelmans, M Chipot, M Grüter and J Saint Jean Paulin
267 Progress in partial differential equations: calculus of variations, applications
C Bandle, J Bemelmans, M Chipot, M Grüter and J Saint Jean Paulin
268 Stochastic partial differential equations and applications
G Da Prato and L Tubaro
269 Partial differential equations and related subjects
M Miranda
270 Operator algebras and topology
W B Arveson, A S Mishchenko, M Putinar, M A Rieffel and S Stratila
271 Operator algebras and operator theory
W B Arveson, A S Mishchenko, M Putinar, M A Rieffel and S Stratila
272 Ordinary and delay differential equations
J Wiener and J K Hale
273 Partial differential equations
J Wiener and J K Hale
274 Mathematical topics in fluid mechanics
J F Rodrigues and A Sequeira

275 Green functions for second order parabolic integro-differential problems
M G Garroni and J F Menaldi
276 Riemann waves and their applications
M W Kalinowski
277 Banach C(K)-modules and operators preserving disjointness
Y A Abramovich, E L Arenson and A K Kitover
278 Limit algebras: an introduction to subalgebras of C*-algebras
S C Power
279 Abstract evolution equations, periodic problems and applications
D Daners and P Koch Medina
280 Emerging applications in free boundary problems
J Chadam and H Rasmussen
281 Free boundary problems involving solids
J Chadam and H Rasmussen
282 Free boundary problems in fluid flow with applications
J Chadam and H Rasmussen
283 Asymptotic problems in probability theory: stochastic models and diffusions on fractals
K D Elworthy and N Ikeda
284 Asymptotic problems in probability theory: Wiener functionals and asymptotics
K D Elworthy and N Ikeda
285 Dynamical systems
R Bamon, R Labarca, J Lewowicz and J Palis
286 Models of hysteresis
A Visintin
287 Moments in probability and approximation theory
G A Anastassiou
288 Mathematical aspects of penetrative convection
B Straughan
289 Ordinary and partial differential equations. Volume IV
B D Sleeman and R J Jarvis
290 K-theory for real C*-algebras
H Schröder
291 Recent developments in theoretical fluid mechanics
G P Galdi and J Necas
292 Propagation of a curved shock and nonlinear ray theory
P Prasad
293 Non-classical elastic solids
M Ciarletta and D Ieşan
294 Multigrid methods
J Bramble
295 Entropy and partial differential equations
W A Day
296 Progress in partial differential equations: the Metz surveys 2
M Chipot
297 Nonstandard methods in the calculus of variations
C Tuckey
298 Barrelledness, Baire-like- and (LF)-spaces
M Kunzinger
299 Nonlinear partial differential equations and their applications. Collège de France Seminar. Volume XI
H Brezis and J L Lions

300 Introduction to operator theory
T Yoshino

301 Generalized fractional calculus and applications
V Kiryakova

302 Nonlinear partial differential equations and their applications. Collège de France Seminar Volume XII
H Brezis and J L Lions

303 Numerical analysis 1993
D F Griffiths and G A Watson

304 Topics in abstract differential equations
S Zaidman

305 Complex analysis and its applications
C C Yang, G C Wen, K Y Li and Y M Chiang

306 Computational methods for fluid-structure interaction
J M Crolet and R Ohayon

307 Random geometrically graph directed self-similar multifractals
L Olsen

308 Progress in theoretical and computational fluid mechanics
G P Galdi, J Málek and J Necas

309 Variational methods in Lorentzian geometry
A Masiello

310 Stochastic analysis on infinite dimensional spaces
H Kunita and H-H Kuo

311 Representations of Lie groups and quantum groups
V Baldoni and M Picardello

312 Common zeros of polynomials in several variables and higher dimensional quadrature
Y Xu

313 Extending modules
N V Dung, D van Huynh, P F Smith and R Wisbauer

314 Progress in partial differential equations: the Metz surveys 3
M Chipot, J Saint Jean Paulin and I Shafrir

315 Refined large deviation limit theorems
V Vinogradov

316 Topological vector spaces, algebras and related areas
A Lau and I Tweddle

317 Integral methods in science and engineering
C Constanda

318 A method for computing unsteady flows in porous media
R Raghavan and E Ozkan

319 Asymptotic theories for plates and shells
R P Gilbert and K Hackl

320 Nonlinear variational problems and partial differential equations
A Marino and M K V Murthy

321 Topics in abstract differential equations II
S Zaidman

322 Diffraction by wedges
B Budaev

323 Free boundary problems: theory and applications
J I Diaz, M A Herrero, A Liñan and J L Vazquez

324 Recent developments in evolution equations
A C McBride and G F Roach

325 Elliptic and parabolic problems: Pont-à-Mousson 1994
C Bandle, J Bemelmans, M Chipot, J Saint Jean Paulin and I Shafrir

326 Calculus of variations, applications and computations: Pont-à-Mousson 1994
C Bandle, J Bemelmans, M Chipot, J Saint Jean Paulin and I Shafrir

327 Conjugate gradient type methods for ill-posed problems
M Hanke

328 A survey of preconditioned iterative methods
A M Bruaset

329 A generalized Taylor's formula for functions of several variables and certain of its applications
J-A Riestra

330 Semigroups of operators and spectral theory
S Kantorovitz

331 Boundary-field equation methods for a class of nonlinear problems
G N Gatica and G C Hsiao

332 Metrizable barrelled spaces
J C Ferrando, M López Pellicer and L M Sánchez Ruiz

333 Real and complex singularities
W L Marar

334 Hyperbolic sets, shadowing and persistence for noninvertible mappings in Banach spaces
B Lani-Wayda

335 Nonlinear dynamics and pattern formation in the natural environment
A Doelman and A van Harten

336 Developments in nonstandard mathematics
N J Cutland, V Neves, F Oliveira and J Sousa-Pinto

337 Topological circle planes and topological quadrangles
A E Schroth

338 Graph dynamics
E Prisner

339 Localization and sheaves: a relative point of view
P Jara, A Verschoren and C Vidal

340 Mathematical problems in semiconductor physics
P Marcati, P A Markowich and R Natalini

341 Surveying a dynamical system: a study of the Gray–Scott reaction in a two-phase reactor
K Alhumaizi and R Aris

342 Solution sets of differential equations in abstract spaces
R Dragoni, J W Macki, P Nistri and P Zecca

343 Nonlinear partial differential equations
A Benkirane and J-P Gossez

344 Numerical analysis 1995
D F Griffiths and G A Watson

345 Progress in partial differential equations: the Metz surveys 4
M Chipot and I Shafrir

346 Rings and radicals
B J Gardner, Liu Shaoxue and R Wiegandt

347 Complex analysis, harmonic analysis and applications
R Deville, J Esterle, V Petkov, A Sebbar and A Yger

348 The theory of quantaloids
K I Rosenthal

349 General theory of partial differential equations and microlocal analysis
Qi Min-you and L Rodino

350 Progress in elliptic and parabolic partial differential equations
A Alvino, P Buonocore, V Ferone, E Giarrusso, S Matarasso, R Toscano and G Trombetti

351 Integral representations for spatial models of mathematical physics
V V Kravchenko and M V Shapiro

352 Dynamics of nonlinear waves in dissipative systems: reduction, bifurcation and stability
G Dangelmayr, B Fiedler, K Kirchgässner and A Mielke

353 Singularities of solutions of second order quasilinear equations
L Véron

354 Mathematical theory in fluid mechanics
G P Galdi, J Málek and J Necas

355 Eigenfunction expansions, operator algebras and symmetric spaces
R M Kauffman

356 Lectures on bifurcations, dynamics and symmetry
M Field

357 Noncoercive variational problems and related results
D Goeleven

358 Generalised optimal stopping problems and financial markets
D Wong

359 Topics in pseudo-differential operators
S Zaidman

360 The Dirichlet problem for the Laplacian in bounded and unbounded domains
C G Simader and H Sohr

361 Direct and inverse electromagnetic scattering
A H Serbest and S R Cloude

362 International conference on dynamical systems
F Ledrappier, J Lewowicz and S Newhouse

363 Free boundary problems, theory and applications
M Niezgódka and P Strzelecki

364 Backward stochastic differential equations
N El Karoui and L Mazliak

365 Topological and variational methods for nonlinear boundary value problems
P Drábek

366 Complex analysis and geometry
V Ancona, E Ballico, R M Mirò-Roig and A Silva

367 Integral expansions related to Mehler–Fock type transforms
B N Mandal and N Mandal

368 Elliptic boundary value problems with indefinite weights: variational formulations of the principal eigenvalue and applications
F Belgacem

369 Integral transforms, reproducing kernels and their applications
S Saitoh

S Saitoh

Gunma University, Japan

Integral transforms, reproducing kernels and their applications

 LONGMAN

Addison Wesley Longman Limited
Edinburgh Gate, Harlow
Essex CM20 2JE, England
and Associated Companies throughout the world.

Published in the United States of America
by Addison Wesley Longman Inc.

First published 1997

AMS Subject Classifications   (Main) 30C40, 44A05, 35A22
                               (Subsidiary) 30B40, 45A05, 41A50

ISSN 0269-3674

ISBN 0 582 31758 4

British Library Cataloguing in Publication Data

A catalogue record for this book is
available from the British Library

Printed and bound in Great Britain
by Biddles Ltd, Guildford and King's Lynn

To Robert Pertsch Gilbert and Toshio Umezawa

Contents

Preface

Acknowledgments

1. **Introduction**   1

2. **Reproducing kernel Hilbert spaces**   20

   §1 Fundamental theorems of linear transforms   20
   §2 Elementary properties of reproducing kernel Hilbert spaces   34
   §3 Generations of reproducing kernel Hilbert spaces   43
   §4 Examples of reproducing kernels   53
   §5 Realizations of reproducing kernel Hilbert spaces   87

3. **Isometrical identities and inversion formulas**   97

   §1 Laplace and Fourier transforms   97
   §2 The heat equation   128
   §3 The wave equation   143
   §4 Analytic and harmonic functions of class L_2 154
   §5 The Meyer wavelets   168

4. **Applications to the approximation of functions**   178

   §1 Best approximations by the functions in a RKHS   178
   §2 Approximations in reproducing kernel Hilbert spaces   189

5. **Applications to analytic extension formulas and real inversion formulas for the Laplace transform**   200

   §1 Representations of the norms in Bergman–Selberg spaces   200
   §2 Real inversion formulas for the Laplace transform   211

6. **Applications to source inverse problems**   218

   §1 Inverse source formulas in Poisson's equation   218
   §2 Inverse source formulas for Helmholtz's equation   229

Appendix 1. Applications to representations of inverse functions **237**

§1 A general approach   237
§2 Reasonable settings   239
§3 Examples   240

Appendix 2. Natural norm inequalities in nonlinear transforms **243**

§1 General principles   243
§2 Concrete examples   248

Appendix 3. Stability of Lipschitz type in determination of initial heat distribution **259**

§1 Introduction and theorem   259
§2 Proof of theorem   262

References   **267**

Books   **267**

Articles   **272**

Preface

This book is an essentially self-contained presentation of the theory of reproducing kernels in connection with integral transforms in the framework of Hilbert spaces. It is a general and fundamental concept and a potentially powerful theory when combined with the integral transforms. A variety of concrete results of its application will be given systematically for isometrical identities and inversion formulas for various typical integral transforms, linear differential equations with variable coefficients, linear integral equations, best approximation theories of functions, analytic extension formulas, real inversion formulas for the Laplace transform, inverse source problems, representations of inverse functions, natural norm inequalities in nonlinear transforms and stability of Lipschitz type in the determination of initial heat distribution. The general theories and results of applications are fairly elementary, but definite values have been obtained. However, almost all mathematicians and scientists will recognize the novelty of this information. These results have been obtained over the last fourteen years, mainly by the author.

In particular, the general theories will give rise to new ideas and methods for the natural inversion formulas for general linear mappings in the framework of Hilbert spaces containing the natural solutions for Fredholm integral equations of the first kind, representing the relations between input (cause) functions and output (effect) functions and typically causing ill-posed problems in the natural sciences.

I would like to thank Professors Hiroaki Aikawa, Nakao Hayashi, Tsuyosi Ando, Masahiro Yamamoto and Dr. D.-W. Byun for their pleasant collaboration. A survey article of the contents of this book was published in [Sa29], following the very kind suggestion of Professor Anatoli P. Prudnikov. I also wish to thank Mrs. Noriko Kimura who, with great competence and patience, typed the manuscript of this book, and the staff at Addison Wesley Longman for their help.

February, 1997
Kiryu

S Saitoh

Acknowledgments

The author thanks, in particular, the following mathematicians for their direct or indirect contributions to the publication of this book: Takehisa Abe, Hiroaki Aikawa, Russell Brown, Constantin Corduneanu, Takayuki Furuta, Robert P. Gilbert, Rudolf Gorenflo, Nakao Hayashi, Masaru Ikehata, Victor Isakov, Yuusuke Iso, Joji Kajiwara, Joe Kamimoto, Zuhair Nashed, Takeo Ohsawa, Kazuyoshi Okubo, Anatoli P. Prudnikov, Juri M. Rappoport, G. F. Roach, Megumi Saigo, Masahiko Taniguchi, Toshiyuki Sugawa, Toshio Umezawa, Wolgang Walter, and Masahiro Yamamoto.

Professors Roach, Nashed and Gilbert encouraged the publication of this book. Dr. Elizabeth Anne Kamei kindly checked the English of the introduction in the first draft. Professor Gilbert checked the whole manuscript of this book as the editor with many and valuable suggestions. The two referees gave valuable comments and suggestions to the first draft of this book.

The research was supported in part by the Japanese Ministry of Education, Science, Sports and Culture; Grant-in-Aid for Scientific Research, General Research (C)03804003, (C)04804003 and (C)066402201.

The following meetings were invaluable to the completion of this book:

The 2nd International Symposium 'Inverse Problems in Engineering Sciences' (July 27–30, 1994, Osaka).

The meeting on 'Integral Transforms and Their Applications' at Kyoto University (August 24–26, 1994) was supported by Kyoto University. The author was the representative of the research meeting.

The NSF and Kentucky NSF-ESPCoR program; and the Gunma University Foundation for Science & Technology (1995) supported participation at the NSF-CBMS Regional Conference 'Nondestructive Evaluation and Inverse Problems' at the University of Kentucky (June 13–17, 1995).

Mathematisches Forschungsinstitut Oberwolfach and the Japanese Ministry of Education, Science, Sports and Culture supported participation at the '7th International Conference on General Inequalities' (November 12–18, 1995, Oberwolfach).

The 4th Seminar on Function Spaces (December 25–27, 1995, Sapporo).

Volterra Centennial Symposium at the University of Texas at Arlington (May 23–25, 1996).

Lectures and a colloquium talk at Nagoya University, based on the manuscript of the second draft of this book (June 23–29, 1996).

INVERSE and ILL-POSED PROBLEMS (IIPP–96) at Moscow Lomonosov State University (September 9–14, 1996).

Kansai Seminar for Complex Analysis at Kyoto University (October 12, 1996).

1. Introduction

In this chapter, the global framework of fundamental theorems with new viewpoints and methods for linear transforms in Hilbert spaces and their miscellaneous applications are stated. At the same time, the contents of this book are described.

1 A generalized isoperimetric inequality

In 1976, the author obtained the generalized isoperimetric inequality in his thesis [Sa1]:

For a bounded regular region G in the complex $z = x + iy$ plane surrounded by a finite number of analytic Jordan curves and for any analytic functions $\varphi(z)$ and $\psi(z)$ on $\overline{G} = G^{\cup} \partial G$,

$$\frac{1}{\pi} \iint_G |\varphi(z)\psi(z)|^2 dx dy \leqq \frac{1}{2\pi} \int_{\partial G} |\varphi(z)|^2 |dz| \frac{1}{2\pi} \int_{\partial G} |\psi(z)|^2 |dz|.$$

The crucial point in this thesis was to determine completely the case when equality holds in the inequality.

In order to prove this simple inequality, surprisingly enough, we must apply the historical results of

G.F.B.Riemann (1826–1866); F.Klein (1849–1925); S.Bergman; G.Szegö;
Z.Nehari; M.M.Schiffer; P.R.Garabedian; D.A.Hejhal (1972, thesis).

In particular, a profound result of D. A. Hejhal, which establishes the fundamental interrelationship between the Bergman and the Szegö reproducing kernels of G ([[He]]), must be applied. Furthermore, we must use the general theory of reproducing kernels by N. Aronszajn ([Ar1]) described in 1950. These circumstances have not changed, since the paper [Sa1] was published about 18 years ago.

This thesis is a milestone in the development of the theory of reproducing kernels. In the thesis, the author realized that miscellaneous applications of the general theory of reproducing kernels are possible for many concrete problems. See also [[Sa]] for the details. It seems that the general theory of reproducing kernels was, in a strict sense, not active in the theory of concrete reproducing kernels until the publication of the thesis. Indeed, after the publication of the thesis, we derived miscellaneous fundamental norm inequalities containing quadratic norm inequalities in matrices which have been described in over 20 papers. Furthermore, we obtained a general idea for linear transforms essentially by using the general theory of reproducing kernels, which is the main tool in this book.

2   Linear transforms of Hilbert spaces

In 1982 and 1983, we published the very simple theorems in [Sa4] and [Sa6], respectively. Certainly the results are very simple mathematically, but they appear to be extremely fundamental and widely applicable for general linear transforms. Moreover, the results gave rise to several new ideas for linear transforms themselves.

We shall formulate a 'linear transform' as follows:

$$F(t) \longrightarrow \quad \boxed{} \quad \longrightarrow f(p)$$

$$h(t,p)$$

$$f(p) = \int_T F(t)\overline{h(t,p)}dm(t), \quad p \in E. \tag{1}$$

Here, the input $F(t)$ (source) is a function on a set T, E is an arbitrary set, $dm(t)$ is a σ-finite positive measure on the dm measurable set T, and $h(t,p)$ is a complex-valued function on $T \times E$ which determines the transform of the system.

This formulation will give a generalized form of a linear transform L:

$$L(aF_1 + bF_2) = aL(F_1) + bL(F_2).$$

Indeed, following the Schwartz kernel theorem (cf. [[Tr2]]), we see that very general linear transforms are realized as integral transforms as in (1) above by using generalized functions as the integral kernels $h(t,p)$.

We shall assume that $F(t)$ is a member of the Hilbert space $L_2(T, dm)$ satisfying

$$\int_T |F(t)|^2 dm(t) < \infty. \tag{2}$$

The space $L_2(T, dm)$ whose norm gives an energy integral will be the most fundamental space as the input function space. In other spaces we shall modify them in order to comply with our situation. As a prototype case, we shall consider primarily or, as the first stage, the linear transform (1) in our situation.

As a natural result of our basic assumption (2), we assume that

$$\text{for any fixed} \quad p \in E, \quad h(\cdot, p) \in L_2(T, dm) \tag{3}$$

for the existence of the integral in (1).

2

We shall consider the following two typical linear transforms:

We take $E = \{1, 2\}$ and let $\{e_1, e_2\}$ be some orthonormal vectors of $\mathbb{R}^2$. Then, we shall consider the linear transform from $\mathbb{R}^2$ to $\{x_1, x_2\}$ as follows:

$$\mathbf{x} \longrightarrow \begin{cases} x_1 = (\mathbf{x}, e_1) \\ x_2 = (\mathbf{x}, e_2). \end{cases} \tag{4}$$

For $F \in L_2(\mathbb{R}, dx)$ we shall consider the integral transform

$$u(x, t) = \frac{1}{\sqrt{4\pi t}} \int_{-\infty}^{\infty} F(\xi) e^{-\frac{(x-\xi)^2}{4t}} \, d\xi, \tag{5}$$

which gives the solution $u(x, t)$ of the heat equation

$$u_{xx}(x, t) = u_t(x, t) \quad \text{on} \quad \mathbb{R} \times \{t > 0\} \tag{6}$$

subject to the initial condition

$$u(x, 0) = F(x) \quad \text{on} \quad \mathbb{R}. \tag{7}$$

3   Identification of the images of linear transforms

We formulated linear transforms as the integral transforms (1) satisfying (2) and (3) in the framework of Hilbert spaces. In this general situation, we can identify the space of output functions $f(p)$ and we can completely characterize the output functions $f(p)$. Regarding this fundamental idea, it seems that the mathematical community still does not realize its importance, since the papers [Sa4] and [Sa6] were published about fourteen years ago.

One reason why we have no idea of the identification of the images of linear transforms is based on the definition of linear transforms themselves. A linear transform is, in general, a linear mapping from a linear space into a linear space, and so the image space of the linear mapping will be considered as an, a priori, given one. For this, our idea will show that the image spaces of linear transforms, in our situation, form the uniquely determined and intuitive ones which are, in general, different from the image spaces stated in the definitions of linear transforms.

Another reason is that the very fundamental theory of reproducing kernels by Aronszajn is still not widely known. The general theory seems to be a very fundamental one in mathematics, as in the theory of Hilbert spaces.

Recall the paper of [Schw] for this fact, which extended globally the theory of Aronszajn. Our basic idea for linear transforms is very simple mathematically, but it was initially deduced from the theory of Schwartz using the direct integrals of reproducing kernel Hilbert spaces. See [[Sa]] for the details.

In order to identify the image space of the integral transform (1), we consider the Hermitian form

$$K(p,q) = \int_T h(t,q)\overline{h(t,p)}dm(t) \quad \text{on} \quad E \times E. \tag{8}$$

The kernel $K(p,q)$ is apparently a positive matrix on E in the sense of

$$\sum_{j=1}^{n}\sum_{j'=1}^{n} C_j \overline{C_{j'}} K(p_{j'}, p_j) \geqq 0$$

for any finite points $\{p_j\}$ of E and for any complex numbers $\{C_j\}$. Then, following the fundamental theorem of Aronszajn–Moore (Theorem 2.2.2), there exists a uniquely determined Hilbert space H_K comprising functions $f(p)$ on E satisfying

for any fixed $q \in E, K(p,q)$ belongs to H_K as a function in p, $\quad$ (9)

and

for any $q \in E$ and for any $f \in H_K$
$$(f(\cdot), K(\cdot, q))_{H_K} = f(q).$$

$$\tag{10}$$

Then, the point evaluation $f(p)(p \in E)$ is continuous on H_K and, conversely, a functional Hilbert space such that the point evaluation is continuous admits the reproducing kernel $K(p,q)$ satisfying (9) and (10). Then, we obtain, in Theorem 2.1.2

Theorem 1 *The images $f(p)$ of the integral transform (1) for $F \in L_2(T, dm)$ form precisely the Hilbert space H_K admitting the reproducing kernel $K(p,q)$ in (8).*

In Example (4), using Theorem 1 we can naturally deduce that

$$\|\{x_1, x_2\}\|_{H_K} = \sqrt{x_1^2 + x_2^2}$$

for the image.

In Example (5), we naturally deduce the very surprising result that the image $u(x,t)$ is extended analytically onto the entire complex $z = x + iy$ plane and when we denote its analytic extension by $u(z,t)$, we have

$$\|u(z,t)\|_{H_K}^2 = \frac{1}{\sqrt{2\pi t}} \iint_{\mathbf{C}} |u(z,t)|^2 e^{-\frac{y^2}{2t}} dxdy \tag{11}$$

4

(Theorem 3.2.1).

The images $u(x,t)$ of (5) for $F \in L_2(\mathbb{R}, dx)$ are characterized by (11); that is, $u(x,t)$ are entire functions in the form $u(z,t)$ with finite integrals (11).

In 1989 ([HS1]), we deduced that (11) equals

$$\sum_{j=0}^{\infty} \frac{(2t)^j}{j!} \int_{-\infty}^{\infty} (\partial_x^j u(x,t))^2 dx \qquad (12)$$

using the property that $u(x,t)$ is the solution of the heat equation (6) with (7) (see Theorem 3.2.2 for a direct proof). Hence, we see that the images $u(x,t)$ of (5) are also characterized by the property that $u(x,t) \in C^\infty$ with finite integrals (12).

4   Relationship between the magnitudes of input and output functions – a generalized Pythagoras theorem

Our second theorem which is given by Theorem 2.1.2 is,

Theorem 2 *In the integral transform (1), we have the inequality*

$$\|f\|_{H_K}^2 \le \int_T |F(t)|^2 dm(t).$$

Furthermore, there exist functions F^* with the minimum norms satisfying (1), and we have the isometrical identity

$$\|f\|_{H_K}^2 = \int_T |F^*(t)|^2 dm(t).$$

In Example (4), we have, surprisingly enough, the Pythagoras theorem

$$\|\mathbf{x}\|^2 = x_1^2 + x_2^2.$$

In Example (5), we have the isometrical identity

$$\int_{-\infty}^{\infty} |F(\xi)|^2 d\xi = \frac{1}{\sqrt{2\pi t}} \int\int_{\mathbf{C}} |u(z,t)|^2 e^{-\frac{y^2}{2t}} dx dy, \qquad (13)$$

whose integrals are independent of $t > 0$ in Theorem 2.2.1. At this moment, we will be able to say that by the general principle (Theorems 1 and 2) for linear transforms we can prove the Pythagoras (572–492 B.C.) theorem apart from the idea of *'orthogonality'*, and we can understand Theorem 2 as a generalized theorem of Pythagoras in our general situation of linear transforms.

5

By using the general principle, we derived miscellaneous Pythagoras type theorems in over 30 papers, whose typical results will be given in Chapter 3. We shall refer to one typical example.

For the solution $u(x,t)$ of the simplest wave equation

$$u_{tt}(x,t) = c^2 u_{xx}(x,t) \quad (c > 0 : \text{constant})$$

subject to the initial conditions

$$u_t(x,t)|_{t=0} = F(x), \quad u(x,0) = 0 \quad \text{on} \quad \mathbb{R}$$

for $F \in L_2(\mathbb{R}, dx)$, we obtain the isometrical identity

$$\frac{1}{2} \int_{-\infty}^{\infty} |F(x)|^2 dx = \frac{2\pi c}{t} \int_{-\infty}^{\infty} \left| l.i.m._{N \to \infty} \int_{-N}^{N} u(x,t) \exp(\frac{ix\xi}{2\pi ct}) dx \right|^2 \frac{d\xi}{\left(\frac{\sin \frac{\xi}{2}}{\frac{\xi}{2}}\right)^2}, \quad (14)$$

whose integrals are independent of $t > 0$.

Recall here the *conservative law of energy*

$$\frac{1}{2} \int_{-\infty}^{\infty} |F(x)|^2 dx = \frac{1}{2} \int_{-\infty}^{\infty} (u_t(x,t)^2 + c^2 u_x(x,t)^2) dx. \quad (15)$$

To compare the two integrals (14) and (15) will be very interesting, because (15) contains the derived functions $u_t(x,t)$ and $u_x(x,t)$; meanwhile (14) contains the values $u(x,t)$ only.

In the viewpoint of the conservative law of energy in (14) and (15), could we give some physical interpretation to the isometrical identities in (13) and (12) whose integrals are independent of $t > 0$ in the heat equation?

5   Inversion formulas for linear transforms

In our **Theorem 3,** we establish the inversion formula

$$f \quad \longrightarrow \quad F^* \quad (16)$$

of the integral transform (1) in the sense of Theorem 2.

The basic idea to derive the inversion formula (16) is, first, to represent $f \in H_K$ in the space H_K in the form

$$f(q) = (f(\cdot), K(\cdot, q))_{H_K},$$

6

secondly, to consider as follows:

$$f(q) = (f(\cdot), \int_T h(t,q)\overline{h(t,\cdot)}dm(t))_{H_K}$$

$$= \int_T (f(\cdot), \overline{h(t,\cdot)})_{H_K}\overline{h(t,q)}dm(t)$$

$$= \int_T F^*(t)\overline{h(t,q)}dm(t)$$

and, finally, to deduce that

$$F^*(t) = (f(\cdot), \overline{h(t,\cdot)})_{H_K}. \tag{17}$$

However, in these arguments the integral kernel $h(t,p)$ does not generally belong to H_K as a function of p and therefore (17) is generally not valid.

For this reason, we shall realize the norm in H_K in terms of a σ-finite positive measure $d\mu$ in the form

$$\|f\|_{H_K}^2 = \int_E |f(p)|^2 d\mu(p).$$

Then, for some suitable exhaustion $\{E_N\}$ of E, we obtain, in general, the inversion formula

$$F^*(t) = s - \lim_{N\to\infty} \int_{E_N} f(p)h(t,p)d\mu(p) \tag{18}$$

in the sense of strong convergence in $L_2(T, dm)$, which will be given by Theorem 2.1.5.

Note that F^* is a member of the visible component of $L_2(T, dm)$ which is the orthocomplement of the null space (the invisible component)

$$\left\{ F_0 \in L_2(T, dm); \int_T F_0(t)\overline{h(t,p)}dm(t) = 0 \quad \text{on} \quad E \right\}$$

of $L_2(T, dm)$. Therefore, our inversion formula (16) will be considered as a very natural one.

By our Theorem 3, for example, in Example (5) we can establish the inversion formulas

$$u(z,t) \longrightarrow F(x)$$

and

$$u(x,t) \longrightarrow F(x)$$

for any fixed $t > 0$, which will be given by Theorem 3.2.2.

Our inversion formula will present a new viewpoint and a new method for Fredholm integral equations of the first kind which are fundamental in the theory of integral equations. The characteristics of our inversion formula are as follows:

(i) Our inversion formula is given in terms of the reproducing kernel Hilbert space H_K which is intuitively determined as the image space of the integral transform (1).

(ii) Our inversion formula gives the visible component F^* of F with the minimum $L_2(T, dm)$ norm.

(iii) The inverse F^* is, in general, given in the sense of strong convergence in $L_2(T, dm)$.

(iv) Our integral equation (1) is, in general, an ill-posed problem, but our solution F^* is given as the solution of a well-posed problem in the sense of Hadamard (1902, 1923) ([H1-2]).

At this point, we can see why we meet ill-posed problems; that is, because we do not consider the problems in the natural image spaces H_K, but in some artificial spaces.

Nowday it is considered that the general theory of integral equations of the first kind has not been formed yet ([[Bi]], Preface and see also [[Gr]]). Our method will give a general theory for the integral equations in the framework of Hilbert spaces.

For a general reference and for a historical background to integral equations, see [[Co]] with its references. See also [[Kon]], [[Tri]] and [[Yo]].

6   Determination of the system by input and output functions

In our **Theorem 4**, we can construct the integral kernel $h(t, p)$ conversely, in terms of the isometrical mapping $\widetilde{L}$ from a reproducing kernel Hilbert space H_K onto $L_2(T, dm)$ and the reproducing kernel $K(p, q)$, in the form

$$h(t, p) = \widetilde{L} K(\cdot, p), \tag{19}$$

which will be discussed in Chapter 2, Section 5.

7   General applications

Our basic assumption for the integral transform (1) is (3). When this assumption is not valid, we will be able to apply the following techniques to comply with our assumption (3).

(a) We restrict the sets E or T, or we exchange the set E.

(b) We multiply a positive continuous function ρ in the form $L_2(T, \rho dm)$.

For example, in the Fourier transform

$$\int_{-\infty}^{\infty} F(t) e^{-itx} dt, \tag{20}$$

we consider the integral transform with the weighted function such that

$$\int_{-\infty}^{\infty} F(t)e^{-itx}\frac{dt}{1+t^2}.$$

(c) We integrate the integral kernel $h(t,p)$.

For example, in the Fourier transform (20), we consider the integral transform

$$\int_{-\infty}^{\infty} F(t)\left(\int_0^{\hat{x}} e^{-itx}dx\right)dt = \int_{-\infty}^{\infty} F(t)\left(\frac{e^{-it\hat{x}}-1}{-it}\right)dt.$$

By these techniques we can apply our general method even for integral transforms with integral kernels of generalized functions. Furthermore, for the integral transforms with the integral kernels of

miscellaneous Green's functions,

Cauchy's kernel,

and

Poisson's kernel

and even for cases of the Fourier transform and the Laplace transform, we were able to derive novel results. We shall give the miscellaneous isometrical identities and inversion formulas in Chapter 3.

Recall the Whittaker–Kotel'nikov–Shannon sampling theorem:

In the integral transform

$$f(t) = \frac{1}{\sqrt{2\pi}}\int_{-\pi}^{\pi} F(\omega)e^{i\omega t}d\omega$$

for functions $F(\omega)$ satisfying

$$\int_{-\pi}^{\pi} |F(\omega)|^2 d\omega < \infty,$$

we have the expression

$$f(t) = \sum_{n=-\infty}^{\infty} f(n)\frac{\sin \pi(n-t)}{\pi(n-t)} \quad \text{on} \quad (-\infty, \infty).$$

All the signals $f(t)$ are expressible in terms of the discrete data $f(n)$ (n: integers) only, therefore many scientists are interested in this theorem. Thus, this theorem is applied in miscellaneous fields. Furthermore, very interesting relationships between

9

fundamental theorems and formulas of signal analysis, of analytic number theory and of applied analysis have been described recently in [Klu].

In our general situation (1), the essence of the sampling theorem is given clearly and simply as follows:

For a sequence of points $\{p_n\}$ of E, if $\{h(t, p_n)\}_n$ is a complete orthonormal system in $L_2(T, dm)$, then for any $f \in H_K$, we have the sampling theorem

$$f(p) = \sum_n f(p_n) K(p, p_n) \quad \text{on} \quad E.$$

These results and related topics will be discussed in Chapter 4, Section 2.1, in detail.

Meanwhile, the theory of wavelets which was created by [MAFG] and [Mo] about fourteen years ago is developing rapidly in both the mathematical sciences and pure mathematics. The theory is applicable to signal analysis, numerical analysis and many other fields, as in Fourier transforms. Since the theory is that of integral transforms in the framework of Hilbert spaces, our general theory for integral transforms will be applicable to wavelet theory, globally, and in particular, our method will give a good unified understanding of the wavelet transform, frames, multiresolution analysis and sampling theory in the theory of wavelets. For the typical Meyer wavelets, we shall examine the isometrical identities and inversion formulas in Chapter 3, Section 5.

8 Analytic extension formulas

The equality of the two integrals (11) and (12) means that a C^∞ function $g(x)$ with a finite integral

$$\sum_{j=0}^{\infty} \frac{(2t)^j}{j!} \int_{-\infty}^{\infty} |\partial_x^j g(x)|^2 dx < \infty,$$

is extended analytically onto $\mathbb{C}$ and when we denote its analytic extension by $g(z)$, we have the identity

$$\sum_{j=0}^{\infty} \frac{(2t)^j}{j!} \int_{-\infty}^{\infty} |\partial_x^j g(x)|^2 dx = \frac{1}{\sqrt{2\pi t}} \iint_{\mathbb{C}} |g(z)|^2 \exp\{-\frac{y^2}{2t}\} dx\, dy. \tag{21}$$

In this way, we have derived miscellaneous analytic extension formulas in over 15 papers with H. Aikawa and N. Hayashi, and the analytic extension formulas are applied to the investigation of analyticity of solutions of nonlinear partial differential equations. See, for example [HS1], [HS2], [Hay1], [Hay2], [HK] and [BHK].

One typical result of another type is obtained from the integral transform

$$v(x, t) = \frac{1}{t} \int_0^t F(\xi) \frac{x \exp[\frac{-x^2}{4(t-\xi)}]}{2\sqrt{\pi}(t-\xi)^{\frac{3}{2}}} \xi\, d\xi$$

10

in connection with the heat equation

$$u_t(x,t) = u_{xx}(x,t) \quad \text{for} \quad u(x,t) = tv(x,t)$$

satisfying the conditions

$$u(0,t) = tF(t) \quad \text{on} \quad t \geq 0$$

and

$$u(x,0) = 0 \quad \text{on} \quad x \geq 0.$$

Then, we obtain:

Let $\Delta(\frac{\pi}{4})$ denote the sector $\{|\arg z| < \frac{\pi}{4}\}$. For any analytic function $f(z)$ on $\Delta(\frac{\pi}{4})$ with a finite integral

$$\iint_{\Delta(\frac{\pi}{4})} |f(z)|^2 dx dy < \infty,$$

we have the identity

$$\iint_{\Delta(\frac{\pi}{4})} |f(z)|^2 dx dy = \sum_{j=0}^{\infty} \frac{2^j}{(2j+1)!} \int_0^{\infty} x^{2j+1} |\partial_x^j f(x)|^2 dx. \tag{22}$$

Conversely, for any smooth function $f(x)$ with a finite integral in (22) on $(0, \infty)$, there exits an analytic extension $f(z)$ onto $\Delta(\frac{\pi}{4})$ satisfying (22).

We shall discuss miscellaneous analytic extension formulas in Chapter 3, Sections 1, 2, 4, 5 and Chapter 5, Section 1.

9 Best approximation formulas

As shown, when we consider linear transforms in the framework of Hilbert spaces, we naturally have the idea of reproducing kernel Hilbert spaces. As a natural extension of our theorems, we have the fundamental theorems for approximations of functions in the framework of Hilbert spaces.

For a function F on a set X, we shall look for a function which is nearest to F among some family of functions $\{f\}$. In order to formulate the 'nearest' precisely, we shall consider F as a member of some Hilbert space $H(X)$ comprising functions on X. Meanwhile, as the family $\{f\}$ of approximation functions, we shall consider some reproducing kernel Hilbert space H_K comprising functions f on, in general, a set E containing the set X. Here the reproducing kernel Hilbert space H_K as a family of approximation functions will be considered as a natural one, since the point evaluation $f(p)$ is continuous on H_K.

11

We shall assume that for the relation between the two Hilbert spaces $H(X)$ and H_K:

for the restriction $f|_X$ of the members f of H_K to the set X,
$f|_X$ belongs to the Hilbert space $H(X)$,

(23)

and

the linear operator $Lf = f|_X$ is continuous from H_K into $H(X)$. (24)

In this natural situation, we can discuss the best approximation problem

$$\inf_{f \in H_K} \|Lf - F\|_{H(X)}$$

for a member F of $H(X)$.

For the sake of the nice properties of the restriction operator L and its adjoint L^*, we can obtain 'algorithms' to decide whether best approximations f^* of F in the sense of

$$\inf_{f \in H_K} \|Lf - F\|_{H(X)} = \|Lf^* - F\|_{H(X)}$$

exist. Further, when the best approximations f^* exist, we can obtain constructive 'algorithms' for them. Moreover, we can give the representations of f^* in terms of the given function F and the reproducing kernel $K(p, q)$. Meanwhile, when the best approximations f^* do not exist, we can construct the minimizing or approximating sequence $\{f_n\}$ of H_K satisfying

$$\inf_{f \in H_K} \|Lf - F\|_{H(X)} = \lim_{n \to \infty} \|Lf_n - F\|_{H(X)}.$$

As an example, for an $L_2(\mathbb{R}, dx)$ function $h(x)$, we shall approximate it by the family of functions $u_F(x, t)$ for any fixed $t > 0$ which are the solutions of the heat equation (6) with (7) for $F \in L_2(\mathbb{R}, dx)$.

Then, we can see that a member F of $L_2(\mathbb{R}, dx)$ exists such that

$$u_F(x, t) = h(x) \quad \text{on} \quad \mathbb{R}$$

if and only if

$$\iint_{\mathbf{C}} \left| \int_{-\infty}^{\infty} h(\xi) \exp\left\{ -\frac{\xi^2}{8t} + \frac{\xi z}{4t} \right\} d\xi \right|^2 \exp\left\{ \frac{-3x^2 + y^2}{12t} \right\} dx\,dy < \infty.$$

12

If this condition is not satisfied for h, then we can construct the sequence $\{F_n\}$ satisfying

$$\lim_{n \to \infty} \int_{-\infty}^{\infty} |u_{F_n}(x,t) - h(x)|^2 dx = 0;$$

that is, for any $L_2(\mathbb{R}, dx)$ function $h(x)$, we can construct the initial functions $\{F_n\}$ whose heat distributions $u_{F_n}(x,t)$ of time t later converge to $h(x)$.

We shall discuss these problems in Chapter 4.

10   Applications to random fields estimations

We assume that the random field is of the form

$$u(x) = s(x) + n(x),$$

where $s(x)$ is the useful signal and $n(x)$ is noise. Without loss of generality, we can assume that the mean values of $u(x)$ and $n(x)$ are zero. We assume that the covariance functions

$$R(x,y) = \overline{u(x)u(y)}$$

and

$$f(x,y) = \overline{u(x)s(y)}$$

are known. We shall consider the general form of a linear estimation $\hat{u}$ of u in the form

$$\hat{u}(x) = \int_T u(t)h(x,t)dm(t)$$

for an $L_2(T, dm)$ space and for a function $h(x,t)$ belonging to $L_2(T, dm)$ for any fixed $x \in E$. For the desired information As for a linear operator A of s, we wish to determine the function $h(x,t)$ satisfying

$$\inf \overline{(\hat{u} - As)^2}$$

which gives the minimum of the variance by the least squares method. Many topics in filtering and estimation theory in signal and image processing, underwater acoustics, geophysics, optical filtering, etc., which were initiated by N. Wiener (1894–1964), will be presented in this framework. Then, we see that the linear transform $h(x,t)$ is given by the integral equation

$$\int_T R(x',t)h(x,t)dm(t) = f(x',x)$$

[[Ra1]]. Therefore, our random fields estimation problems will be reduced to finding the inversion formula

$$f(x',x) \longrightarrow h(x,t)$$

13

in our framework. So, our general method for integral transforms will be applied to these problems. For this situation and other topics and methods for the inversion formulas, see [[Ra1]] for details.

11 Applications to scattering and inverse problems

Scattering and inverse problems will be considered as the problems of determining unobservable quantities from observable quantities. These problems are miscellaneous and are, in general, difficult. In many cases, the problems are reduced to certain Fredholm integral equations of the first kind and then our method will be applicable to these equations. Meanwhile, in many cases, the problems will be reduced to determining the inverse F^* from the data $f(p)$ on some subset of E in our integral transform (1). See, for example, [[Ra2]] and [[Gr]].

In each case, we shall state a typical example.

We shall consider the Poisson equation

$$\Delta u = -\rho(\mathbf{r}) \quad \text{on} \quad \mathbb{R}^3 \tag{25}$$

for a real-valued $L_2(\mathbb{R}^3, d\mathbf{r})$ source function ρ whose support is contained in a sphere $r < a(|\mathbf{r}| = r)$. By using our method for the integral transform

$$u(\mathbf{r}) = \frac{1}{4\pi} \int_{r' < a} \frac{1}{|\mathbf{r} - \mathbf{r}'|} \rho(\mathbf{r}') d\mathbf{r}',$$

we can obtain the characteristic property and natural representation of the potential u on the outside of the sphere $\{r < a\}$. Furthermore, we can obtain the surprisingly simple representation of ρ^* in terms of u on any sphere (a', θ', φ') $(a < a')$, which has the minimum $L_2(\mathbb{R}^3, d\mathbf{r})$ norm among ρ satisfying (25) on $r > a$, in the form:

$$\rho^*(r, \theta, \varphi) = \frac{1}{4\pi} \sum_{n=0}^{\infty} \frac{(2n + 1)^2 (2n + 3)}{a^{2n+3}} r^n a'^{n+1}$$

$$\times \sum_{m=0}^{n} \frac{\varepsilon_m (n - m)!}{(n + m)!} P_n^m(\cos\theta) \int_0^{\pi} \int_0^{2\pi} u(a', \theta', \varphi')$$

$$\times P_n^m(\cos\theta') \cos m(\varphi' - \varphi) \sin\theta' d\theta' d\varphi'.$$

Here, ε_m is the Neumann factor $\varepsilon_m = 2 - \delta_{m0}$.

In Chapter 6, we shall examine inverse source problems in the Poisson and the Helmholtz equations.

Next, we shall consider an analytical real inversion formula for the Laplace transform

14

$$f(p) = \int_0^\infty e^{-pt} F(t)dt, \quad p > 0;$$

$$\int_0^\infty |F(t)|^2 dt < \infty.$$

For the polynomial of degree $2N + 2$

$$P_N(\xi) = \sum_{0 \le \nu \le n \le N} \frac{(-1)^{\nu+1}(2n)!}{(n+1)!\nu!(n-\nu)!(n+\nu)!} \xi^{n+\nu}$$

$$\cdot \left\{ \frac{2n+1}{n+\nu+1} \xi^2 - \left(\frac{2n+1}{n+\nu+1} + 3n + 1 \right) \xi + n(n+\nu+1) \right\},$$

we set

$$F_N(t) = \int_0^\infty f(p) e^{-pt} P_N(pt) dp.$$

Then, we have

$$\lim_{N \to \infty} \int_0^\infty |F(t) - F_N(t)|^2 dt = 0.$$

Furthermore, the estimation of the error of $F_N(t)$ will also be given.

In Chapter 5, we shall derive this formula in a more general form.

Compare our formula with ([BW], 1940 and [[Wi]], Chapter VII), and with ([[Ra2]], p.221, 1992):

For the Laplace transform

$$\int_0^b e^{-pt} F(t)dt = f(p),$$

we have

$$F(t) = \frac{2tb^{-1}}{\pi} \frac{d}{du} \int_0^u \frac{G(v)}{(u-v)^{\frac{1}{2}}} dv \bigg|_{u=t^2 b^{-2}};$$

$$G(v) = v^{-\frac{1}{2}} \frac{2}{\pi} \int_0^\infty dy \cos(y \cosh^{-1} v^{-1}) \cosh \pi y$$

$$\times \int_0^\infty dz \cos(zy)(\cosh z)^{-\frac{1}{2}} \int_0^\infty dp f(p) J_0 \left(p \frac{b}{(\cosh z)^{\frac{1}{2}}} \right).$$

Unfortunately, in this very complicated formula, the characteristic properties of both the functions F and f making the inversion formula hold are not given.

15

12   Nonharmonic transforms

In our general transform (1), suppose that $\varphi(t,p)$ is near to the integral kernel $h(t,p)$ in the following sense:

For any $F \in L_2(T, dm)$,

$$\left\| \int_T F(t)\overline{(h(t,p) - \varphi(t,p))}dm(t) \right\|_{H_K}^2 \leq \omega^2 \int_T |F(t)|^2 dm(t)$$

where $0 < \omega < 1$ and ω is independent of $F \in L_2(T, dm)$.

Then, we can see that for any $f \in H_K$, there exists a function F_φ^* belonging to the visible component of $L_2(T, dm)$ in (1) such that

$$f(p) = \int_T F_\varphi^*(t)\overline{\varphi(t,p)}dm(t) \quad \text{on} \quad E \tag{26}$$

and

$$(1 - \omega)^2 \int_T |F_\varphi^*(t)|^2 dm(t) \leq \|f\|_{H_K}^2$$

$$\leq (1 + \omega)^2 \int_T |F_\varphi^*(t)|^2 dm(t).$$

The integral kernel $\varphi(t,p)$ will be considered as a perturbation of the integral kernel $h(t,p)$. When we look for the inversion formula of (26) following our general method, we must calculate the kernel form

$$K_\varphi(p,q) = \int_T \varphi(t,q)\overline{\varphi(t,p)}dm(t) \quad \text{on} \quad E \times E.$$

We will, however, in general, not be able to calculate this kernel.

Suppose that the image $f(p)$ of (26) belongs to the known space H_K. Then, we can construct the inverse F_φ^* by using our inversion formula in H_K repeatedly and by constructing some approximation of F_φ^* by our inverses.

In particular, for the reproducing kernel $K(p,q) \in H_K(q \in E)$ we construct (or we obtain, by some other method or directly) the function $\hat{\varphi}(t,p)$ satisfying

$$K(p,q) = \int_T \hat{\varphi}(t,q)\overline{\varphi(t,p)}dm(t) \quad \text{on} \quad E \times E,$$

where $\hat{\varphi}(t,q)$ belongs to the visible component of $L_2(T, dm)$ in (26) for any fixed $q \in E$. Then, we have the idea of a *'nonharmonic integral transform'* and we can formulate the inversion formula of (26) in terms of the kernel $\hat{\varphi}(t,q)$ and the space H_K, globally ([[Sa]], Chapter 7).

16

13 Nonlinear transforms

Our generalized isoperimetric inequality will mean that for an analytic function $\varphi(z)$ on $\overline{G}$ satisfying

$$\int_{\partial G} |\varphi(z)|^2 |dz| < \infty,$$

the image of the simplest nonlinear transform

$$\varphi(z) \longrightarrow \varphi(z)^2$$

belongs to the space of analytic functions satisfying

$$\iint_G |\varphi(z)|^4 dx\, dy < \infty$$

and we have the norm inequality

$$\frac{1}{\pi} \iint_G |\varphi(z)^2|^2 dx\, dy \leqq \left\{ \frac{1}{2\pi} \int_{\partial G} |\varphi(z)|^2 |dz| \right\}^2.$$

We established in Theorem 1 the method of identification of the images of linear transforms, and we will also be able to look for some Hilbert spaces containing the image spaces of nonlinear transforms. In these cases, however, the spaces will be too large for the image spaces, as in the generalized isoperimetric inequality. So, the inversion formulas for nonlinear transforms will be, in general, extremely involved. However, in many nonlinear transforms of reproducing kernel Hilbert spaces, some norm inequalities exist, as in the generalized isoperimetric inequality. We shall state one example in the strongly nonlinear transform

$$f(x) \longrightarrow e^{f(x)}$$

which was recently obtained in [Sa27]:

For an absolutely continuous real-valued function f on $[a, b)$ $(a > 0)$ satisfying

$$f(a) = 0,$$

$$\int_a^b f'(x)^2 x\, dx < \infty,$$

we obtain the inequality

$$1 + a \int_a^b |(e^{f(x)})'|^2 dx \leqq e^{\int_a^b f'(x)^2 x\, dx}.$$

17

Here, we should note that the equality holds for many functions $f(x)$.

For nonlinear transforms, we shall need other treatments different from linear transforms and so we shall refer to nonlinear transforms in Appendix 2 based on [Sa30].

14 Representations of inverse functions

By considering transforms of reproducing kernels we can obtain some general principles for representations of the inverse function φ^{-1} for an arbitrary mapping $p = \varphi(\hat{p})$

$$\varphi : \widehat{E} \longrightarrow E$$

from an abstract set $\widehat{E}$ into an abstract set E. Of course, the inverse φ^{-1} is, in general, multivalued. We can, however, obtain representations of φ^{-1} in terms of φ for many concrete mappings φ by a unified principle. For example, we have the expression

$$\sqrt[n]{x} = \frac{2}{\pi} \int_0^\infty \int_0^\infty \frac{\cos(\xi^n t) \sin xt}{t} dt d\xi$$

which will be given in Appendix 1, (31).

The construction of this book is as follows:

In Chapter 2, we establish the fundamental theorems 1 − 4 of linear transforms in the framework of Hilbert spaces and furthermore we introduce the basic general theory for reproducing kernel Hilbert spaces which is essentially needed for the proofs and applications of the fundamental theorems, in a self-contained manner. This Chapter 2 will form the main and essential body of this book, as all other chapters are direct applications of this chapter.

In Chapter 3, we establish miscellaneous typical isometrical identities and inversion formulas in various integral transforms which typically appear in analysis.

In Chapters 4, 5 and 6, we shall give concrete applications to approximations of functions, analytic extension formulas and analytical real inversion formulas for the Laplace transform, and inverse source problems, respectively.

In Appendix 1, we establish a general principle to represent the inverse functions and give concrete examples.

In Appendix 2, we shall discuss nonlinear transforms in some general situations and give natural norm inequalities in concrete nonlinear transforms.

18

In Appendix 3, we examine Lipschitz-type stability in the determination of the initial heat distribution, as an application of the real inversion formulas for the Laplace transform.

References to formulas are given as (a) within one section, (a, b) within one chapter; a being the number of the section, b being the number of the formula in this section, (a, b, c) for a formula in Chapter a, Section b, formula c.

2. Reproducing kernel Hilbert spaces

In this chapter, we establish the fundamental theorems of linear transforms. The theorems will be naturally related to the idea of reproducing kernel Hilbert spaces and therefore we essentially need the properties of the spaces. So, we shall introduce the general properties of reproducing kernel Hilbert spaces which are needed in concrete applications of the fundamental theorems. This chapter will present a unified and self-contained theory for reproducing kernel Hilbert spaces.

§1 Fundamental theorems of linear transforms

In this section, we give the fundamental theorems of linear transforms and applications to general linear transforms by smooth functions and general linear integro-differential equations with variable coefficients.

1 Identification of the images of linear transforms and the interrelationship between the domains and the ranges

As shown in Chapter 1, we shall formulate linear transforms as the integral transform Chapter 1, (1) in the framework of Hilbert spaces. For its importance and simplicity, we shall formulate linear transforms in the following general and abstract form.

Let $\mathcal{F}(E)$ be the linear space comprising all complex-valued functions on an abstract set E. Let $\mathcal{H}$ be a Hilbert (possibly finite-dimensional) space equipped with inner product $(\cdot, \cdot)_{\mathcal{H}}$. Let

$$\mathbf{h} : E \longrightarrow \mathcal{H}$$

be a Hilbert space $\mathcal{H}$-valued function on E. Then, we shall consider the linear mapping L from $\mathcal{H}$ into $\mathcal{F}(E)$ defined by

$$f(p) = (L\mathbf{f})(p) = (\mathbf{f}, \mathbf{h}(p))_{\mathcal{H}}. \tag{1}$$

The fundamental problems in the linear mapping (1) will be firstly the indentification (characterization) of the images $f(p)$ and secondly the relationship between $\mathbf{f}$ and $f(p)$.

The key which solves these fundamental problems is to form the function $K(p, q)$ on $E \times E$ defined by

$$K(p, q) = (\mathbf{h}(q), \mathbf{h}(p))_{\mathcal{H}}. \tag{2}$$

We let $\mathcal{R}(L)$ denote the range of L for $\mathcal{H}$ and we introduce the inner product in $\mathcal{R}(L)$ induced from the norm

$$\|f\|_{\mathcal{R}(L)} = \inf\{\|\mathbf{f}\|_{\mathcal{H}}; f = L\mathbf{f}\}. \tag{3}$$

Then, we obtain

Theorem 1 *For the function $K(p,q)$ defined by (2), the space $[\mathcal{R}(L), (\cdot, \cdot)_{\mathcal{R}(L)}]$ is a Hilbert (possibly finite-dimensional) space satisfying the properties that*

(i) for any fixed $q \in E$, $K(p,q)$ belongs to $\mathcal{R}(L)$ as a function in p, and
(ii) for any $f \in \mathcal{R}(L)$ and for any $q \in E$,

$$f(q) = (f(\cdot), K(\cdot, q))_{\mathcal{R}(L)}.$$

Further, the function $K(p,q)$ satisfying (i) and (ii) is uniquely determined by $\mathcal{R}(L)$. Furthermore, the mapping L is an isometry from $\mathcal{H}$ onto $\mathcal{R}(L)$ if and only if

$$\{\mathbf{h}(p); p \in E\} \quad \text{is complete in} \quad \mathcal{H}. \tag{4}$$

Proof : From (1) we see that $N(L)$ (the null space or the kernel of L) is a closed subspace in $\mathcal{H}$. Hence, for $f = L\mathbf{f}$ we have

$$\|f\|_{\mathcal{R}(L)} = \inf\{\|\mathbf{f} - \mathbf{g}\|_{\mathcal{H}}; \mathbf{g} \in N(L)\} = \|P_{[N(L)]^{\perp}}\mathbf{f}\|_{\mathcal{H}}. \tag{5}$$

Here, $P_{[N(L)]^{\perp}}$ is the orthogonal projection from $\mathcal{H}$ onto the orthocomplement of $N(L)$ in $\mathcal{H}$. If we restrict L to $[N(L)]^{\perp}$, then the restriction $L|_{[N(L)]^{\perp}}$ is an isometry between $[[N(L)]^{\perp}, (\cdot, \cdot)_{\mathcal{H}}]$ and $[\mathcal{R}(L), (\cdot, \cdot)_{\mathcal{R}(L)}]$, which implies that $[\mathcal{R}(L), (\cdot, \cdot)_{\mathcal{R}(L)}]$ is a Hilbert space.

Clearly $K(p,q)$ satisfies (i). In order to see the property (ii) of $K(p,q)$, note that for $\mathbf{f}_0 \in N(L)$,

$$(\mathbf{f}_0, \mathbf{h}(p)) = 0 \quad \text{on} \quad E. \tag{6}$$

Hence, for any $q \in E$, $\mathbf{h}(q) \in [N(L)]^{\perp}$. Therefore, for any $f = L\mathbf{f}$

$$\begin{aligned}
(f, K(\cdot, q))_{\mathcal{R}(L)} &= (L\mathbf{f}, L\mathbf{h}(q))_{\mathcal{R}(L)} \\
&= (P_{[N(L)]^{\perp}}\mathbf{f}, P_{[N(L)]^{\perp}}\mathbf{h}(q))_{\mathcal{H}} \\
&= (\mathbf{f}, \mathbf{h}(q))_{\mathcal{H}} \\
&= f(q), \tag{7}
\end{aligned}$$

which shows (ii).

21

Next, suppose that $K^+(p, q)$ has the properties (i) and (ii). Then, we have, for any $q \in E$

$$\|K(\cdot, q) - K^+(\cdot, q)\|_{\mathcal{R}(L)}^2$$
$$= (K(\cdot, q) - K^+(\cdot, q), K(\cdot, q))_{\mathcal{R}(L)}$$
$$- (K(\cdot, q) - K^+(\cdot, q), K^+(\cdot, q))_{\mathcal{R}(L)}$$
$$= 0,$$

by using the property (ii) for $K(\cdot, q)$ and $K^+(\cdot, q)$. Hence, we have the desired result

$$K(p, q) = K^+(p, q) \quad \text{on} \quad E \times E.$$

Finally, (4) is valid if and only if $\mathcal{H} = [N(L)]^\perp$ and then, for $f = L\mathbf{f}$

$$\|f\|_{\mathcal{R}(L)} = \|\mathbf{f}\|_{\mathcal{H}}.$$

In Theorem 1, the properties (i) and (ii) of the function $K(p, q)$ will be called the '*reproducing property*' of $K(p, q)$ in (or for) the Hilbert space $\mathcal{R}(L)$, and then the kernel $K(p, q)$ is called '*reproducing kernel*'. A Hilbert space admitting a reproducing kernel will be called a '*reproducing kernel Hilbert space*' (RKHS).

For Theorem 1 itself, see [Schw], [D1], [P], [NW1] and [Sa4], [Sa6] and [[Sa]], pages 84–85 for detailed comments.

Theorem 1 itself will not become a fundamental theorem of linear transforms. In order to realize Theorem 1 as a fundamental theorem of linear transforms we will need the idea of reproducing kernel Hilbert spaces. As we shall see in the next section, since a reproducing kernel Hilbert space is uniquely determined by the reproducing kernel $K(p, q)$, conversely, we shall write it by H_K; that is,

$$\mathcal{R}(L) = H_K$$

in Theorem 1. We shall consider Theorem 1 as follows: The range $\mathcal{R}(L)$ of L for $\mathcal{H}$ forms precisely the Hilbert space H_K admitting the reproducing kernel $K(p, q)$ defined by (2) and the RKHS H_K admits an intuitively determined inner product $(\cdot, \cdot)_{H_K}$ for the members of H_K which are functions on E, apart from the linear transform (1) and, of course, apart from the space $\mathcal{H}$. Then, we will be able to realize Theorem 1 as a fundamental theorem of linear transforms. Then, Theorem 1 will be stated in the form

Theorem 2 *The images $f(p)$ of the linear transform (1) for $\mathcal{H}$ form precisely the functional Hilbert space H_K admitting the reproducing kernel $K(p, q)$ defined by (2)*

which is uniquely determined by the reproducing kernel $K(p,q)$. Then, we have the inequality

$$\|f\|_{H_K} \leqq \|\mathbf{f}\|_{\mathcal{H}}. \tag{8}$$

Furthermore, for any $f \in H_K$, there exists a uniquely determined member $\mathbf{f}^*$ of $\mathcal{H}$ such that

$$f(p) = (\mathbf{f}^*, \mathbf{h}(p))_{\mathcal{H}} \quad on \quad E$$

and

$$\|f\|_{H_K} = \|\mathbf{f}^*\|_{\mathcal{H}}. \tag{9}$$

In our fundamental Theorem 2, the crucial point is that the image space $\mathcal{R}(L)$ is determined and is characterized as a RKHS H_K which is uniquely determined by the reproducing kernel $K(p,q)$ defined by (2). Therefore, in Theorem 2, it is very important to realize the RKHS H_K concretely. So, we will need some techniques to realize the RKHS H_K from $K(p,q)$ and miscellaneous general properties for reproducing kernel Hilbert spaces.

When we investigate the linear transform (1) in the framework of Hilbert spaces, we will naturally need the theory of reproducing kernel Hilbert spaces. In this viewpoint, it seems that the theory of reproducing kernels is fundamental in mathematics.

Let $\{\mathbf{v}_j\}$ be a complete orthonormal basis for $\mathcal{H}$. Then, for

$$v_j(p) = (\mathbf{v}_j, \mathbf{h}(p))_{\mathcal{H}}, \tag{10}$$

we have

$$\mathbf{h}(p) = \sum_j (\mathbf{h}(p), \mathbf{v}_j)_{\mathcal{H}} \mathbf{v}_j = \sum_j \overline{v_j(p)} \mathbf{v}_j.$$

Hence, by setting

$$\overline{\mathbf{h}}(p) = \sum_j v_j(p) \mathbf{v}_j,$$

$$\overline{\mathbf{h}}(\cdot) = \sum_j v_j(\cdot) \mathbf{v}_j.$$

Thus, we shall define

$$(f, \overline{\mathbf{h}}(p))_{H_K} = \sum_j (f, v_j)_{H_K} \mathbf{v}_j.$$

Then, we have

Theorem 3   *We assume that for $f \in H_K$*

$$(f, \overline{\mathbf{h}})_{H_K} \in \mathcal{H} \tag{11}$$

and for all $p \in E$,

$$(f, (\mathbf{h}(p), \mathbf{h}(\cdot))_{\mathcal{H}})_{H_K} = ((f, \overline{\mathbf{h}})_{H_K}, \mathbf{h}(p))_{\mathcal{H}}. \tag{12}$$

Then, we have

$$\|f\|_{H_K} \leqq \|(f, \overline{\mathbf{h}})_{H_K}\|_{\mathcal{H}}. \tag{13}$$

If $\{\mathbf{h}(p); p \in E\}$ is complete in $\mathcal{H}$, then equality always holds.

Proof: For any $p \in E$, we have formally using the assumptions (11) and (12)

$$
\begin{aligned}
f(p) &= (f(\cdot), K(\cdot, p))_{H_K} \\
&= (f(\cdot), (\mathbf{h}(p), \mathbf{h}(\cdot))_{\mathcal{H}})_{H_K} \\
&= ((f, \overline{\mathbf{h}})_{H_K}, \mathbf{h}(p))_{\mathcal{H}}.
\end{aligned}
$$

This impies that

$$f = L(f, \overline{\mathbf{h}})_{H_K}.$$

Hence, by Theorem 2 we have the desired results.

2   Inversion formula for linear transforms

We shall establish the inversion formula for the linear transform (1). Formally, the inversion formula will be given by using the RKHS H_K in the following way, as we see from the proof of Theorem 3.

Theorem 4   *We assume (11), (12) and*

$$(\mathbf{f}_0, (f, \overline{\mathbf{h}})_{H_K})_{\mathcal{H}} = ((\mathbf{f}_0, \overline{\mathbf{h}})_{\mathcal{H}}, f)_{H_K} \quad for \quad \mathbf{f}_0 \in N(L). \tag{14}$$

Then, we have, for $\mathbf{f}^*$ in Theorem 2

$$\mathbf{f}^* = (f, \overline{\mathbf{h}})_{H_K}. \tag{15}$$

In Theorem 4, the assumption (11) will, however, generally not be valid, in particular, in the case of infinite dimensional Hilbert spaces H_K. In order to establish inversion formulas which are widely applicable in analysis, we shall assume that the norms in $\mathcal{H}$ and H_K are realized in terms of σ-finite positive measures dm and $d\mu$ on measurable spaces T and E in the following ways:

$$\mathcal{H} = L_2(T, dm), \quad H_K \subset L_2(E, d\mu).$$

24

So, we shall consider the integral transform

$$f(p) = \int_T F(t)\overline{h(t,p)}dm(t) \tag{16}$$

where $h(t,p)$ is a complex-valued function on $T \times E$ with

$$h(\cdot,p) \in L_2(T,dm) \quad for \quad p \in E \tag{17}$$

and for the functions F satisfying

$$F \in L_2(T,dm). \tag{18}$$

The corresponding reproducing kernel is

$$K(p,q) = \int_T h(t,q)\overline{h(t,p)}dm(t) \quad on \quad E \times E. \tag{19}$$

The RKHS H_K is isometrically contained in $L_2(E,d\mu)$. Under these situations we shall give a polished version by [[By]] of Theorem VI.7.3 in [[Sa]]

Theorem 5 *Suppose that there exists an exhaustion $\{E_N\}_{N=1}^\infty$ of $d\mu$-measurable sets of E satisfying*

(a) $E_1 \subset E_2 \subset \cdots \subset \cdots$,
(b) $\bigcup_{N=1}^\infty E_N = E$,
(c) $\int_{E_N} K(p,p)d\mu(p) < \infty$.

For a member f of H_K, if

$$\int_{E_N} f(p)h(t,p)d\mu(p) \in L_2(T,dm) \quad for\ any \quad N,$$

then the sequence

$$\left\{ \int_{E_N} f(p)h(t,p)d\mu(p) \right\}_{N=1}^\infty$$

converges strongly to F^* in Theorem 2 for (16); that is,

$$F^*(t) = s - \lim_{N\to\infty} \int_{E_N} f(p)h(t,p)d\mu(p) \tag{20}$$

for the function F^* satisfying

$$f(p) = \int_T F^*(t)\overline{h(t,p)}dm(t), \tag{21}$$

and

$$\|f\|_{H_K} = \|F^*\|_{L_2(T,dm)}. \qquad (22)$$

Proof : The linear operator L defined by (16); that is,

$$(LF)(p) = (F, h(\cdot, p))_{L_2(T,dm)} \quad for \quad F \in L_2(T, dm),$$

is an isometry from $[N(L)]^{\perp}$ onto H_K. Let χ_{E_N} be the characteristic function of E_N, that is

$$\chi_{E_N}(p) = \begin{cases} 1 & \text{on} & E_N \\ 0 & \text{on} & E \backslash E_N. \end{cases}$$

Let L^* be the adjoint of L. Then, for any $G \in L_2(T, dm)$ we have

$$\begin{aligned} (L^*(f\chi_{E_N}), G)_{L_2(T,dm)} &= (f\chi_{E_N}, LG)_{L_2(E,d\mu)} \\ &= \int_{E_N} f(p)\overline{(LG)(p)}d\mu(p) \\ &= \int_{E_N} f(p)d\mu(p) \int_T \overline{G(t)}h(t,p)dm(t) \\ &= \int_T \overline{G(t)}dm(t) \int_{E_N} f(p)h(t,p)d\mu(p) \\ &= \left(\int_{E_N} f(p)h(t,p)d\mu(p), G(t) \right)_{L_2(T,dm)}, \end{aligned}$$

by using Fubini's theorem.

Indeed, from the Schwarz inequality, we have

$$\begin{aligned} \int_{E_N} \int_T |G(t)||f(p)||h(t,p)|d\mu(p)dm(t) \\ \leqq \int_{E_N} \|G\|_{L_2(T,dm)} K(p,p)^{\frac{1}{2}}|f(p)|d\mu(p) \\ \leqq \|G\|_{L_2(T,dm)} \left\{ \int_{E_N} K(p,p)d\mu(p) \right\}^{\frac{1}{2}} \|f\|_{L_2(E,d\mu)} \\ < \infty. \end{aligned}$$

Hence, we obtain

$$L^*(f\chi_{E_N})(t) = \int_{E_N} f(p)h(t,p)d\mu(p).$$

26

Since the sequence $\{f\chi_{E_N}\}_{N=1}^{\infty}$ converges to f in $L_2(E, d\mu)$ and $L^*f = F^*$, we have the desired result.

In miscellaneous concrete inversion problems, the assumptions in Theorem 5 will be automatically satisfied (see the following Chapter 3) and so the inversion formula in Theorem 5 will be widely applicable in analysis.

Meanwhile, in terms of a complete orthonormal system $\{\mathbf{v}_j\}$ of $\mathcal{H}$, we have

Theorem 6   *In Theorem 2, we have the inversion formula*

$$\mathbf{f}^* = \sum_j (f(\cdot), (\mathbf{v}_j, \mathbf{h}(\cdot))_{\mathcal{H}})_{H_K} \mathbf{v}_j.$$

Proof :   For the adjoint L^* of the isometry L between $N(L)^{\perp}$ and H_K, $L^{-1} = L^*$. Hence, from Parseval's identity we obtain

$$L^{-1}f = \mathbf{f}^* = \sum_j (\mathbf{f}^*, \mathbf{v}_j)_{\mathcal{H}} \mathbf{v}_j$$

$$= \sum_j (L^*f, \mathbf{v}_j)_{\mathcal{H}} \mathbf{v}_j = \sum_j (f, L\mathbf{v}_j)_{H_K} \mathbf{v}_j$$

$$= \sum_j (f, (\mathbf{v}_j, \mathbf{h}(\cdot))_{\mathcal{H}})_{H_K} \mathbf{v}_j.$$

Compare this with the inversion formula in terms of a singular system for the mapping L from $\mathcal{H}$ onto H_K. See, for example, [[Gr]].

For inverses, regularization, discretization and approximation of linear operators in reproducing kernel Hilbert spaces, see also [NW1-3] and [W1-4].

3   Determination of the linear system

In Theorem 2, conversely by using an isometrical mapping $\widetilde{L}$ from a Hilbert space H_K admitting a reproducing kernel $K(p, q)$ on E onto a Hilbert space $\mathcal{H}$ and by using the reproducing kernel $K(p, q)$, we can determine the linear system function $\mathbf{h}(p)$ which is a Hilbert $\mathcal{H}$-valued function on E as follows:

$$\mathbf{g}_{\widetilde{L}}(q) = \widetilde{L}K(\cdot, q), \tag{23}$$

which is called the *'generating vector'* of $\widetilde{L}$. See [SS] and [Bu] for many concrete examples. Then, we have

$$K(p, q) = (K(\cdot, q), K(\cdot, p))_{H_K}$$

$$= (\widetilde{L}K(\cdot, q), \widetilde{L}K(\cdot, p))_{\mathcal{H}}$$

$$= (\mathbf{g}_{\widetilde{L}}(q), \mathbf{g}_{\widetilde{L}}(p))_{\mathcal{H}}. \tag{24}$$

27

Then, we obtain

Theorem 7   *For the linear mapping*

$$f(p) = (\mathbf{f}, \mathbf{g}_{\widetilde{L}}(p))_{\mathcal{H}}, \quad for \quad \mathbf{f} \in \mathcal{H}, \tag{25}$$

we have the isometrical identity

$$\|f\|_{H_K} = \|\mathbf{f}\|_{\mathcal{H}}. \tag{26}$$

Furthermore, the linear mapping (25) gives the inverse for the isometry $\widetilde{L}$ and the family of vectors $\{\mathbf{g}_{\widetilde{L}}(p); p \in E\}$ is complete in $\mathcal{H}$.

Proof :   We note that in (25), $\widetilde{L}f = \mathbf{f}$ for any $\mathbf{f} \in \mathcal{H}$. Indeed, since $\widetilde{L}$ is an isometry, for $\widetilde{L}\tilde{f} = \mathbf{f}$ we have

$$\begin{aligned}
f(p) &= (\mathbf{f}, \mathbf{g}_{\widetilde{L}}(p))_{\mathcal{H}} \\
&= (\widetilde{L}\tilde{f}, \widetilde{L}K(\cdot, p))_{\mathcal{H}} \\
&= (\tilde{f}, K(\cdot, p))_{H_K} \\
&= \tilde{f}(p).
\end{aligned}$$

Hence, $f = \tilde{f}$ and so (25) gives the inverse of the isometry $\widetilde{L}$. Of course, we have (26) and we see that $\{\mathbf{g}_{\widetilde{L}}(p); p \in E\}$ is complete in $\mathcal{H}$.

4   Identification of the images of linear transforms in terms of the adjoint transforms

By Theorem 2, we see that the linear transform (1) maps from $\mathcal{H}$ onto the RKHS H_K precisely. However, we know frequently that the linear transform (1) maps from $\mathcal{H}$ into some Hilbert space H comprising functions on E. Then, for a given $f \in H$ we wish to know whether $f \in H_K$ or not. Of course, if the RKHS H_K is realized reasonably, then there is no problem in this. However, even if we do not know the RKHS H_K, there exists the following simple theorem which characterizes the members of H_K in terms of the adjoint operator L^*. The theorem is powerful, in particular, for L^2 solvability of the $\bar{\partial}$-equation. See, [Hö], [[Hö]] and [Oh2].

Theorem 8   *For any $f \in H$,*

$$f \in H_K \quad with \quad L\mathbf{f} = f \quad and \quad \|\mathbf{f}\|_{\mathcal{H}} \leqq c$$

if and only if

$$|(g, f)_H| \leqq c \|L^* g\|_{\mathcal{H}} \quad for \ all \quad g \in Dom L^*.$$

Proof :   If the inequality holds, the linear map

$$L^* g \longrightarrow (g, f)_H$$

is well defined and bounded. Hence, by the Riesz representation theorem there exists $\mathbf{f} \in Dom L$ such that

$$(L^* g, \mathbf{f})_{\mathcal{H}} = (g, f)_H, \quad \|\mathbf{f}\|_{\mathcal{H}} \leq c,$$

which implies the desired result $L\mathbf{f} = f$. The other implication is trivial.

Of course, for the mapping L in Theorem 8, it is sufficient to assume that L is a densely defined closed linear operator from $\mathcal{H}$ into H.

5   Linear transforms by smooth functions

For a general linear transform (1) in the framework of Hilbert spaces we shall consider (1) for some smooth functions $\mathbf{f}$. Here, for 'some smooth' functions we shall consider them as members of some reproducing kernel Hilbert space.

For a typical example, compare the usual Hilbert space $L_2(a, b)$ which does not admit a reproducing kernel with the Sobolev Hilbert space $H^1(a, b)$ comprising absolutely continuous functions on (a, b) and equipped with the inner product

$$(f, g)_{H^1(a,b)} = \int_a^b f(x)\overline{g(x)}dx + \int_a^b f'(x)\overline{g'(x)}dx,$$

admitting a reproducing kernel.

For various reproducing kernel Hilbert spaces, see §3.

Here, we shall examine the linear transform (1) for a reproducing kernel Hilbert space (as $\mathcal{H}$) and we wish to establish the corresponding isometrical identity and inversion formula.

We shall consider a reproducing kernel Hilbert space $H_K(E)$ on E as an input function space satisfying (i) and (ii) in Theorem 1. We shall consider a linear transform of $H_K(E)$ in the form

$$\hat{f}(\hat{p}) = (f(\cdot), h(\cdot, \hat{p}))_{H(E)}, \quad f \in H_K(E) \tag{27}$$

by some functional Hilbert space $H(E)$-valued function $h(p, \hat{p})$ on E from an abstract set $\hat{E} = \{\hat{p}\}$ containing the reproducing kernel Hilbert space $H_K(E)$ as members $H_K(E) \subset H(E)$. Note here that the linear transform (27) is, in general, of a type of different from the linear transform (1), since, in general, $H(E) \neq H_K(E)$.

For the linear transform (27) we shall give a general method obtaining a natural isometrical identity and inversion formula, based on [SY2].

Theorem 9   *For the reproducing kernel $K(p,q)$ we construct a Hilbert space $\mathcal{H}$-valued function $\mathbf{h}(p)$ satisfying (2). We assume that*

$$((\mathbf{f}, \mathbf{h}(p))_{\mathcal{H}}, h(p, \hat{p}))_{H(E)} = (\mathbf{f}, (h(p, \hat{p}), \overline{\mathbf{h}(p)})_{H(E)})_{\mathcal{H}}$$

$$\text{on} \quad E \times \hat{E} \quad \text{for} \quad \mathbf{f} \in \mathcal{H}, \tag{28}$$

and

$$\{\mathbf{h}(p), p \in E\} \quad \text{and} \quad \{h(p, \hat{p}); \hat{p} \in \hat{E}\} \tag{29}$$

are complete in $\mathcal{H}$ and $H(E)$, respectively.
   Then, we form the positive matrix

$$\hat{K}(\hat{p}, \hat{q}) = \Big((h(\cdot, \hat{q}), \overline{\mathbf{h}(\cdot)})_{H(E)}, (h(\cdot, \hat{p}), \overline{\mathbf{h}(\cdot)})_{H(E)}\Big)_{\mathcal{H}} \quad \text{on} \quad \hat{E} \times \hat{E}. \tag{30}$$

Then, in the linear transform from $\mathcal{H}$ onto the reproducing kernel Hilbert space $H_{\hat{K}}$ admitting the reproducing kernel (30) on $\hat{E}$

$$\hat{f}(\hat{p}) = (\mathbf{f}, (h(\cdot, \hat{p}), \overline{\mathbf{h}(\cdot)})_{H(E)})_{\mathcal{H}}, \tag{31}$$

we have the inversion formula in the form

$$\hat{f} \longrightarrow \mathbf{f} \tag{32}$$

in the space $\mathcal{H}$ by Theorem 4 or Theorem 5.
   Then, we obtain the isometrical identity

$$\|\hat{f}\|_{H_{\hat{K}}(\hat{E})} = \|f\|_{H_K(E)} \tag{33}$$

and the inversion formula for the mapping (27) from $H_K(E)$ onto $H_{\hat{K}}(\hat{E})$ using (32) and

$$f(p) = (\mathbf{f}, \mathbf{h}(p))_{\mathcal{H}}. \tag{34}$$

Proof :   By (1), (27) and (28), we have

$$\hat{f}(\hat{p}) = (f(\cdot), h(\cdot, \hat{p}))_{H(E)}$$
$$= ((\mathbf{f}, \mathbf{h}(\cdot))_{\mathcal{H}}, h(\cdot, \hat{p}))_{H(E)}$$
$$= \Big(\mathbf{f}, (h(\cdot, \hat{p}), \overline{\mathbf{h}(\cdot)})_{H(E)}\Big)_{\mathcal{H}}.$$

30

By the assumptions (29), the mapping from $\mathbf{f}$ to $\hat{f}$ is one to one and so we have the isometrical identity

$$\|\hat{f}\|_{H_{\hat{K}(\hat{E})}} = \|\mathbf{f}\|_{\mathcal{H}}.$$

Hence we have the desired isometrical identity (33)

$$\|\hat{f}\|_{H_{\hat{K}(\hat{E})}} = \|\mathbf{f}\|_{\mathcal{H}} = \|f\|_{H_K(E)}.$$

Since the linear mappings (1) and (27) are isometrical, respectively, we obtain the desired inversion formulas (32) and (34).

Note that the assumption (28) corresponds to Fubini's theorem on exchanging the order of integrals if the norms in the Hilbert spaces $H(E)$ and $\mathcal{H}$ are given in terms of some L_2 spaces on E and $\hat{E}$, respectively, and so the assumption will be a natural one.

Theorem 9 is very simple, but it will work for various concrete integral transforms for some smooth functions which are considered as members of suitable reproducing kernel Hilbert spaces. So Theorem 9 will create a new field connecting special functions with integral transforms and reproducing kernels. For some concrete applications, see [SY2].

6   Linear integro-differential equations and reproducing kernels

We shall formulate Volterra-type integral equations of the first kind in the following way:

$$\int_a^t F(\xi)h(\xi,t)d\xi = f(t). \tag{35}$$

We shall assume that

$$F \in L_2(a,\infty) \tag{36}$$

and

$$h(\cdot,t) \in L_2(a,\infty) \quad for \quad t > a, \tag{37}$$

in the framework of Hilbert spaces. Then, note that by using the characteristic function

$$\chi(\xi;(a,t)) = \begin{cases} 1 & on \quad (a,t) \\ 0 & on \quad t < \xi, \end{cases}$$

the integral equation (35) of Volterra type can be reduced to the integral equation (16) of Fredholm type as follows:

$$\int_a^\infty F(\xi)h(\xi,t)\chi(\xi;(a,t))d\xi = f(t). \tag{38}$$

31

As examples, see Theorem 3.3.4 and Theorem 3.3.10.

In general, integral equations of Volterra type can be transformed to those of Fredholm type by using the characteristic function as in (38) from the viewpoint of our fundamental theorems of linear transforms.

Next, we shall formulate Fredholm-type integral equations of the second kind in the form

$$F(t) + \int_T F(\xi)\overline{h(\xi, t)}dm(\xi) = f(t), \tag{39}$$

where we shall assume that F belongs to some reproducing kernel Hilbert space $H_{\mathbb{K}}$ on E and for the reproducing kernel $\mathbb{K}(t, t')$

$$\mathbb{K}(t, t') = \int_{\hat{E}} \hat{h}(\hat{\xi}, t')\overline{\hat{h}(\hat{\xi}, t)}d\hat{m}(\hat{\xi}), \quad \text{on} \quad E \times E \tag{40}$$

and

$$\left\{\hat{h}(\hat{\xi}, t); t \in E\right\} \quad \text{is complete in} \quad L_2(\hat{E}, d\hat{m}), \tag{41}$$

as in (19). Further we assume that

$$h(\cdot, t) \in L_2(T, dm) \quad \text{on} \quad E$$

and

$$H_{\mathbb{K}} \subset L_2(T, dm).$$

Then, any member $F \in H_{\mathbb{K}}$ is expressible in the form

$$F(t) = \int_{\hat{E}} \hat{F}(\hat{\xi})\overline{\hat{h}(\hat{\xi}, t)}d\hat{m}(\hat{\xi}) \tag{42}$$

and we have the isometrical identity

$$\|F\|_{H_{\mathbb{K}}}^2 = \int_{\hat{E}} |\hat{F}(\hat{\xi})|^2 d\hat{m}(\hat{\xi}). \tag{43}$$

Now, from (39) and (42) we have

$$\int_{\hat{E}} \hat{F}(\hat{\xi}) \left\{\overline{\hat{h}(\hat{\xi}, t)} + \int_T \overline{\hat{h}(\hat{\xi}, \xi)}\,\overline{h(\xi, t)}dm(\xi)\right\} d\hat{m}(\hat{\xi}) = f(t), \tag{44}$$

where we assume that the order of integrals is exchangeable. If

$$\int_T \mathbb{K}(t, t)dm(t) < \infty,$$

then this assumption will be satisfied by Fubini's theorem.

We thus see that following this procedure integral equations of the second kind can be transformed to those of the first kind. The circumstances are similar for integral equations of Fredholm type of the third kind. We shall state the procedure in the following more general integro-differential equations.

We shall formulate integro-differential equations as follows. For real intervals T and E,

$$a_0(t)F(t) + a_1(t)F'(t) + \cdots + a_n(t)F^{(n)}(t)$$
$$+ \int_T F(\xi)\overline{h(\xi,t)}dm(\xi) = f(t), \quad \text{on} \quad E \qquad (45)$$

where

$$h(\cdot,t) \in L_2(T,dm) \quad \text{for} \quad t \in E,$$

and $\{a_j\}_{j=0}^n$ are *arbitrary* complex-valued functions on E.

From the form (45), we shall assume that F belongs to some reproducing kernel Hilbert space $H_{\mathbb{K}}$ on E satisfying (40), (41) (and so, satisfying (42) and (43)). From (45), we shall assume furthermore that

$$\frac{\partial^{j+j'}\mathbb{K}(t,t')}{\partial t^j \partial t'^{j'}} \quad (j,j' = 0,1,2,\cdots,n) \quad \text{on} \quad E \times E \qquad (46)$$

are continuously differentiable. Then, we see that any member F of $H_{\mathbb{K}}$ belongs to the C^n-class and we have the expression

$$F^{(j)}(t) = \int_{\hat{E}} \hat{F}(\hat{\xi})\overline{\frac{\partial^j}{\partial t^j}\hat{h}(\hat{\xi},t)}d\hat{m}(\hat{\xi}), \quad \text{on} \quad E. \qquad (47)$$

See Theorem 2.9.

From (45) and (47) we have

$$\int_{\hat{E}} \hat{F}(\hat{\xi})\left\{a_0(t)\overline{\hat{h}(\hat{\xi},t)} + \cdots + a_n(t)\overline{\partial_t^n \hat{h}(\hat{\xi},t)}\right.$$
$$+ \left.\int_T \overline{\hat{h}(\hat{\xi},\xi)}\ \overline{h(\xi,t)}dm(\xi)\right\}d\hat{m}(\hat{\xi}) = f(t), \quad \text{on} \quad E. \qquad (48)$$

Here we assume that the order of the integrals is exchangeable. Again, if

$$\int_E \mathbb{K}(t,t)dm(t) < \infty,$$

it is assured by Fubini's theorem. Our procedure implies that the integro-differential equation (45) can be transformed to the Fredholm-type integral equation of the first kind ([Sa32]).

The difficulty of solving the integro-differential equation (45) with variable co-efficients will be transformed to that of the complicated form in the integral kernel in (48). In the integral equation (48) of the first kind we can apply also various numerical constructions of the solutions. See [[Gr]], [[TA]], [NW1-3], [Stg1-2] and [W1-2].

§2   Elementary properties of reproducing kernel Hilbert spaces

As shown in Section 1, when we consider linear transforms in the framework of Hilbert spaces, we naturally have the idea of reproducing kernel Hilbert spaces. Therefore, we shall review the basic properties of reproducing kernel Hilbert spaces. Basic references are [Ar1], [Schw] and [Kre]. See also [[M]] as a book for the theory of reproducing kernel Hilbert spaces.

We will not be able to refer to reproducing kernel Hilbert spaces for vector-valued functions. See, for example, [AD], [ABDS1-2], [[Bru]], [[BuM]] and [[Dy]]. See [GH] and its references for Hilbert function modules admitting reproducing kernels.

1   Positive matrices and reproducing kernel Hilbert spaces

Let H be a (possibly finite-dimensional) Hilbert space composed of complex-valued functions on a set E admitting a reproducing kernel $K(p, q)$ on E; that is,

$$\text{for any fixed} \quad q \in E, K(p, q) \quad \text{belongs to} \quad H \quad \text{as a function in} \quad p, \qquad (1)$$

and

$$\text{for any fixed} \quad q \in E \quad \text{and for any member} \quad f \quad \text{of} \quad H,$$
$$f(q) = (f(\cdot), K(\cdot, q))_H. \qquad (2)$$

As shown in Theorem 1.1, the reproducing kernel $K(p, q)$ having the properties (1) and (2) is uniquely determined by the Hilbert space H.

Theorem 1   *For a Hilbert space H comprising functions $f(p)$ on E, there exists a reproducing kernel $K(p, q)$ for H if and only if for any point q of E, the point evaluation $f(q)$ is a bounded linear functional on H.*

Proof :   If there exists a reproducing kernel $K(p, q)$ for H satisfying (1) and (2), then from (2) we have, for any $q \in E$

$$|f(q)| \leqq \|f\|_H \|K(\cdot, q)\|_H = \|f\|_H K(q, q)^{\frac{1}{2}};$$

34

that is, $f(q)$ is a bounded linear functional on H.

Conversely, we assume that $f(q)$ is a bounded linear functional on H. Then, by the Riesz theorem, there exists a uniquely determined member K_q of H satisfying

$$f(q) = (f, K_q)_H \quad \text{for} \quad f \in H. \tag{3}$$

Since K_q is a function on E as a member of H, we can represent it as $K_q(p)$ for $p \in E$. Then, the function

$$K(p, q) = K_q(p) = (K_q, K_p)_H \quad \text{on} \quad E \times E \tag{4}$$

is the desired reproducing kernel for H having the properties (1) and (2).

A complex-valued function $k(p, q)$ on $E \times E$ is called a '*positive matrix*' on E if for any complex-valued function X_p on E which vanishes except for a finite number of points of E

$$\sum_{p,q} \overline{X_p} X_q k(p, q) \geq 0. \tag{5}$$

Note that for a positive matrix $k(p, q)$ on E, we have the elementary properties:

$$k(p, p) \geq 0 \quad \text{for} \quad p \in E, \tag{6}$$
$$k(p, q) = \overline{k(q, p)} \quad \text{on} \quad E \times E, \tag{7}$$

and

$$|k(p, q)|^2 \leq k(p, p) k(q, q) \quad \text{on} \quad E \times E. \tag{8}$$

For any reproducing kernel $K(p, q)$ in Theorem 1, we see from the expression (4) that $K(p, q)$ is a positive matrix on E, because

$$\sum_{p,q} \overline{X_p} X_q K(p, q)$$

$$= \left(\sum_q X_q K_q, \sum_p X_p K_p \right)_H$$

$$= \left\| \sum_p X_p K_p \right\|_H^2 \geq 0. \tag{9}$$

Conversely we have the fundamental theorem of Moore–Aronszajn ([Ar1]):

Theorem 2 *For any positive matrix $K(p, q)$ on E, there exists a uniquely determined functional Hilbert space H_K admitting the reproducing kernel $K(p, q)$.*

Proof :   We shall consider the functions on E defined by

$$K_p(\cdot) = K(\cdot, p) \quad \text{for any} \quad p \in E \tag{10}$$

and the linear span H_0 by (10). We introduce an inner product in H_0 by

$$\left(\sum_q Y_q K_q, \sum_p X_p K_p \right)_{H_0} = \sum_{p,q} \overline{X_p} Y_q K(p, q). \tag{11}$$

This inner product is well defined, and so H_0 forms a pre-Hilbert space, as we see from (9).

Note that for the pre-Hilbert space H_0, we have the reproducing property (i) and (ii) in Theorem 1. Hence, every point evaluation is a bounded linear functional on H_0 as in the proof in Theorem 1. We now consider the completion of H_0 by including the limiting functions of Cauchy sequences in H_0. Then, the completion H is realized as a space of functions on E and the reproducing property of $K(p, q)$ holds for the completion H, for the sake of the reproducing property of $K(p, q)$ for H_0. Hence, the completion H is the desired reproducing kernel Hilbert space.

The uniqueness of the reproducing kernel Hilbert space H is deduced from the fact that the family (10) of functions is complete in any functional Hilbert space admitting the reproducing kernel $K(p, q)$.

For a complete proof of Theorem 2, see [Ar1], [[M]] or ([[Sa]], pages 6–10).

Note the very interesting fact that for any positive matrix $K(p, q)$ on E we can construct a Hilbert space $\mathcal{H}$ and a Hilbert space valued function $\mathbf{h}(p)$ from E into $\mathcal{H}$ satisfying (1.2), by using an n dimensional complex Gaussian probability distribution $\mu^{p_1 \cdots p_n}$ with mean value zero and covariance matrix $\|K(p_j, p_{j'})\|$ and by using Kolmogorov's theorem on measures. See [[PS]].

If we apply this fact, then the fundamental Theorem 2 can be deduced from Theorem 2.1 or Theorem 2.2.

Following Theorem 2, we shall denote the reproducing kernel Hilbert space (RKHS)H by H_K because H is uniquely determined by the positive matrix (or the reproducing kernel) $K(p, q)$.

Theorem 3   *A sequence of functions $\{f_n\}$ converging to a function f in H_K converges also to f in the ordinary sense as the functions on E. Further, this convergence is uniform on every subset on which $K(p, p)$ is bounded.*

Proof :   The results follow directly from

$$
\begin{aligned}
|f(q) - f_n(q)| &= |(f - f_n, K_q)_{H_K}| \\
&\leqq \|f - f_n\|_{H_K} \|K_q\|_{H_K} \\
&= \|f - f_n\|_{H_K} K(q, q)^{\frac{1}{2}}.
\end{aligned}
$$

This nice property will show why Hilbert spaces admitting reproducing kernels are used in, in particular, approximation theory. See, for example, [[Da]], [[Sh]], [NW1-3], [W1-4] and [KW].

The following two theorems will be clear.

Theorem 4   *If a RKHS H_K is a subspace of a Hilbert space $\mathcal{H}$, then*

$$f(p) = (f(\cdot), K(\cdot, p))_{\mathcal{H}}$$

gives a projection from $\mathcal{H}$ onto H_K.

Conversely, we have

Theorem 5   *For any subspace H_0 of a RKHS H_K, there exists the reproducing kernel $K_{H_0}(p, q)$ for H_0 and it is given by*

$$K_{H_0}(p, q) = (P_{H_0} K_q, K_p)_{H_K}$$

for the orthogonal projection P_{H_0} from H_K onto H_0.

For two positive matrices $K^{(1)}(p, q)$ and $K^{(2)}(p, q)$ on E, if $K^{(2)}(p, q) - K^{(1)}(p, q)$ is a positive matrix on E, then we shall write it as follows:

$$K^{(1)} << K^{(2)}.$$

If $K^{(1)} << K^{(2)}$ and $K^{(2)} << K^{(1)}$, then we have $K^{(1)}(p, q) = K^{(2)}(p, q)$ on $E \times E$.

Indeed, then, we have, in particular, $K^{(1)}(p, p) \leq K^{(2)}(p, p)$ and $K^{(2)}(p, p) \leq K^{(1)}(p, p)$, and so $K^{(1)}(p, p) = K^{(2)}(p, p)$ on E. Then, since $K^{(2)} - K^{(1)}$ is a positive matrix on E, we have, for any $p, q \in E$,

$$|K^{(2)}(p, q) - K^{(1)}(p, q)|^2$$
$$\leq [K^{(2)}(p, p) - K^{(1)}(p, p)][K^{(2)}(q, q) - K^{(1)}(q, q)]$$
$$= 0,$$

that is, $K^{(1)}(p, q) = K^{(2)}(p, q)$ on $E \times E$. We thus see that the symboll $<<$ is a partial ordering in the class of all positive matrices on E.

Theorem 6   *For two positive matrices $K^{(1)}(p, q)$ and $K^{(2)}(p, q)$ on E,*

$$H_{K^{(1)}} \subset H_{K^{(2)}} \quad \text{(as members)} \tag{12}$$

if and only if there exists a positive constant γ such that

$$K^{(1)}(p, q) << \gamma^2 K^{(2)}(p, q). \tag{13}$$

37

Here, the minimum of such constants γ coincides with the norm of the inclusion map J from $H_{K^{(1)}}$ into $H_{K^{(2)}}$.

Proof :   First we shall show that the inclusion map J becomes a closed operator. Suppose that

$$f_n \longrightarrow f \quad \text{in} \quad H_{K^{(1)}} \quad \text{and} \quad f_n \longrightarrow g \quad \text{in} \quad H_{K^{(2)}}.$$

Then, by Theorem 3 we have the desired result $f = g$ on E. Hence, by the closed graph theorem, we see that J is continuous.

For any $f \in H_{K^{(1)}}$ and $q \in E$,

$$(Jf, K_q^{(2)})_{H_{K^{(2)}}} = (Jf)(q) = f(q) = (f, K_q^{(1)})_{H_{K^{(1)}}}.$$

Hence, we have

$$J^* K_q^{(2)} = K_q^{(1)} \quad \text{for} \quad q \in E.$$

Then, we have, for any $\gamma \geqq \|J\|$

$$\left\| \sum_q X_q K_q^{(1)} \right\|_{H_{K^{(1)}}}^2 = \left\| J^* \sum_q X_q K_q^{(2)} \right\|_{H_{K^{(1)}}}^2 \leqq \gamma^2 \left\| \sum_q X_q K_q^{(2)} \right\|_{H_{K^{(2)}}}^2,$$

which implies (13).

Next, note that

$$\|f\|_{H_{K^{(j)}}}^2 = \sup_{\{X_q\}} \frac{|\sum_q X_q f(q)|^2}{\sum_{p,q} \overline{X_p} X_q K^{(j)}(p, q)},$$

as we see from the Schwarz inequality and from (11). Hence, from (13) we have

$$\|f\|_{H_{K^{(2)}}} \leqq \gamma \|f\|_{H_{K^{(1)}}} \quad \text{for} \quad f \in H_{K^{(1)}}$$

and so (12) and $\|J\| \leqq \gamma$.

For a Hilbert space H comprising functions on E, if we know the existence of the reproducing kernel for H by Theorem 1, then we are basically interested in the construction of the reproducing kernel. For this fundamental problem, we have

Theorem 7   *For any Hilbert space $\mathcal{H}$ and for a complete orthonormal system $\{v_j\}$ for $\mathcal{H}$, we consider the linear transform from $\mathcal{H}$ into the complex numbers $\{f_j\}_j$ defined by*

$$f_j = (\mathbf{f}, \mathbf{v}_j)_{\mathcal{H}} \quad \text{for} \quad \mathbf{f} \in \mathcal{H}. \tag{14}$$

38

Then, we have the Parseval identity

$$\sum_j |f_j|^2 = \|\mathbf{f}\|_{\mathcal{H}}^2. \tag{15}$$

If a Hilbert space H comprising functions on E admits a reproducing kernel, then the reproducing kernel $K(p,q)$ is represented by a complete orthonormal system $\{v_j(p)\}_j$ in the form

$$K(p,q) = \sum_j v_j(p)\overline{v_j(q)} \tag{16}$$

which converges absolutely on $E \times E$.

Proof :   In the linear transform (14), the corresponding reproducing kernel is the Kronecker delta on $\mathbb{N} \times \mathbb{N}$, for the set $\mathbb{N}$ of natural numbers,

$$\delta_{jj'} = (\mathbf{v}_{j'}, \mathbf{v}_j)_{\mathcal{H}}. \tag{17}$$

The Kronecker delta is apparently the reproducing kernel for the usual l^2 space as we see from its reproducing property in the form $(\mathbf{c}, \delta_j) = c_j$, for any $\mathbf{c} = \{c_j\} \in l^2$ and for $\delta_j = (0, 0, \cdots, 0, 1, 0, \cdots)$.

Hence, from Theorem 1.2 we have, immediately, the Parseval identity.

Next, for $\mathbf{v}_j = v_j(p)$ we have, for any $p \in E$

$$(K_p, v_j)_{H_K} = \overline{(v_j, K_p)_{H_K}} = \overline{v_j(p)} \in l^2.$$

Hence, we have, from the Parseval identity

$$\begin{aligned} K(p,q) &= (K_q, K_p)_{H_K} \\ &= \sum_j (K_q, v_j)_{H_K} \overline{(K_p, v_j)_{H_K}} \\ &= \sum_j v_j(p)\overline{v_j(q)} \end{aligned}$$

which converges, of course, absolutely on $E \times E$.

In Theorem 7, a complete orthonormal system for H_K can be constructed by the Gram–Schmidt orthogonalization procedure from some independent complete system for H_K. In this sense, reproducing kernels are computable. One fundamental idea of Bergman–Schiffer will be that reproducing kernels are computable and so representations in terms of reproducing kernels are valuable. The typical example was the construction of the Riemann mapping function. See Appendix 1, **4.1.** Therefore, many important domain functions which are defined on the domain were expressed by the classical Bergman reproducing kernels, which will be introduced in Section 4.6. See also [[Ber]] and [[Ne]] for the details, in particular, for the Bergman and the Szegö reproducing kernels and their miscellaneous applications to complex analysis.

2 Smoothness of reproducing kernel Hilbert spaces

In order to examine some smoothness properties of a reproducing kernel $K(p,q)$ on E and its RKHS H_K, we will need a topology on E. When we need a topology on E, we shall, for simplicity, consider E as a domain on a Euclidean space. We shall use the expression

$$K(p,q) = (\mathbf{h}(q), \mathbf{h}(p))_{\mathcal{H}} \quad \text{on} \quad E \times E \tag{18}$$

in terms of a Hilbert space $\mathcal{H}$-valued function $\mathbf{h}(p)$ on E appeared in Theorem 1.2. Note that some similarity exists between this expression (18) and the covariance for a time series $\{X(t); t \in T\}$

$$R(t_1, t_2) = Cov[X(t_1)\overline{X(t_2)}].$$

See, for example, [Kre], ([[Lo]], Chapter XI) and ([[Pa]], Chapter 9).
    First, from the identity

$$\|\mathbf{h}(p) - \mathbf{h}(q)\|_{\mathcal{H}}^2 = K(p,p) - 2Re K(p,q) + K(q,q),$$

we have

Theorem 8 *If*

(i) $K(p,p)$ is continuous on E
and
(ii) for any fixed $p \in E$, $Re K(p,q)$ is continuous at $q = p$,

or more simply, if $Re K(p,q)$ is continuous on the diagonal $p = q$ of $E \times E$, then $K(p,q)$ is continuous on $E \times E$ and $\mathbf{h}(p)$ is a continuous $\mathcal{H}$-valued function on E in the strong topology of $\mathcal{H}$. In particular, all the members of H_K are continuous functions on E.

Corollary 1 *If $K(p,q)$ is continuous on $E \times E$, then for any expression*

$$K(p,q) = \sum_j g_j(p)\overline{g_j(q)} \tag{19}$$

which converges absolutely on $E \times E$, all the functions $g_j(p)$ are continuous on E, and the convergence of (19) is uniform on $E_0 \times E_0$, for any compact subset E_0 of E.

Proof : In the expression (19), consider l^2 as $\mathcal{H}$ in Theorem 8. Then, we have the first statement. The uniform convergence can be deduced from the expression

$$K(p,p) = \sum_j |g_j(p)|^2 \quad \text{on} \quad E$$

and from Dini's theorem.

Next, we shall consider the differentiabiliy of the reproducing kernel $K(p,q)$ and of the members of H_K. For this purpose, for simplicity, we shall assume that $E = I$, an open interval on $\mathbb{R}$. Then, we have

Theorem 9 *If $K(p,q)$ has continuous derivatives*

$$K_{mn}(p,q) = \frac{\partial^{m+n} K(p,q)}{\partial p^m \partial q^n} \quad (m,n = 0,1,\cdots,r) \quad on \quad I \times I, \qquad (20)$$

then $\mathbf{h}(p)$ has continuous strong derivatives in $\mathcal{H}$

$$\mathbf{h}^{(m)}(p) = \frac{d^m \mathbf{h}(p)}{dp^m} \quad (m = 0,1,\cdots,r) \quad on \quad I, \qquad (21)$$

and for any expression (19) on $I \times I$, all the functions $g_j(p)$ have continuous derivatives

$$g_j^{(m)}(p) \quad (m = 0,1,\cdots,r) \quad on \quad I$$

and we have

$$K_{mn}(p,q) = \sum_j g_j^{(m)}(p)\overline{g_j^{(n)}(q)} \quad on \quad I \times I \qquad (22)$$

where the convergence is uniform on $I_0 \times I_0$, for any compact subset I_0 of I.

Proof : The proof will now be almost routine. Note, for example, for $m = 1, n = 0$, the identity

$$\left(\mathbf{h}(q), \frac{\mathbf{h}(p+\varepsilon) - \mathbf{h}(p)}{\varepsilon} \right)_{\mathcal{H}} = \frac{K(p+\varepsilon,q) - K(p,q)}{\varepsilon}$$

and repeat this procedure until $m, n = r$.

Next, we shall examine the analyticity of the members of RKHS H_K. We assume that $K(p,q)$ is analytic at the point $p = q = p_0$; that is,

$$\sum_{m,n=0}^{\infty} \frac{K_{mn}(p_0,p_0)}{m!n!}(z - p_0)^m(\zeta - p_0)^n \qquad (23)$$

41

converges on some four-dimensional bicylinder, for complex variables z and ζ

$$|z - p_0| < R, \quad |\zeta - p_0| < R. \tag{24}$$

Then, for any positive $r(< R)$ and for a positive $r_0(< R - r)$, there exists a constant M such that

$$|K_{mn}(p, q)| \leqq m!n! \frac{M^{m+n}}{r^{m+n}} \quad \text{for} \quad |r - p_0| < r_0 \tag{25}$$

$$(m, n = 0, 1, 2, \cdots).$$

Hence,

$$\|\mathbf{h}^{(n)}(p)\|_{\mathcal{H}} \leqq K_{nn}(p, p)^{\frac{1}{2}} \leqq n! \frac{M^n}{r^n} \quad \text{for} \quad |r - p_0| < r_0$$

and so, in

$$f(p) = (\mathbf{f}, \mathbf{h}(p))_{\mathcal{H}}, \mathbf{f} \in \mathcal{H} \tag{26}$$

we have

$$|f^{(n)}(p)| \leqq n! \frac{M^n}{r^n} \|\mathbf{f}\|_{\mathcal{H}} \quad \text{for} \quad |r - p_0| < r_0 \tag{27}$$

$$(n = 0, 1, 2, \cdots).$$

Hence, $f(p)$ is analytic on $|z - p_0| < R$.

Now, for any $z(|z - p_0| < R)$ and for $C_r = \{|z - p_0| = r\}$, we have, for $|z - p_0| < r < R$

$$f(z) = (\mathbf{f}, \mathbf{h}(z))_{\mathcal{H}} = \frac{1}{2\pi i} \int_{C_r} \frac{(\mathbf{f}, \mathbf{h}(\zeta))_{\mathcal{H}}}{\zeta - z} d\zeta$$

$$= \left(\mathbf{f}, \frac{1}{2\pi i} \int_{C_r} \frac{\mathbf{h}(\zeta)}{\zeta - z} d\zeta \right)_{\mathcal{H}};$$

that is,

$$\mathbf{h}(z) = \frac{1}{2\pi i} \int_{C_r} \frac{\mathbf{h}(\zeta)}{\zeta - z} d\zeta. \tag{28}$$

Therefore, we have

Theorem 10   *If $K(p, q)$ is analytic in the bicylinder (24), then the $\mathcal{H}$-valued function $\mathbf{h}(p)$ is analytic on $|z - p_0| < R$. Further, all the members of H_K are also analytic on $|z - p_0| < R$.*

Furthermore, from the expression (18), we see that if $K(p, q)$ is analytic on the bicylinder (24) and a bicylinder

$$|z - q_0| < R, \quad |\zeta - q_0| < R_2$$

42

then $K(p, q)$ is analytic on the bicylinder

$$|z - p| < R, \quad |\zeta - q| < R_2,$$

for any point (p, q) in the bicylinder.

Next, we shall consider a region D on the complex plane as E. Then, we must consider the analytic positive matrix $K(p, q)$ on D in the form $K(z, \overline{u})$ on $D \times \overline{D}$, for $\overline{D} = \{\overline{z}; z \in D\}$. Note that if $K(z, u)$ is not identically constant on D, then $K(z, u)$ is not an Hermitian positive matrix on D.

§3 Generations of reproducing kernel Hilbert spaces

For any positive matrix $K(p, q)$ on E, there exists a uniquely determined RKHS H_K admitting the reproducing kernel $K(p, q)$. We can, in principle, consider the relationship among the reproducing kernel Hilbert spaces induced from the relationship among positive matrices. In this section, we shall examine such relationships induced from restrictions, sums, products and transforms of positive matrices. Furthermore, we shall construct the reproducing kernels for Hilbert spaces induced from some larger norms and smaller norms for a RKHS H_K, respectively.

1 Restrictions, sums and products of reproducing kernels

In order to examine the construction of reproducing kernel Hilbert spaces, we need the idea of the generation of reproducing kernel Hilbert spaces from restrictions, sums and products of reproducing kernels. For complete and detailed arguments, see also [Ar1]. For more general formulations, see [Schw]. For compact arguments for these, we shall apply here the arguments by [[An]] in the constructions of the generated reproducing kernel Hilbert spaces based on Theorem 1.2.

For a positive matrix $K(p, q)$ on E and for any subset E_1 of E, the restriction $K^{(1)}(p, q) = K(p, q)|_{E_1}$ to $E_1 \times E_1$ is, of course, a positive matrix on E_1. Hence, for the restriction $K^{(1)}(p, q)$ to E_1, there exists a RKHS $H_{K^{(1)}}$ admitting the reproducing kernel $K^{(1)}(p, q)$. The interrelationship between $H_{K^{(1)}}$ and H_K is given by

Theorem 1 *Any member $f^{(1)}$ of $H_{K^{(1)}}$ is the restriction $f^{(1)} = f|_{E_1}$ of a member f of H_K and its norm in $H_{K^{(1)}}$ is given by*

$$\|f^{(1)}\|_{H_{K^{(1)}}} = \min\left\{ \|f\|_{H_K}; f|_{E_1} = f^{(1)}, f \in H_K \right\}.$$

43

Proof : By using the reproducing property we have

$$(K_q, K_p)_{H_K} = K(p,q) = K^{(1)}(p,q)$$
$$= (K_q^{(1)}, K_p^{(1)})_{H_{K^{(1)}}} \quad \text{for} \quad p, q \in E_1.$$

Hence, the mapping

$$S : K_p \longrightarrow K_p^{(1)} \quad (p \in E_1)$$

is uniquely extended to an isometry from the closed linear span H_0 of $\{K_p; p \in E_1\}$ onto the closed linear span $\{K_p^{(1)}; p \in E_1\}$, that is, the RKHS $H_{K^{(1)}}$. For the orthogonal projection P_{H_0} of H_K onto H_0, we define the mapping $T = SP_{H_0}$. Since $(TK_q)(p) = K_q^{(1)}(p) = K_q(p)$ for $p, q \in E_1$,

$$(Tf)(p) = f(p) \quad \text{for} \quad f \in H_0 \quad \text{and for} \quad p \in E_1$$

and

$$(Tf)(p) = 0 = (f, K_p)_{H_K} = f(p)$$
$$\text{for} \quad f \in H_0^{\perp} \quad \text{and} \quad p \in E_1;$$

that is, T is just the restriction map of H_K to E_1 and

$$(Tf)(p) = (Tf, K_p^{(1)})_{H_{K^{(1)}}}$$
$$= (f, T^* K_p^{(1)})_{H_K} \quad \text{for} \quad p \in E_1. \tag{1}$$

By Theorem 1.2, the reproducing kernel which determines the images of (1) for H_K is, by using the isometry of S,

$$(T^* K_q^{(1)}, T^* K_p^{(1)})_{H_K} = K^{(1)}(p,q).$$

Hence, by Theorem 1.2 we have the desired result.

Next, for two positive matrices $K^{(1)}(p,q)$ and $K^{(2)}(p,q)$ on E, their usual sum as functions on E

$$K(p,q) = K^{(1)}(p,q) + K^{(2)}(p,q)$$

is also a positive matrix. The interrelationship among the corresponding reproducing kernel Hilbert spaces H_K, $H_{K^{(1)}}$ and $H_{K^{(2)}}$ is given by

Theorem 2 *Any member f of H_K is expressible in the form*

$$f(p) = f^{(1)}(p) + f^{(2)}(p) \quad on \quad E$$

44

for some function $f^{(1)} \in H_{K^{(1)}}$ and for some function $f^{(2)} \in H_{K^{(2)}}$, and its norm in H_K is given by

$$\|f\|_{H_K}^2 = \min \{\|f^{(1)}\|_{H_{K^{(1)}}}^2 + \|f^{(2)}\|_{H_{K^{(2)}}}^2;$$
$$f(p) = f^{(1)}(p) + f^{(2)}(p) \quad on \quad E, f^{(1)} \in H_{K^{(1)}}, f^{(2)} \in H_{K^{(2)}}\}.$$

Proof :   We shall consider the usual direct sum $H = H_{K^{(1)}} \oplus H_{K^{(2)}}$ of the Hilbert spaces, and the mapping

$$h_p = K_p^{(1)} \oplus K_p^{(2)}$$

from E into H. Then, we have

$$K(p, q) = (K_q^{(1)}, K_p^{(1)})_{H_{K^{(1)}}} + (K_q^{(2)}, K_p^{(2)})_{H_{K^{(2)}}} = (h_q, h_p)_H.$$

In the linear transform

$$L(f^{(1)} \oplus f^{(2)})(p) = (f^{(1)} \oplus f^{(2)}, h_p)_H \quad for \quad H,$$

we have

$$N(L) = \{f \oplus -f; f \in H_{K^{(1)}} \cap H_{K^{(2)}}\}.$$

Hence, from Theorem 1.2 we have the desired results.

The following two corollaries follow from Theorem 2 or Theorem 2.6.

Corollary 1   *For two positive matrices $K_1(p,q)$ and $K_2(p,q)$ on E, if $K_1 << K_2$ on E, then for the corresponding reproducing kernel Hilbert spaces H_{K1} and H_{K2}, we have*

$$H_{K1} \subset H_{K2} \quad (as\ members)$$

and

$$\|f\|_{H_{K1}} \geqq \|f\|_{H_{K2}} \quad for\ any \quad f \in H_{K1}.$$

Corollary 2   *A function f on E belongs to H_K if and only if there exists a constant γ such that*

$$f(p)\overline{f(q)} << \gamma^2 K(p,q) \quad on \quad E.$$

The minimum of all such γ coincides with $\|f\|_{H_K}$.

For two positive matrices $K^{(1)}(p_1, q_1)$ and $K^{(2)}(p_2, q_2)$ on E_1 and E_2, respectively, the usual product

$$K^{(1)}(p_1, q_1)K^{(2)}(p_2, q_2)$$

45

is the reproducing kernel for the tensor product

$$H = H_{K^{(1)}} \otimes H_{K^{(2)}},$$

because for $f^{(1)} \in H_{K^{(1)}}, f^{(2)} \in H_{K^{(2)}}$

$$(f^{(1)} \otimes f^{(2)})(q_1, q_2) = f^{(1)}(q_1)f^{(2)}(q_2) = \left(f^{(1)} \otimes f^{(2)}, K_{q_1}^{(1)} \otimes K_{q_2}^{(2)} \right)_H.$$

In particular, for two positive matrices $K^{(1)}(p,q)$ and $K^{(2)}(p,q)$ on E, the product

$$K(p,q) = K^{(1)}(p,q)K^{(2)}(p,q)$$

is the restriction to the diagonal set of $E \times E$. In particular, $K(p,q)$ is a positive matrix on E and by Theorem 1 we have the following

Theorem 3   *For the product $K(p,q) = K^{(1)}(p,q)K^{(2)}(p,q)$ on $E \times E$, the corresponding RKHS H_K consists of all the functions f on E which are expressible in the form*

$$f(p) = \sum_j f_j^{(1)}(p)f_j^{(2)}(p) \quad on \quad E \tag{2}$$

which converge absolutely for some functions

$$f_j^{(1)} \in H_{K^{(1)}}, f_j^{(2)} \in H_{K^{(2)}} \tag{3}$$

satisfying

$$\sum_j \|f_j^{(1)}\|_{H_{K^{(1)}}}^2 \|f_j^{(2)}\|_{H_{K^{(2)}}}^2 < \infty. \tag{4}$$

Furthermore, the norm in H_K is given by

$$\|f\|_{H_K}^2 = \min \sum_j \|f_j^{(1)}\|_{H_{K^{(1)}}}^2 \|f_j^{(2)}\|_{H_{K^{(2)}}}^2$$

where the minimum is taken over all the functions $\{f_j^{(1)}\}$ and $\{f_j^{(2)}\}$ satisfying (2), (3) and (4).

This theorem was used to derive our generalized isoperimetric inequality in Chapter 1, (1) and, furthermore, we shall use it to derive miscellaneous norm inequalities in some general nonlinear transforms which will be discussed in Appendix 2 (see also ([[Sa]]) and [Sa30]).

In particular, we have

Corollary 3 *For a RKHS H_K on E and for any nonvanishing complex-valued function $s(p)$ on E,*

$$K_s(p,q) = s(p)\overline{s(q)}K(p,q) \quad on \quad E \times E$$

is a reproducing kernel for the Hilbert space H_{K_s} comprising all the functions $f_s(p)$ on E which are expressible in the form

$$f_s(p) = f(p)s(p) \quad for \quad f \in H_K$$

and which is equipped with the inner product

$$(f_s, g_s)_{H_{K_s}} = \left(\frac{f_s}{s}, \frac{g_s}{s}\right)_{H_K}.$$

2 Transforms of reproducing kernels

We shall consider transforms of reproducing kernels and examine the relationships between the reproducing kernel Hilbert spaces admitting the reproducing kernels and the transformed reproducing kernels. The material is taken from [Sa25].

We shall start with the expression

$$K(p,q) = (\mathbf{h}(q), \mathbf{h}(p))_{\mathcal{H}} \quad on \quad E \times E \tag{5}$$

in terms of an $\mathcal{H}$-valued function $\mathbf{h}(p)$ on E. We further assume that

$$\{\mathbf{h}(p); p \in E\} \quad \text{is complete in} \quad \mathcal{H}; \tag{6}$$

and so we have, by Theorem 1.2, for the linear transform

$$f(p) = (\mathbf{f}, \mathbf{h}(p))_{\mathcal{H}}, \mathbf{f} \in \mathcal{H}, \tag{7}$$

$$\|f\|_{H_K} = \|\mathbf{f}\|_{\mathcal{H}}. \tag{8}$$

We first consider any operator A from $\mathcal{H}$ into $\mathcal{H}$ and form the positive matrix $K_A(p,q)$ on E defined by

$$K_A(p,q) = (A\mathbf{h}(q), A\mathbf{h}(p))_{\mathcal{H}} \quad on \quad E \times E. \tag{9}$$

We shall consider the linear transform induced by the operator

$$f_A(p) = (\mathbf{f}, A\mathbf{h}(p))_{\mathcal{H}} \quad on \quad E \tag{10}$$

47

from $\mathcal{H}$ onto the RKHS H_{K_A}. Then, we shall define a linear transform

$$A : f \longrightarrow f_A \tag{11}$$

from H_K onto H_{K_A}. For this transform, we have

Theorem 4   *In the linear transform (11) we have*

$$\|f\|_{H_K} \geqq \|f_A\|_{H_{K_A}}. \tag{12}$$

Isometry holds here if and only if the mapping (11) is one to one; that is,

$$\{A\mathbf{h}(p); p \in E\} \quad \text{is complete in} \quad \mathcal{H}. \tag{13}$$

In particular, we have

$$K(q, q) \geqq \|K_A(\cdot, q)\|^2_{H_K} \quad on \quad E. \tag{14}$$

Proof :   In (10), we have, by Theorem 1.2

$$\|f_A\|_{H_{K_A}} \leqq \|\mathbf{f}\|_{\mathcal{H}}$$

and so, from (8) we have (12).

If A is one to one, then we have (13) and so (12) with the equality. The converse is also true.

Theorem 4 is very simple, but we can apply it to many kinds of operators on $\mathcal{H}$, for example, using Theorem 2.9. This theorem may be stated as the *decreasing principle* for operators in reproducing kernel Hilbert spaces.

We shall consider the inversion formula for the mapping A in (11). For this purpose, for any $f_A \in H_{K_A}$, we take the function $f^* \in H_K$ such that

$$A f^* = f_A \quad \text{and} \quad \|f^*\|_{H_K} = \|f_A\|_{H_{K_A}} \tag{15}$$

in Theorem 4. Then, f^* is determined by

$$f_A(p) = (\mathbf{f}^*, A\mathbf{h}(p))_{\mathcal{H}}, \quad \mathbf{f}^* \in [N(A)]^{\perp} \tag{16}$$

and

$$f^*(p) = (\mathbf{f}^*, \mathbf{h}(p))_{\mathcal{H}}. \tag{17}$$

48

Here $[N(A)]^\perp$ denotes the orthogonal complement in $\mathcal{H}$ of the null space of A defined by the mapping (10).

Then, we have, by using the isometry of (10) from $\mathcal{H}$ onto H_{K_A},

$$f^*(p) = (\mathbf{f}^*, \mathbf{h}(p))_{\mathcal{H}} = (f_A, AK_p)_{H_{K_A}}. \qquad (18)$$

We thus obtain

Theorem 5 *In the linear mapping (15) we have the inversion formula (18).*

Next, we shall consider any mapping $p = \phi(\hat{p})$

$$\phi : \widehat{E} \longrightarrow E$$

from an abstract set $\widehat{E}$ into an abstract set E and the family $f(\phi(\hat{p}))$ of functions on $\widehat{E}$. Then, we shall define the positive matrix defined by

$$K_\phi(\hat{p}, \hat{q}) = (\mathbf{h}(\phi(\hat{q})), \ \mathbf{h}(\phi(\hat{p})))_{\mathcal{H}} \quad \text{on} \quad \widehat{E} \times \widehat{E}$$

and the linear mapping

$$f(\phi(\hat{p})) = (\mathbf{f}, \mathbf{h}(\phi(\hat{p})))_{\mathcal{H}}, \quad \mathbf{f} \in \mathcal{H} \qquad (19)$$

in (7). Then we obtain, as in Theorem 4

Theorem 6 *In (7) and (19) we consider the mapping ϕ defined by*

$$\phi : f \longrightarrow f(\phi) \qquad (20)$$

from H_K onto H_{K_ϕ}. Then, we obtain

$$\|f\|_{H_K} \geqq \|f(\phi)\|_{H_{K_\phi}}. \qquad (21)$$

Isometry holds here if and only if $f(p) = 0$ on $\phi(\widehat{E})$ implies $f(p) = 0$ on E.
In particular, we obtain

$$K(q, q) \geqq \|K(\phi(\cdot), q)\|^2_{H_{K_\phi}} \quad \text{on} \quad \widehat{E}.$$

In Theorem 6, we note that if $p = \phi(\hat{p})$ is a mapping from $\widehat{E}$ onto E, the RKHS H_{K_ϕ} can be regarded as a faithful representation of H_K on $\widehat{E}$, even when ϕ is not one to one. See [Sa25] for a concrete example.

49

Theorem 6 will be stated as the *decreasing principle* for pull backs of reproducing kernel Hilbert spaces.

As in Theorem 5, we obtain the inversion formula for the mapping ϕ in (20)

Theorem 7   *In the mapping ϕ in (20), for any $\hat{f} \in H_{H_\phi}$ we take the inverse $f^* \in H_K$ satisfying*

$$f^*(\phi) = \hat{f} \quad and \quad \|\hat{f}\|_{H_{K_\phi}} = \|f^*\|_{H_K}. \tag{22}$$

Then, we have the inversion formula

$$f^*(p) = \left(\hat{f}(\cdot), K(\phi(\cdot), p) \right)_{H_{K_\phi}}. \tag{23}$$

In Appendix 1, as an application of Theorem 7 we shall establish a general principle representing of the inverse ϕ^{-1} in terms of ϕ and we shall give concrete expressions.

3   Squeezes and balloons of a RKHS

Let T denote a bounded linear operator from a RKHS H_K on E into a Hilbert space $\mathcal{H}$. Then, by introducing the inner products

$$(f, g)_{H_K(\mathcal{H})} = (f, g)_{H_K} + (Tf, Tg)_{\mathcal{H}} \tag{24}$$

and

$$(f, g)_{H_K^-(\mathcal{H})} = (f, g)_{H_K} - (Tf, Tg)_{\mathcal{H}}, \tag{25}$$

we shall construct the reproducing kernels for the Hilbert spaces $H_K(\mathcal{H})$ and $H_K^-(\mathcal{H})$ such that

$$H_K(\mathcal{H}) \subset H_K \subset H_K^-(\mathcal{H}) \quad \text{(as subspaces)}, \tag{26}$$

in some constructive ways. The material is taken from [AS].

First we have the elementary results directly. For Lemma 3, use Theorem 2.6.

Lemma 1   *For the linear mapping from $\mathcal{H}$ into the functions on E defined by*

$$f(p) = (\mathbf{f}, TK_p)_{\mathcal{H}} \quad for \quad \mathbf{f} \in \mathcal{H}, \tag{27}$$

we have

$$f = T^* \mathbf{f} \tag{28}$$

50

and

$$\|f\|_{H_K}^2 = (\mathbf{f}, T_p(\mathbf{f}, TK_p)_{\mathcal{H}})_{\mathcal{H}}. \qquad (29)$$

Lemma 2 *The following items are equivalent:*

(i) $\quad K(p,q) >> (TK_q, TK_p)_{\mathcal{H}},$

(ii) $\quad \|T\| \leqq 1$

and

(iii) $\quad \|f\|_{H_K} \leqq \|\mathbf{f}\|_{\mathcal{H}} \quad in \quad (27).$

In the linear mapping (27), for the positive matrix

$$\mathbb{K}(p,q) = (TK_q, TK_p)_{\mathcal{H}} \quad \text{on} \quad E \times E,$$

we obtain, by Theorem 1.2

$$\|f\|_{H_{\mathbb{K}}} \leqq \|\mathbf{f}\|_{\mathcal{H}}.$$

Hence, if an item in Lemma 2 is valid, then we have the sharp inequalities in (iii)

$$\|f\|_{H_K} \leqq \|f\|_{H_{\mathbb{K}}} \leqq \|\mathbf{f}\|_{\mathcal{H}}. \qquad (30)$$

Lemma 3 *The mapping $f = T^*\mathbf{f}$ from $\mathcal{H}$ into H_K is onto H_K if and only if there exists a positive constant γ such that*

$$K(p,q) << \gamma^2 \quad \mathbb{K}(p,q) \quad on \quad E.$$

By definition, we note that for $f \in H_K(\mathcal{H})$

$$\|f\|_{H_K(\mathcal{H})} \geqq \|f\|_{H_K} \qquad (31)$$

and so there exists a reproducing kernel $K_{\mathcal{H}}(p,q)$ for the space $H_K(\mathcal{H})$ such that

$$K_{\mathcal{H}}(p,q) << K(p,q) \quad \text{on} \quad E \qquad (32)$$

by Corollary 1. Note that here

$$H_K(\mathcal{H}) = H_K \quad \text{(as sets of functions)}.$$

For a construction of the reproducing kernel $K_{\mathcal{H}}(p,q)$, we obtain

51

Theorem 8   *The reproducing kernel* $K_\mathcal{H}(p, q)$ *is characterized as the solution* $\tilde{K}(p, q)$ *of the functional equation*

$$K(p, q) = \tilde{K}(p, q) + (T\tilde{K}_q, TK_p)_\mathcal{H} \tag{33}$$

satisfying

$$\tilde{K}_q = \tilde{K}(\cdot, q) \in H_K \quad for \quad q \in E. \tag{34}$$

Proof :   We have directly: for any $q \in E$,

$$(K_q, K_p)_{H_K} = (\tilde{K}_q, K_p)_{H_K} + (T\tilde{K}_q, TK_p)_\mathcal{H} \quad \text{for all} \quad p \in E$$

$\Longleftrightarrow$

$$K_q = \tilde{K}_q + T^* T\tilde{K}_q$$

$\Longleftrightarrow$

$$(f, K_q)_{H_K} = (f, \tilde{K}_q)_{H_K} + (f, T^* T\tilde{K}_q)_{H_K} \quad \text{for all} \quad f \in H_K$$

$\Longleftrightarrow$

$$f(q) = (f, \tilde{K}_q)_{H_K(\mathcal{H})} \quad \text{for all} \quad f \in H_K.$$

If, for any $f \in H_K$

$$\|f\|_{H_K} \geqq \|Tf\|_\mathcal{H} \tag{35}$$

and equality holds if and only if $f = 0$, then we can introduce the pre-Hilbert space H' equipped with the inner product

$$(f, g)_{H'} = (f, g)_{H_K} - (Tf, Tg)_\mathcal{H}. \tag{36}$$

Note the following fact which was used in the proof of the fundamental Theorem 2.2.

In order that there exists a functional completion of H' if and only if

(a)   for any fixed $p \in E$, the point evaluation $f(p)$ is bounded on H'
and

(b)   for any Cauchy sequence $\{f_n\}$ in H', the condition $f_n(p) \to 0$ on E implies $\|f_n\|_{H'} \to 0$.

If the functional completion is possible, it is unique and admits the reproducing kernel.

We shall denote the functional completion of H' by $H_K^-(\mathcal{H})$, if the conditions (a) and (b) are satisfied. Then, for any $f \in H_K$

$$\|f\|_{H_K^-(\mathcal{H})} \leqq \|f\|_{H_K} \tag{37}$$

and so H_K is a subspace of $H_K^-(\mathcal{H})$. We shall denote the reproducing kernel for $H_K^-(\mathcal{H})$ by $K_{\mathcal{H}}^-(p,q)$. For this kernel, we obtain a similar result by exchanging $+$ and $-$ in (33).

We shall give typical concrete examples of the reproducing kernels $K_{\mathcal{H}}(p,q)$ and $K_{\mathcal{H}}^-(p,q)$ in Section 4.16.

§4 Examples of reproducing kernels

Since all the positive matrices are reproducing kernels for some uniquely determined functional Hilbert spaces, we have many concrete reproducing kernels. In those positive matrices we will, however, be interested in and we will need essentially in Theorem 1.2 the realizations (characterizations and norms) of the Hilbert spaces admitting the reproducing kernels. We shall consider some general methods for the realizations in Section 5. In this section, we shall give typical and concrete reproducing kernels and their reproducing kernel Hilbert spaces. These examples will show the importance and fruitful world of the theory of reproducing kernels. At the same time, concrete reproducing kernels will be valuable, because reproducing kernels have the fundamental information for the reproducing kernel Hilbert spaces. See also Appendix 1 and Appendix 2 for the importance of concrete reproducing kernels.

1 One dimensional space

The simplest reproducing kernel is the constant

$$1 \tag{1}$$

for the complex space $\mathbb{C}$ equipped with the usual inner product

$$(\alpha, \beta)_{\mathbb{C}} = \alpha\bar{\beta}. \tag{2}$$

Any positive number

$$\varpi(> 0) \tag{3}$$

is the reproducing kernel on $\mathbb{C}(\varpi)$ equipped with the inner product

$$(\alpha, \beta)_{\mathbb{C}(\varpi)} = \frac{1}{\varpi}\alpha\bar{\beta}. \tag{4}$$

53

The results are, of course, similarly valid on the real space $\mathbb{R}$.

For any abstract set E and for any complex-valued function $f(p)$ ($|f(p)| < \infty$) on E, the function

$$f(p)\overline{f(q)} \quad \text{on} \quad E \times E \tag{5}$$

is the reproducing kernel for the one-dimensional space generated by $f(p)$; that is, $\{\alpha f(p); \alpha \in \mathbb{C}\}$ and equipped with the inner product

$$(\alpha f(p), \beta f(p)) = \alpha\overline{\beta}. \tag{6}$$

Conversely, Theorem 2.7 shows that any reproducing kernel $K(p, q)$ for a separable Hilbert space comprising functions on E is expressible in the form

$$K(p, q) = \sum_j f_j(p)\overline{f_j(q)}$$

which converges absolutely on $E \times E$.

2 Finite dimensional spaces

For any n-dimensional space H_n comprising complex-valued functions on a set E, let $\{W_j(p)\}_{j=1}^n$ be any basis on H_n. Then, any member $f(p)$ of H_n is uniquely expressible in the form

$$f(p) = \sum_{j=1}^n c_j W_j(p) \quad \text{on} \quad E, \quad c_j \in \mathbb{C}.$$

The most general metric in H_n will be introduced by a positive definite Hermitian matrix $\|\alpha_{jk}\|$ in the form

$$\|f\|_{H_n}^2 = \sum_{j,j'=1}^n \alpha_{jj'} c_j \overline{c_{j'}}. \tag{7}$$

Then, we have

$$\alpha_{jj'} = (W_j, W_{j'})_{H_n};$$

that is, $\|\alpha_{jj'}\|$ is the Gram matrix of the system $\{W_j\}_{j=1}^n$.

We take the complex conjugate inverse $\|\beta_{jj'}\|$ of $\|\alpha_{jj'}\|$; that is,

$$\sum_{j=1}^n \alpha_{\nu j} \overline{\beta_{j\mu}} = \delta_{\nu\mu}, \quad \text{the Kronecker} \quad \delta.$$

54

Then, the function

$$K(p,q) = \sum_{j,j'=1}^{n} \beta_{jj'} W_j(p)\overline{W_{j'}(q)} \quad \text{on} \quad E \times E \tag{8}$$

is the reproducing kernel for the space H_n equipped with the metric (7).

In particular, for any positive definite $n \times n$ Hermitian matrix $A = \|a_{\nu\mu}\|$,

$$K(\nu,\mu) = a_{\nu\mu} \tag{9}$$

is the reproducing kernel for the vector space $\mathbb{C}^n$ equipped with the inner product

$$(\mathbf{x}, \mathbf{y})_{\mathbb{C}^n(A)} = \mathbf{y}^* \widetilde{A} \mathbf{x}, \tag{10}$$

where $*$ denotes complex conjugate transpose and $\widetilde{A}$ is the complex conjugate inverse of A.

In this setting, we can connect the theory of reproducing kernels to that of positive definite Hermitian matrices. See [Sa3] and ([[Sa]], pages 12–14; 104–105).

For systematic applications of finite dimensional reproducing kernels to a constructive theory of approximations to multivariate functions by polynomial restrictions, see [[Re]]. There, we find several concrete reproducing kernels in terms of special functions.

3   On the half line

Let H denote the Hilbert space composed of all complex-valued functions on $[0,\infty)$ such that

$$f(x) \quad \text{are absolutely continuous on} \quad [0,\infty),$$
$$f(0) = 0,$$
$$f'(x) \in L_2(0,\infty)$$

and equipped with the inner product

$$(f,g) = \int_0^\infty f'(x)\overline{g'(x)}dx. \tag{11}$$

Then, we have the reproducing kernel, as we can see directly:

$$K(x,y) = \min(x,y) = \frac{2}{\pi}\int_0^\infty \frac{\sin xt \sin yt}{t^2}dt. \tag{12}$$

A general form of (12) is given, for a positive continuous function ρ on (a,b)

55

$(-\infty \leq a < b \leq \infty)$, by

$$K_\rho(x, y) = \int_a^{\min(x,y)} \rho(t)dt. \tag{13}$$

Then, $K_\rho(x, y)$ is the reproducing kernel for H_{K_ρ} comprising all complex-valued functions $f(x)$ on $[a, b)$ such that

$$f(x) \quad \text{are absolutely continuous on} \quad [a, b),$$
$$f(a) = 0,$$
$$f'(x) \in L_2((a, b), \rho(x)^{-1}dx),$$

and equipped with the inner product

$$(f, g)_{H_{K_\rho}} = \int_a^b f'(x)\overline{g'(x)}\frac{dx}{\rho(x)}. \tag{14}$$

Furthermore, note that

$$K_\rho(x, y)^r \quad (r > 1) \tag{15}$$

is also the reproducing kernel for the Hilbert space $H_{(K_\rho)^r}$ composed of all complex-valued functions $f(x)$ on $[a, b)$ such that

$$f(x) \quad \text{are absolutely continuous on} \quad [a, b),$$
$$f(a) = 0,$$
$$f'(x) \in L_2\left((a, b), \left(\int_a^x \rho(t)dt\right)^{1-r}\rho(x)dx\right),$$

and equipped with the inner product

$$(f, g)_{H_{(K_\rho)^r}} = \frac{1}{r}\int_a^b f'(x)\overline{g'(x)}\frac{dx}{(\int_a^x \rho(t)dt)^{r-1}\rho(x)} \tag{16}$$

([[Sa]]). See also Theorem 3.3 for two reproducing kernels (13) and (15).

As a modification of (12), we see that

$$K_a(x, y) = (1 - a) + \min(x, y) \tag{17}$$

is the reproducing kernel of the Hilbert space H_{K_a} composed of all complex-valued functions $f(x)$ on $[a, b)$ $(-\infty < a < b \leq \infty)$ such that

$$f(x) \quad \text{are absolutely continuous on} \quad [a, b),$$
$$f'(x) \in L_2(a, b),$$

and equipped with the inner product

$$(f,g)_{HK_a} = f(a)\overline{g(a)} + \int_a^b f'(x)\overline{g'(x)}dx. \tag{18}$$

The positive matrix on $(0, \infty)$

$$\int_0^\infty \frac{\sin xt \sin yt}{t^2+1}dt = \frac{\pi}{4}\left(e^{-|x-y|} - e^{-x}e^{-y}\right) \tag{19}$$

$$\left(\text{resp. } \int_0^\infty \frac{\cos xt \cos yt}{t^2+1}dt = \frac{\pi}{4}\left(e^{-|x-y|} + e^{-x}e^{-y}\right)\right) \tag{20}$$

is the reproducing kernel for the Sobolev Hilbert space

$$\|f\|^2 = \frac{1}{\pi}\int_0^\infty (|f'(x)|^2 + |f(x)|^2)dx \tag{21}$$

composed of all odd (resp. even) functions $f(x)$ with respect to the origin such that

$$f(x) \quad \text{are absolutely continuous on} \quad (-\infty, \infty),$$
$$f'(x) \in L_2(-\infty, \infty),$$

as we can see directly, using Fourier integrals.

4   On the whole line

Recall the identity

$$\lim_{L\to\infty} \frac{1}{2L}\int_{-L}^L e^{iyt}e^{-ixt}dt = \delta(x-y) = \begin{cases} 0 & \text{for} \quad x \neq y \\ 1 & \text{for} \quad x = y. \end{cases} \tag{22}$$

Then, the function $\delta(x-y)$ is the reproducing kernel for the Hilbert space AP comprising all almost periodic functions on $(-\infty, \infty)$ with finite norms

$$\left\{\lim_{L\to\infty} \frac{1}{2L}\int_{-L}^L |f(x)|^2dx\right\}^{\frac{1}{2}} < \infty. \tag{23}$$

For this space, see, for example, ([[AG]], page 29). This reproducing kernel $\delta(x-y)$ will be considered as a continuous version of the Kronecker $\delta_{jj'}$ stated in Theorem 2.4.

On the Sobolev Hilbert space $H^1(\mathbb{R}; a, b)$ on $(-\infty, \infty)$ comprising all complex-valued and absolutely continuous functions $f(x)$ with finite norms

$$\left\{ \int_{-\infty}^{\infty} (a^2 |f'(x)|^2 + b^2 |f(x)|^2) dx \right\}^{\frac{1}{2}} \quad (a, b > 0), \tag{24}$$

the function

$$G_{a,b}(x, y) = \frac{1}{2ab} e^{-\frac{b}{a}|x-y|} = \frac{1}{2\pi} \int_{-\infty}^{\infty} \frac{e^{i\xi(x-y)}}{a^2 \xi^2 + b^2} d\xi \tag{25}$$

is the reproducing kernel. See [Sa24] for an application of this fact to norm inequalities.

Similarly, for the Sobolev Hilbert space on $(-\infty, \infty)$ with finite norms

$$\left\{ \int_{-\infty}^{\infty} (|f''(x)|^2 + 2|f'(x)|^2 + |f(x)|^2) dx \right\}^{\frac{1}{2}} < \infty, \tag{26}$$

we see that using Fourier integrals

$$\frac{1}{4} e^{-|x-y|} \{1 + |x - y|\} = \frac{1}{2\pi} \int_{-\infty}^{\infty} \frac{e^{i\xi(x-y)}}{(1 + \xi^2)^2} d\xi \tag{27}$$

is the reproducing kernel for the space.

In general, the relationship among differential equations, Green's functions, Sobolev Hilbert spaces and reproducing kernels will be given as follows.

We consider the family $L_2^{(n)}(a, b)$ of real-valued (for simplicity) functions $F(x)$ on $[a, b)$ $(-\infty \leqq a < b \leqq \infty)$ which are expressible in the form

$$F(x) = \int_a^x dx \overset{n-times}{\cdots} \int_a^x f(x) dx$$

$$\text{for} \quad f \in L_2(a, b).$$

We introduce the inner product in $L_2^{(n)}(a, b)$ $(n \geqq 1)$

$$(F_1, F_2) = \int_a^b \left\{ \sum_{j=0}^{n} a_j(x) F_1^{(j)}(x) F_2^{(j)}(x) \right\} dx;$$

$$a_0(x) > 0, a_n(x) > 0 \quad \text{and} \quad a_j(x) \geqq 0 \quad \text{for} \quad j = 1, 2, \cdots, n - 1. \tag{28}$$

Then, the Euler–Lagrange equation is given by

$$\sum_{j=0}^{n} (-1)^j (a_j(x) F^{(j)}(x))^{(j)} = 0. \tag{29}$$

58

Clearly, the Green's function $G(x, y)$ satisfying the $2n$ boundary conditions

$$\sum_{j=m}^{n} (-1)^j (a_j(x) F^{(j)}(x))^{(j)} = 0 \quad \text{at} \quad x = a \quad \text{and} \quad b,$$

$$\text{for} \quad m = 1, 2, \cdots, n, \tag{30}$$

is the reproducing kernel for the Hilbert space $L_2^{(n)}(a, b)$ equipped with the inner product (28) ([[Da]], pages 322–323).

Meanwhile,

$$K_n(x, \xi; \alpha) = \sum_{j=0}^{n-1} \frac{(x - \alpha)^j}{j!} \frac{(\xi - \alpha)^j}{j!} + \int_a^b g_n(x, t) g_n(\xi, t) dt; \tag{31}$$

$$g_n(x, t) = \frac{(x - t)^{n-1}}{(n - 1)!} \psi(\alpha, t, x)$$

for

$$\psi(\alpha, t, x) = \begin{cases} 1 & \alpha \leqq t < x \\ -1 & x \leqq t < \alpha, \\ 0 & otherwise \end{cases}$$

is the reproducing kernel for the Hilbert space $\mathcal{F}_n(a, b)$ equipped with the inner product

$$(f_1, f_2)_{\mathcal{F}_n(a,b)} = \sum_{j=0}^{n-1} f_1^{(j)}(\alpha) f_2^{(j)}(\alpha) + \int_a^b f_1^{(n)}(x) f_2^{(n)}(x) dx \tag{32}$$

([BL]).

In connection with these reproducing kernels, the following reproducing kernel will be very instructive.

For the Hilbert space comprising all real-valued and absolutely continuous functions on $[0, 1]$ whose derivatives are in $L_2(0, 1)$ and equipped with the inner product

$$(f_1, f_2) = \int_0^1 f_1(x) f_2(x) dx + \int_0^1 f_1'(x) f_2'(x) dx, \tag{33}$$

the reproducing kernel $k(x, y)$ is given by

$$k(x, y) = \begin{cases} a_y e^x + b_x e^{-x}, & x \leqq y \\ c_y e^x + d_y e^{-x}, & x \geqq y \end{cases} \tag{34}$$

59

where

$$a_y = b_y = \frac{e^y + e^2 e^{-y}}{2(e^2 - 1)},$$

$$c_y = \frac{e^y + e^{-y}}{e^2 - 1},$$

and

$$d_y = e^2 c_y,$$

([Ag]).

Much more general reproducing kernel Hilbert spaces were investigated by G. Kimeldorf and G. Wahba ([KW]) on the interval $[a, b]$ from the viewpoints of miscellaneous extremal problems, differential equations and Green's functions. They solved explicitly the generalized Birkhoff interpolation and smoothing problems involving Tchebycheffian spline functions.

5 Sobolev Hilbert spaces on $\mathbb{R}^n$

We set

$$G_\alpha(x - y) = (2\pi)^{-\frac{n}{2}} |x - y|^{\frac{2-n}{2}} \int_0^\infty \frac{\rho^{n/2}}{(1 + \rho^2)^{\alpha/2}} J_{\frac{n-2}{2}}(\rho|x - y|) d\rho$$

$$= \frac{1}{2^{\frac{n+\alpha-1}{2}} \pi^{\frac{n-1}{2}} \Gamma(\frac{n}{2}) \Gamma(\frac{n-\alpha+1}{2})}$$

$$\cdot e^{-|x-y|} \int_0^\infty e^{-|x-y|t} t^{\frac{n-\alpha-1}{2}} \left(1 + \frac{t}{2}\right)^{\frac{n-\alpha-1}{2}} dt. \qquad (35)$$

Then, for $\alpha > \frac{n}{2}, G_{2\alpha}(x - y)$ is the reproducing kernel for the Hilbert space equipped with the following norm $\|u\|_\alpha$ and is also the fundamental solution of the differential equation

$$(1 - \Delta)^\alpha u = 0 \quad \text{on} \quad \mathbb{R}^n. \qquad (36)$$

First,

$$\|u\|_0^2 = \int_{\mathbb{R}^n} |u|^2 dx,$$

for $0 < \alpha < 1$,

$$\|u\|_\alpha^2 = \frac{1}{C(n+1, \alpha)} \int_{-\infty}^\infty \int_{\mathbb{R}^n} \int_{\mathbb{R}^n} \frac{\left|e^{\frac{1}{2} i z_0} u(x) - e^{-\frac{1}{2} i z_0} u(y)\right|^2}{[|x - y|^2 + z_0^2]^{\frac{n+1+2\alpha}{2}}} dx dy dz_0;$$

60

if m is the greatest integer $\leq \alpha$, and, in general

$$\|u\|_\alpha^2 = \sum_{k=0}^{m} \binom{m}{k} \sum_{|j|=k} \|D_j u\|_{\alpha-m}^2. \tag{37}$$

Here,

$$C(n, \alpha) = \frac{2^{-2\alpha+1} \pi^{\frac{n+2}{2}}}{\Gamma(\alpha+1)\Gamma(\alpha+\frac{n}{2}) \sin \pi\alpha}$$

and

$$D_j u = \frac{\partial^m u}{\partial x_{j_1} \cdots \partial x_{j_m}}$$

for $j = (j_1, \cdots, j_m)$ $(1 \leq j_m \leq n)$ and $|j| = m$.

For the relationship among the Bessel potentials, the fundamental solutions for the operators $(1 - \Delta)^\alpha$ and the reproducing kernels, see [ArSm].

6   On the whole complex plane

For the Hilbert space $\mathcal{F}(A)$ $(A > 0)$ comprising all entire functions $f(z)$ with finite norms

$$\left\{ \frac{A}{\pi} \int\!\!\int_{\mathbf{C}} |f(z)|^2 e^{-A|z|^2} dx dy \right\}^{\frac{1}{2}} < \infty, \tag{38}$$

the function

$$e^{Az\overline{u}} \tag{39}$$

is the reproducing kernel. This space will be called the *Fischer space*, or the *Fock space*, or the *Bargmann–Fock space*. See, for example, [B1-2], [JPR], [Pe] and [Se1-2] for various applications.

As a much further generalization, for the Hilbert space comprising all entire functions $f(z)$ with finite norms

$$\|f\|_{\alpha,\beta}^2 = \frac{\beta^\alpha}{\pi\Gamma(\alpha)} \int\!\!\int_{\mathbf{C}} |f(z)|^2 |z|^{2(\alpha-1)} e^{-\beta|z|^2} dx dy, \tag{40}$$
$$(\alpha \geq 1, \beta > 0)$$

the function

$$k_{\alpha,\beta}(z, \overline{u}) = F_1(1; 2; \beta z\overline{u}) \tag{41}$$

61

is the reproducing kernel. Note that

$$k_{1,\beta}(z,\overline{u}) = e^{\beta z \overline{u}}, \tag{42}$$

$$k_{2,\beta}(z,\overline{u}) = (\beta z \overline{u})^{-1}(e^{\beta z \overline{u}} - 1), \tag{43}$$

and

$$k_{\alpha,\beta}(z,\overline{u}) = (\alpha - 1) \int_0^1 e^{\beta z \overline{u} t}(1-t)^{\alpha-\beta} dt \tag{44}$$

$$(\alpha > 1).$$

These facts will be confirmed by using the Taylor expansion, see [Bu].
In connection with these reproducing kernels, note that

$$\frac{b^2}{2\pi} \frac{\sinh b\sqrt{z\overline{u}}}{\sqrt{z\overline{u}}} \tag{45}$$

is the reproducing kernel for the Hilbert space comprising all entire functions $f(z)$ with finite norms

$$\left\{ \int\int_{\mathbf{C}} |f(z)|^2 e^{-b|z|} dx dy \right\}^{\frac{1}{2}} < \infty \tag{46}$$

([Lue]).

Meanwhile, from the identity

$$K_a(z,\overline{u}) = \frac{1}{2\pi} \int_{-a}^a e^{t(z+\overline{u})} dt$$

$$= \frac{\sinh a(z+\overline{u})}{\pi(z+\overline{u})} \quad \text{on} \quad \mathbb{C} \times \overline{\mathbb{C}}, \tag{47}$$

we see that the images of the integral transform for $L_2(-a,a)$ functions F

$$f(z) = \frac{1}{2\pi} \int_{-a}^a F(t) e^{tz} dt \tag{48}$$

form the Hilbert space H_{K_a} admitting the reproducing kernel (47) equipped with the inner product

$$(f,g)_{H_{K_a}} = \int_{-\infty}^\infty f(iy)\overline{g(iy)} dy. \tag{49}$$

Similarly, for

$$\widehat{K}_a(z,\overline{u}) = \frac{1}{2\pi} \int_{-a}^a e^{i\omega z} e^{-i\omega \overline{u}} d\omega$$

$$= \frac{\sin a(z-\overline{u})}{\pi(z-\overline{u})} \quad \text{on} \quad \mathbb{C} \times \overline{\mathbb{C}}, \tag{50}$$

62

the images of the integral transform for $L_2(-a, a)$ functions

$$f(z) = \frac{1}{2\pi} \int_{-a}^{a} e^{i\omega z} F(\omega) d\omega \qquad (51)$$

form the RKHS $H_{\widehat{K}_a}$ comprising all entire functions of exponential type a equipped with the inner product

$$(f, g)_{H_{\widehat{K}_a}} = \int_{-\infty}^{\infty} f(x)\overline{g(x)} dx. \qquad (52)$$

L. de Branges ([[Br]]) examined a much greater generalization of these spaces. Let $E(z)$ be an entire function satisfying

$$|E(\bar{z})| < |E(z)|$$

for $y > 0$. We set

$$E(z) = A(z) - iB(z)$$

where $A(z)$ and $B(z)$ are entire functions which are real for real z, and we set

$$K(z, \bar{u}) = \frac{B(z)\overline{A(u)} - A(z)\overline{B(u)}}{\pi(z - \bar{u})}. \qquad (53)$$

Let $H(E)$ be the Hilbert space comprising all entire functions $f(z)$ satisfying

$$\|f\|_{H(E)}^2 = \int_{-\infty}^{\infty} \frac{|f(x)|^2}{|E(x)|^2} dx < \infty \qquad (54)$$

and

$$|f(z)|^2 \leq \|f\|_{H(E)}^2 K(z, \bar{z}) \quad \text{on} \quad \mathbb{C}. \qquad (55)$$

Then, (53) is the reproducing kernel for the Hilbert space $H(E)$.
    If

$$E(z, a) = \exp(-iaz) \quad (a > 0),$$

then we have the famous Theorem X of Paley–Wiener ([[PW]]).

    In the following three subsections, we shall introduce the Bergman kernel, the Szegö kernel, and the Hardy H_2 kernel and the related kernel as classical and typical reproducing kernels in one complex variable analysis. The corresponding reproducing kernels in several complex variables will be omitted, because we have too many concrete reproducing kernels to state them in this book. See, for example, [[AY]], [[Fu1-2]] and [[Hu]]. The Bergman kernel and Szegö kernel have been studied by

S.Bergman, M.Schiffer, P.R.Garabedian, Z.Nehari, D.A.Hejhal, J.D.Fay and many others, and many beautiful and fundamental results have been obtained. See [[Ber]], [[Ne]], [[M]], [[He]], [[Fa1-2]] and [Sch1-2]. For the interrelationship among the solutions, boundary value problems, and reproducing kernels in partial differential equations, see [[BS]], [[G]] and [GW]. For reproducing kernels for higher order operators and polyanalytic functions, see [[BG1]] and for higher-order systems and a complex first-order equations, see [[BG2]] with their references. The properties of the Hardy H_2 kernel and the related kernel were systematically examined by the author. See [[Sa]].

7   The Bergman kernel

Let D denote a finitely connected bounded regular region surrounded by disjoint analytic Jordan curves on the complex plane. Let $AL_2(D)$ denote the *Bergman space* defined as the family of all analytic functions $f(z)$ on D with finite norms

$$\|f\|_{AL_2(D)} = \left\{ \iint_D |f(z)|^2 dx dy \right\}^{\frac{1}{2}} < \infty.$$

As we see from the mean value theorem for analytic functions, $f(z)$ $(z \in D)$ on $AL_2(D)$ is a bounded linear functional. Hence, there exists the reproducing kernel $K(z, u)$ for the space $AL_2(D)$ such that

$$f(u) = \iint_D f(z)\overline{K(z, u)} dx dy \qquad (56)$$

for all $u \in D$ and for all $f \in AL_2(D)$. Since $K(z, u) = \overline{K(u, z)}$, the kernel $K(z, u)$ is analytic in $\overline{u}$ (complex conjugate). Hence, we will denote it by $K(z, \overline{u})$. This is called the *Bergman kernel* for (or, of) D.

Let $G(z, u)$ denote Green's function for the Laplace equation on D with pole at u; that is,

(i) $G(z, u)$ is harmonic in z on $D \cup \partial D - \{u\}$,
(ii) $G(z, u) = \log \frac{1}{|z-u|} +$ regular terms, around u,
and
(iii) $G(z, u) = 0$ on $z \in \partial D$.

We set

$$\frac{\partial}{\partial z} = \frac{1}{2}\left(\frac{\partial}{\partial x} - i\frac{\partial}{\partial y} \right) \quad \text{and} \quad \frac{\partial}{\partial \overline{z}} = \frac{1}{2}\left(\frac{\partial}{\partial x} + i\frac{\partial}{\partial y} \right).$$

Then, Schiffer found the fundamental identity

$$K(z, \overline{u}) = -\frac{2}{\pi} \frac{\partial^2 G(z, u)}{\partial z \partial \overline{u}}, \qquad (57)$$

64

and introduced the *adjoint L kernel* for the Bergman kernel $K(z, \overline{u})$

$$L(z, u) = -\frac{2}{\pi} \frac{\partial^2 G(z, u)}{\partial z \partial u}.$$

The adjoint L-kernel $L(z, u)$ has one double pole at u as follows:

$$L(z, u) = \frac{1}{\pi(z - u)^2} + \text{regular terms.}$$

We also have the relation

$$\overline{K(z, \overline{u})dz} = -L(z, u)dz \quad \text{along} \quad \partial D.$$

This important relation implies that the Bergman kernel $K(z, \overline{u})dz$ as an analytic differential can be extended analytically onto the reflection of D, and the extension has a double pole at the symmetric point u of $\overline{u}$. See [[SS]] for a unified theory of the Bergman kernel on Riemann surfaces.

As a characteristic property of $L(z, u)$, we have the orthogonality property

$$(P) \iint_D f(z)\overline{L(z, u)}dxdy = 0 \quad \text{for all} \quad f \in AL_2(D) \tag{58}$$

among the class of meromorphic functions on D with the same singularity as $L(z, u)$ and with finite integrals (58). Here, (P) means that the integral is taken in the sense of Cauchy's principal value; that is,

$$\lim_{\varepsilon \to 0} \iint_{D-\{|z-u|<\varepsilon\}} f(z)\overline{L(z, u)}dxdy.$$

For generalizations of (57), see [G2] and [[BG1-2]].

On the unit disc $\Delta = \{|z| < 1\}$, from the expression

$$G(z, u) = -\log \left| \frac{z - u}{1 - \overline{u}z} \right|,$$

we obtain the explicit expressions

$$K(z, \overline{u}) = \frac{1}{\pi(1 - \overline{u}z)^2}, \quad L(z, u) = \frac{1}{\pi(z - u)^2}. \tag{59}$$

In Theorem 3.4, by considering the transform

$$\int_0^z \int_0^{\overline{u}} K(\xi, \overline{\eta})d\xi\overline{d\eta} = \frac{1}{\pi} \log \frac{1}{1 - \overline{u}z} \tag{60}$$

65

or directly we can see that this is the reproducing kernel for the Hilbert space comprising all analytic functions $f(z)$ on Δ with finite norms

$$\left\{ \iint_\Delta |f'(z)|^2 \, dx\, dy \right\}^{\frac{1}{2}} < \infty$$

and satisfying

$$f(0) = 0.$$

Similarly,

$$\frac{1}{\pi} \log \frac{1}{1 - \frac{1}{\bar{u}z}} \tag{61}$$

is the reproducing kernel for the Hilbert space comprising all analytic functions $f(z)$ on $|z| > 1$ with finite norms

$$\left\{ \iint_{|z|>1} |f'(z)|^2 \, dx\, dy \right\}^{\frac{1}{2}} < \infty$$

and satisfying

$$f(\infty) = 0.$$

In connection with Theorem 3.2, note that

$$\frac{1 + \bar{u}z}{(1 - \bar{u}z)^2} \tag{62}$$

is the reproducing kernel for the Hilbert space comprising all analytic functions $f(z)$ on Δ with finite norms

$$\left\{ \frac{1}{2\pi} \iint_\Delta |f(z)|^2 \frac{dx\, dy}{|z|} \right\}^{\frac{1}{2}} < \infty, \tag{63}$$

as we can see directly. See [Bu].

8   The Szegö kernel

Let $H_2(D)$ denote the *analytic Hardy class* on D defined as the family of all analytic functions $f(z)$ on D such that the subharmonic functions $|f(z)|^2$ have harmonic majorants $U(z)$:

$$|f(z)|^2 \leq U(z) \quad \text{on} \quad D.$$

Then, each function $f(z) \in H_2(D)$ has Fatou's nontangential boundary values a.e. on ∂D belonging to $L_2(\partial D)$ (see, for example, [[Du]]).

66

Let $AL_2(\partial D)$ denote the *Szegö space* defined as the family of all $H_2(D)$ analytic functions $f(z)$ on D with finite norms

$$\left\{\int_{\partial D} |f(z)|^2 |dz|\right\}^{\frac{1}{2}} < \infty,$$

where $f(z)$ mean Fatou's nontangential boundary values at ∂D. For $f \in AL_2(\partial D)$, from the Cauchy integral formula

$$f(u)^2 = \frac{1}{2\pi i} \int_{\partial D} \frac{f(z)^2}{z - u} dz,$$

we have

$$|f(u)| \leqq \left\{\frac{1}{2\pi \text{dis}(\partial D, u)}\right\}^{\frac{1}{2}} \left\{\int_{\partial D} |f(z)|^2 |dz|\right\}^{\frac{1}{2}}.$$

Hence, there exists the reproducing kernel $\hat{K}(z, u)$ such that

$$f(u) = \int_{\partial D} f(z)\overline{\hat{K}(z, u)}|dz| \tag{64}$$

for all $u \in D$ and for all $f \in AL_2(\partial D)$. Since the kernel $\hat{K}(z, u)$ is analytic in $\overline{u}$, we shall denote it by $\hat{K}(z, \overline{u})$, which is the *Szegö kernel* on (or, of) D.

A very important notion is the *adjoint L kernel* for $\hat{K}(z, \overline{u})$. That is, there exists a uniquely determined meromorphic function $\hat{L}(z, u)$ on $D \cup \partial D$ with one simple pole at u such that

$$\hat{L}(z, u) = \frac{1}{2\pi(z - u)} + \text{regular terms}, \quad \text{around} \quad z = u \tag{65}$$

which satisfies the relation

$$\overline{\hat{K}(z, \overline{u})}|dz| = \frac{1}{i}\hat{L}(z, u)dz \quad \text{along} \quad \partial D. \tag{66}$$

Note that

$$\hat{L}(z, u) = -\hat{L}(u, z) \quad \text{on} \quad D \times D, \tag{67}$$

as we see from the relation (66).

The adjoint L kernel $\hat{L}(z, u)$ is characterized by the minimum property:

$$\min \int_{\partial D} |h(z, u)|^2 |dz| = \int_{\partial D} |\hat{L}(z, u)|^2 |dz|,$$

where the minimum is taken over all meromorphic functions $h(z, u)$ on D having one simple pole at u as in (65) and finite norms

$$\left\{\int_{\partial D} |h(z, u)|^2 |dz|\right\}^{\frac{1}{2}} < \infty.$$

Another characterization for the adjoint L kernel is the following orthogonality property:

$$\int_{\partial D} f(z)\overline{\hat{L}(z, u)}|dz| = 0 \quad \text{for all} \quad f \in AL_2(\partial D) \tag{68}$$

in the above class, as we see directly from (66).

A deep and fundamental result for the adjoint L kernel is that the function $\hat{L}(z, u)$ does not vanish on $D \cup \partial D$. From this property and the relation (66), we can easily see that the function

$$f_0(z; u) = \frac{\hat{K}(z, \overline{u})}{\hat{L}(z, u)}$$

is the Ahlfors function on D with respect to u; that is, the function $f_0(z; u)$ is the uniquely determined extremal function such that in the class of analytic functions $f(z; u)$ on D satisfying

$$f(u; u) = 0, f'(u; u) \geq 0 \quad \text{and} \quad |f(z; u)| \leq 1 \quad \text{on} \quad D,$$

the function $f_0(z; u)$ satisfies

$$f_0'(u; u) = \max f'(u; u).$$

Then, as we see from (66) and the argument principle, the function $f_0(z; u)$ maps D onto an N sheeted unit disc, if D is N-ply connected.

From (66), we have

$$\overline{\hat{K}(z, \overline{u})^2 dz} = -\hat{L}(z, u)^2 dz \quad \text{along} \quad \partial D. \tag{69}$$

This relation means that the kernels $\hat{K}(z, \overline{u})$ and $\hat{L}(z, u)$ are, in fact, half order differentials when we consider them on Riemann surfaces or when we consider them for conformal mappings of D ([G1]). In general, therefore their multivaluedness is a serious problem on Riemann surfaces. A detailed and profound analysis for the Szegö kernel on Riemann surfaces was given by D.A.Hejhal ([[He]]) and J.D.Fay ([[Fa1-2]]) in terms of the Riemann theta function and the Klein prime form.

On the unit disc $\{|z| < 1\}$, we have the explicit expressions

$$\hat{K}(z, \overline{u}) = \frac{1}{2\pi(1 - \overline{u}z)}, \quad \hat{L}(z, u) = \frac{1}{2\pi(z - u)}, \tag{70}$$

since the family

$$\left\{\frac{z^j}{(2\pi)^{\frac{1}{2}}}\right\}_{j=0}^{\infty}$$

is a complete orthonormal system for $AL_2(\partial\Delta)$.

In connection with the arguments in Section 3.3, note that

$$\frac{\pi(1+\overline{u}z)}{1-\overline{u}z} \tag{71}$$

and

$$\frac{4\pi(2-\overline{u}z)}{1-\overline{u}z} \tag{72}$$

are the reproducing kernels for the Hilbert spaces comprising all analytic functions $f(z)$ on Δ with finite norms

$$\frac{1}{2\pi}\left\{\int_{\partial\Delta}|f(z)|^2|dz|+2\pi|f(0)|^2\right\}^{\frac{1}{2}}<\infty \tag{73}$$

and

$$\left[\frac{1}{8\pi^2}\left\{\int_{\partial\Delta}|f(z)|^2|dz|-\pi|f(0)|^2\right\}\right]^{\frac{1}{2}}<\infty, \tag{74}$$

respectively.

Note that for $\alpha \geq 0$

$$K_\alpha(z,\overline{u}) = \frac{\Gamma(\alpha+1)}{(1-\overline{u}z)^\alpha} \tag{75}$$

are the *Bergman–Selberg* reproducing kernels for the Hilbert spaces comprising all analytic functions $f(z)$ on Δ with finite norms

$$\|f\|_\alpha^2 = \frac{1}{\pi\Gamma(\alpha)}\iint_\Delta|f(z)|^2(1-|z|^2)^{\alpha-1}dxdy \tag{76}$$

and

$$\|f\|_0^2 = \frac{1}{2\pi}\int_{\partial\Delta}|f(z)|^2|dz|, \tag{77}$$

respectively. See, for example, [AFP], [Bu], [[DS]], [I] and [Z].

Furthermore, note that

$$\frac{1}{z\overline{u}}\log\frac{1}{1-\overline{u}z} \tag{78}$$

69

is the reproducing kernel for the Hilbert space comprising all analytic functions $f(z)$ on Δ with finite norms

$$\left\{ \frac{1}{\pi} \iint_\Delta |f'(z)|^2 \, dx dy + \frac{1}{2\pi} \int_{\partial\Delta} |f(z)|^2 |dz| \right\}^{\frac{1}{2}} < \infty. \tag{79}$$

These results are also confirmed by using the Taylor expansion. See [Bu].

The kernel

$$\log \left[\frac{1}{z\overline{u}} \log \frac{1}{1 - \overline{u}z} \right]$$

is also the reproducing kernel on Δ comprising analytic functions and relating to Bernoulli numbers and combinatorial theory, but its reproducing kernel Hilbert space is not known ([R]).

Further, for concrete reproducing kernels on Δ, see [I] and for reproducing kernels for automorphic forms on Δ, see for example, [Se 1, 2], [E] and [[Kr]] with their references.

Meanwhile, in the Hilbert space $PA_n^2(\Delta)$ comprising all polyanalytic functions $f(z)$ satisfying

$$f(z) = \varphi_0(z) + \overline{z}\varphi_1(z) + \cdots + \overline{z}^{n-1}\varphi_{n-1}(z);$$
$$\varphi_j(z) \quad \text{are analytic on} \quad \Delta$$

and equipped with the norm

$$\left\{ \iint_\Delta |f(z)|^2 \, dx dy \right\}^{\frac{1}{2}} < \infty,$$

$$K_n(z, \overline{u}) = \frac{2}{\pi} \frac{\sum_{j=1}^n (-1)^{j-1} {}_n C_j \, {}_n C_{n+j-1} |1 - \overline{u}z|^{2(n-j)}}{(1 - \overline{u}z)^{2n}}$$

$$= \sum_{j=1}^{n-1} \sum_{m=0}^\infty e_{mj}(z) \overline{e_{mj}(u)};$$

$$e_{mj}(z) = \sqrt{\frac{m+j+1}{\pi}} \frac{1}{(m+j)!} \frac{\partial^{m+j}}{\partial z^j \, \partial \overline{z}^m} (|z|^2 - 1)^{m+j}$$

is the reproducing kernel, because the system

$$\{e_{mj}(z)\}_{m=0}^\infty \quad (j = 0, 1, \cdots, m-1)$$

is complete orthonormal in $PA_n^2(\Delta)$ ([Kos]).

70

9 The Hardy H_2 kernel and its conjugate kernel

Let $H_2(D)$ (resp. $\hat{H}_2(D)$) denote the analytic (resp. conjugate) Hardy space on D defined as the family of all $H_2(D)$ analytic functions $f(z)$ on D with finite norms

$$\left\{ \frac{1}{2\pi} \int_{\partial D} |f(z)|^2 \frac{\partial G(z,t)}{\partial \nu_z} |dz| \right\}^{\frac{1}{2}} \tag{80}$$

$$\left(\text{resp. } \left\{ \frac{1}{2\pi} \int_{\partial D} |f(z)|^2 \left(\frac{\partial G(z,t)}{\partial \nu_z} \right)^{-1} |dz| \right\}^{\frac{1}{2}} \right) \tag{81}$$

where $f(z)$ mean Fatou's nontangential boundary values and $\partial/\partial \nu_z$ denotes the inner normal derivative with respect to D. Since $\partial G(z,t)/\partial \nu_z$ is a positive continuous function on ∂D, there exists, as in the Szegö kernel, the reproducing kernel $K_t(z, \overline{u})$ (resp. $\hat{K}_t(z, \overline{u})$) such that

$$f(u) = \frac{1}{2\pi} \int_{\partial D} f(z) \overline{K_t(z, \overline{u})} \frac{\partial G(z,t)}{\partial \nu_z} |dz| \tag{82}$$

$$\left(\text{resp. } f(u) = \frac{1}{2\pi} \int_{\partial D} f(z) \overline{\hat{K}_t(z, \overline{u})} \left(\frac{\partial G(z,t)}{\partial \nu_z} \right)^{-1} |dz| \right) \tag{83}$$

for all $u \in D$ and for all $f \in H_2(D)$ (resp. $\hat{H}_2(D)$). The reproducing kernels $K_t(z, \overline{u})$ and $\hat{K}_t(z, \overline{u})$ will be called the *Hardy H_2 kernel* and the *conjugate Hardy H_2 kernel* on (or, of) D, respectively.

For an arbitrary domain S in the complex plane and, more generally, for any open Riemann surface S, we can define the Hardy space $H_2(S)$. The space $H_2(S)$ is composed of all analytic functions $f(z)$ on S such that for the subharmonic functions $|f(z)|^2$, there exist harmonic majorants $U(z)$ satisfying

$$|f(z)|^2 \leq U(z) \quad \text{on} \quad S.$$

We take a regular exhaustion $\{D_n\}_{n=1}^{\infty}$ of S containing the point $t \in S$ defined by

(i) ∂D_n is composed of a finite number of disjoint analytic Jordan curves,
(ii) $D_n \cup \partial D_n$ is compact,
(iii) $D_n \subset D_{n+1} \subset S$,
and
(iv) $\cup_{n=1}^{\infty} D_n = S$.

Let $G_n(z,t)$ denote the Green's function of D_n with pole at t. Then, the sequence

$$\left\{ \frac{1}{2\pi} \int_{\partial D_n} |f(z)|^2 \frac{\partial G_n(z,t)}{\partial \nu_z} |dz| \right\}_{n=1}^{\infty}$$

increases as n increases, and furthermore the limit is equal to the least harmonic majorant of $|f(z)|^2$ on S. Then, the norm of $f(z)$ in $H_2(S)$ is given by

$$\|f\|_{H_2(S)} = \lim_{n \to \infty} \left\{ \frac{1}{2\pi} \int_{\partial D_n} |f(z)|^2 \frac{\partial G_n(z,t)}{\partial \nu_z} |dz| \right\}^{\frac{1}{2}}. \tag{84}$$

Therefore, for any open Riemann surface S, there exists the Hardy H_2 kernel $K_t^{(S)}(z, \overline{u})$ on S which satisfies the reproducing property

$$f(u) = \lim_{n \to \infty} \frac{1}{2\pi} \int_{\partial D_n} f(z) \overline{K_t^{(S)}(z, \overline{u})} \frac{\partial G_n(z,t)}{\partial \nu_z} |dz|$$

for all $u \in S$ and for all $f \in H_2(S)$.

Meanwhile, the conjugate norm (81) is, in general, complicated. For the regular exhaustion $\{D_n\}$ of S, the sequence

$$\frac{1}{2\pi} \int_{\partial D_n} |f(z)|^2 \left(\frac{\partial G_n(z,t)}{\partial \nu_z} \right)^{-1} |dz| \tag{85}$$

is, in general, not monotone with respect to n. But note that the integrals have their senses on the compact bordered Riemann surfaces D_n.

Indeed, let $G_n^*(z, t)$ denote a conjugate harmonic function of $G_n(z, t)$, and define the multivalued meromorphic function

$$W_n(z, t) = G_n(z, t) + i G_n^*(z, t).$$

Then, the meromorphic differential $i dW_n(z, t)$ is single valued, positive along ∂D_n, and satisfies

$$\frac{\partial G_n(z,t)}{\partial \nu_z} |dz| = i dW_n(z, t) \quad \text{along} \quad \partial D_n.$$

Therefore, the integrals (85) are expressible in the form

$$\frac{1}{2\pi} \int_{\partial D_n} \frac{|f(z) dz|^2}{i dW_n(z, t)},$$

which exist on compact bordered Riemann surfaces D_n. This fact means that the conjugate Hardy space $\hat{H}_2(D_n)$ is composed of analytic differentials $f(z) dz$ on D_n.

In particular, from the uniqueness of reproducing kernels, we see that

$$K_t(z, \overline{t}) \equiv 1, \quad K_t(t, \overline{u}) = 1. \tag{86}$$

72

The *adjoint L kernel* $L_t(z, u)$ (resp. $\hat{L}_t(z, u)$) for $K_t(z, \overline{u})$ (resp. $\hat{K}_t(z, \overline{u})$) is characterized by the boundary relation

$$\overline{K_t(z, \overline{u})}idW(z, t) = \frac{1}{i}L_t(z, u)dz \quad \text{along} \quad \partial D \tag{87}$$

$$\left(\text{resp.} \ \overline{\hat{K}_t(z, \overline{u})}dz = \frac{1}{i}\hat{L}_t(z, u)idW(z, t) \quad \text{along} \quad \partial D\right) \tag{88}$$

in the class of meromorphic functions $h(z, u)$ on $D \cup \partial D$ having one simple pole at u with residue 1. Here, from (87) and (88), we have the relation

$$L_t(z, u) = -\hat{L}_t(u, z) \quad \text{on} \quad D \times D. \tag{89}$$

The relations (87), (88) and (89) mean that

$$L_t(z, u)dz$$

is a meromorphic differential with respect to z and a meromorphic function in u on $D \times D$, with one simple pole along $z = u$; that is, it is a *Cauchy kernel* on D.

For a representation of $L_t(z, u)$ in terms of the Riemann theta function and the Klein prime form on compact bordered Riemann surfaces, see [[Fa1]].

On the unit disc Δ, we have the explicit expressions:

$$W'(z, t) = -\frac{1 - |t|^2}{(z - t)(1 - \overline{t}z)},$$

$$K_t(z, \overline{u}) = \frac{(1 - t\overline{u})(1 - \overline{t}z)}{(1 - |t|^2)(1 - \overline{u}z)}, \tag{90}$$

$$L_t(z, u) = \frac{1 - \overline{t}u}{(1 - \overline{t}z)(z - u)},$$

$$\hat{K}_t(z, \overline{u}) = \frac{1 - |t|^2}{(1 - t\overline{u})(1 - \overline{t}z)(1 - \overline{u}z)}, \tag{91}$$

and

$$\hat{L}_t(z, u) = \frac{1 - \overline{t}z}{(1 - \overline{t}u)(z - u)}.$$

By using the conformal mapping from the unit disc onto the strip $Sa = \{|\text{Im} z| < a\}$, we have the concrete expressions of the Szegö kernel $\hat{K}_a(z, \overline{u})$, the Bergman

73

kernel $K_a(z,\overline{u})$, the Hardy H_2 kernel $K_t^{(a)}(z,\overline{u})$ and its conjugate kernel $\hat{K}_t^{(a)}(z,\overline{u})$ as follows:

$$\hat{K}_a(z,\overline{u}) = \frac{1}{8a}\frac{1}{\cosh\frac{\pi(z-\overline{u})}{4a}},$$

$$K_a(z,\overline{u}) = \frac{\pi}{16a^2}\frac{1}{[\cosh\frac{\pi(z-\overline{u})}{4a}]^2},$$

$$K_t^{(a)}(z,\overline{u}) = \frac{\cosh\frac{\pi(\overline{u}-t)}{4a}\cosh\frac{\pi(z-\overline{t})}{4a}}{\cosh\frac{\pi(t-\overline{t})}{4a}\cosh\frac{\pi(z-\overline{u})}{4a}},$$

and

$$\hat{K}_t^{(a)}(z,\overline{u}) = \frac{\pi^2}{16a^2}\frac{\cosh\frac{\pi(t-\overline{t})}{4a}}{\cosh\frac{\pi(\overline{u}-t)}{4a}\cosh\frac{\pi(z-\overline{t})}{4a}\cosh\frac{\pi(z-\overline{u})}{4a}}.$$

For explicit representations of the Bergman and Szegö kernels on domains with hyperelliptic double, see [Ba].

10 On the complex half plane

For $q > \frac{1}{2}$,

$$K_q(z,\overline{u}) = \int_0^\infty e^{-tz}e^{-t\overline{u}}t^{2q-1}dt = \frac{\Gamma(2q)}{(z+\overline{u})^{2q}} \qquad (92)$$

is the Bergman–Selberg reproducing kernel on the half plane $R^+ = \{\mathrm{Re}\,z > 0\}$ comprising all analytic functions $f(z)$ on R^+ with finite norms

$$\|f\|_{HK_q}^2 = \frac{1}{\pi\Gamma(2q-1)}\iint_{R^+}|f(z)|^2[2\mathrm{Re}\,z]^{2q-2}dxdy. \qquad (93)$$

For $q = \frac{1}{2}$ in (92), $K_{\frac{1}{2}}(z,\overline{u})$ is the Szegö reproducing kernel on the half plane R^+ comprising all analytic functions $f(z)$ on R^+ with finite norms

$$\|f\|_{HK_{1/2}}^2 = \frac{1}{2\pi}\sup_{x>0}\int_{-\infty}^\infty|f(x+iy)|^2dy < \infty. \qquad (94)$$

Then, a member $f(z)$ of $H_{K1/2}$ has nontangential boundary values on the imaginary axis belonging to $L_2((-\infty,\infty),dy)$ and we have

$$\|f\|_{HK_{1/2}}^2 = \frac{1}{2\pi}\int_{-\infty}^\infty|f(iy)|^2dy.$$

On the upper half plane $U = \{\operatorname{Im} z > 0\}$, we have the corresponding reproducing kernels

$$K_{q,U}(z, \overline{u}) = \Gamma(2q) \left(\frac{i}{z - \overline{u}} \right)^{2q}. \tag{95}$$

We can obtain the above results using Fourier integrals. See, for example, [[PW]], [Bu] and [DD].

The kernel

$$k(z, \overline{u}) = \sum_{n=0}^{\infty} \binom{z-1}{n} \overline{\binom{u-1}{n}} = \frac{\Gamma(z + \overline{u} - 1)}{\Gamma(z)\Gamma(\overline{u})}, \quad \operatorname{Re} z, \operatorname{Re} \overline{u} > \frac{1}{2} \tag{96}$$

is the reproducing kernel for the Hilbert space comprising all analytic functions $f(z)$ satisfying

$$\sup_{\frac{1}{2} < x} \int_{-\infty}^{\infty} |f(x + iy)|^2 |\Gamma(x + iy)|^2 dy < \infty$$

and equipped with the inner product

$$\int_{-\infty}^{\infty} f(\frac{1}{2} + iy) \overline{g(\frac{1}{2} + iy)} |\Gamma(\frac{1}{2} + iy)|^2 dy$$

where $f(\frac{1}{2} + iy)$ and $g(\frac{1}{2} + iy)$ mean Fatou's nontangential boundary values ([K]).

More generally, for any $h > 0$

$$\frac{\Gamma(h)\Gamma(z + \overline{u} + h)}{\Gamma(z + h)\Gamma(\overline{u} + h)}$$

and

$$\log \frac{\Gamma(h)\Gamma(z + \overline{u} + h)}{\Gamma(z + h)\Gamma(\overline{u} + h)}$$

are the reproducing kernels for the *Newton space* and the *Euler space* comprising analytic functions on $\{\operatorname{Re} z > -\frac{1}{2}h\}$, respectively ([MRR] and [R]).

11 Reproducing kernels for harmonic functions

We can consider reproducing kernels for harmonic functions as for analytic functions, but concrete representations of reproducing kernels are more involved.

$$\frac{1}{\pi} \operatorname{Re} \left[\frac{2}{(1 - \overline{u}z)^2} - 1 \right] \tag{97}$$

is the reproducing kernel for the Hilbert space comprising all harmonic functions $u(z)$ on Δ with finite norms

$$\left\{ \iint_{\Delta} |u(z)|^2 dx dy \right\}^{\frac{1}{2}} < \infty. \tag{98}$$

75

Meanwhile,

$$\frac{1}{\pi} \operatorname{Re}\ \log \frac{1}{1 - \bar{u}z} \tag{99}$$

is the reproducing kernel for the Hilbert space comprising all harmonic functions $u(z)$ on Δ with finite norms

$$\left\{ \iint_\Delta \left((\frac{\partial u}{\partial x})^2 + (\frac{\partial u}{\partial y})^2 \right) dx\,dy \right\}^{\frac{1}{2}} < \infty \tag{100}$$

and satisfying

$$u(0) = 0.$$

See [Ar2] and [NS].

On the other hand,

$$\frac{1}{2\pi} \operatorname{Re} \frac{1 + \bar{u}z}{1 - \bar{u}z} \tag{101}$$

is the reproducing kernel for the harmonic Hardy H_2 space on Δ comprising all harmonic functions $u(z)$ having harmonic majorants $U(z)$ such that

$$|u(z)|^2 \leqq U(z) \quad \text{on} \quad \Delta.$$

Here, the norm $\|u\|_{H_2}$ is given by

$$\left\{ \int_{\partial \Delta} |u(z)|^2 |dz| \right\}^{\frac{1}{2}} < \infty, \tag{102}$$

where $u(z)$ ($z \in \partial \Delta$) means the Fatou nontangential boundary value. See (3.4.50).

On the upper half plane U,

$$\frac{y + y'}{(x - x') + (y + y')^2} \tag{103}$$

is the reproducing kernel for the harmonic Hardy H_2 space with finite norms

$$\left\{ \int_{-\infty}^{\infty} |u(x,0)|^2 dx \right\}^{\frac{1}{2}} < \infty. \tag{104}$$

See (3.4.54).

12 Reproducing kernels of subspaces

For any reproducing kernel $K(p,q)$ on E, the simplest subspace of H_K will be made by the functions $f \in H_K$ satisfying

$$f(p_0) = 0 \quad \text{for a fixed} \quad p_0 \in E.$$

Then, for the subspace $H_K(p_0)$ we have the reproducing kernel

$$K_{p_0}(p, q) = K(p, q) - \frac{K(p, p_0)\overline{K(q, p_0)}}{K(p_0, p_0)}, \tag{105}$$

as we can see directly. Here, note that

$$K(p_0, p_0) = 0$$

if and only if

$$f(p_0) = 0 \quad \text{for all} \quad f \in H_K.$$

Hence, in the above formula we assume that $K(p_0, p_0) \neq 0$.

For a finite point set $\{p_j\}_{j=1}^n$ of E, we can consider the similar reproducing kernel for the subspace of H_K satisfying

$$f(p_j) = 0, j = 1, 2, \cdots, n.$$

For general properties of reproducing kernels for subspaces of a RKHS, see [C].

13 Restrictions of reproducing kernels

The relation in Theorem 3.1 between a RKHS H_K on E and the RKHS $H_{K|_{E_0}}$ induced from the restriction $K|_{E_0}$ of $K(p, q)$ to a subset E_0 of E will have, in general, interesting properties for miscellaneous concrete reproducing kernels.

As stated in (24) and (25),

$$G(x, y) = \frac{1}{2} e^{-|x-y|}$$

is the reproducing kernel for the Sobolev Hilbert space H_G on $(-\infty, \infty)$ equipped with the norm

$$\|f\|_{H_G}^2 = \int_{-\infty}^{\infty} (|f'(x)|^2 + |f(x)|^2) dx.$$

As we can confirm directly, the restriction

$$G\big|_{(c,d)} \tag{106}$$

is the reproducing kernel for the Hilbert space $H_{G|_{(c,d)}}$ comprising all absolutely continuous functions on $[c, d]$ with finite norms

$$\|f\|_{H_{G|_{(c,d)}}}^2 = \int_c^d (|f'(x)|^2 + |f(x)|^2) dx + |f(c)|^2 + |f(d)|^2. \tag{107}$$

Note that all the members of $H_{G|_{(c,d)}}$ are obtained by the restriction to (c, d) of the members of H_G.

The restrictions of a RKHS comprising analytic functions to subsets will sometimes have interesting and deep results which can be connected with analytic extension problems.

For example, we obtain the identity for analytic functions $f(z)$ on the strip $S_a = \{|\text{Im} z| < a\}$

$$\iint_{S_a} |f(z)|^2 dx dy = \sum_{j=0}^{\infty} \frac{(2a)^{2j+1}}{(2j+1)!} \int_{-\infty}^{\infty} |f^{(j)}(x)|^2 dx. \tag{108}$$

The corresponding reproducing kernels are

$$K_a(z, \overline{u}) = \frac{\pi}{16a^2} \frac{1}{\cosh^2 \frac{\pi(z-\overline{u})}{4a}} \tag{109}$$

and

$$K_a(x, x') = \frac{\pi}{16a^2} \frac{1}{\cosh^2 \frac{\pi(x-x')}{4a}}, \tag{110}$$

respectively (Theorem 5.1.1).

Similar results are valid in Chapter 1, (22); that is,

$$\frac{4}{\pi} \frac{z\overline{u}}{(z^2 + \overline{u}^2)^2} \tag{111}$$

is the Bergman kernel for the Hilbert space comprising all analytic functions $f(z)$ on the sector $\Delta(\frac{\pi}{4}) = \{|\arg z| < \frac{\pi}{4}\}$ with finite norms

$$\left\{ \iint_{\Delta(\frac{\pi}{4})} |f(z)|^2 dx dy \right\}^{\frac{1}{2}} < \infty,$$

meanwhile, the restriction to the real axis

$$\frac{4}{\pi} \frac{x x'}{(x^2 + x'^2)^2} \tag{112}$$

is the reproducing kernel for the infinite order Sobolev space comprising C^{∞}-functions on $(0, \infty)$ with finite norms

$$\left\{ \sum_{j=0}^{\infty} \frac{2^j}{(2j+1)!} \int_0^{\infty} x^{2j+1} |f^{(j)}(x)|^2 dx \right\}^{\frac{1}{2}} < \infty \tag{113}$$

78

(Theorem 5.1.3).

Also for the identity in Chapter 1, (21), we obtain similar results. See [HS1] and the later (115) in this section.

14 Product of reproducing kernels

As shown in (38) and (39),

$$e^{z\bar{u}}$$

is the reproducing kernel for the Hilbert space comprising all entire functions with finite norms

$$\left\{ \frac{1}{\pi} \iint_{\mathbf{C}} |f(z)|^2 e^{-|z|^2} dx dy \right\}^{\frac{1}{2}} < \infty.$$

Hence, by Corollary 3.3

$$e^{z\bar{u}} e^{-\frac{z^2}{2}} e^{-\frac{\bar{u}^2}{2}} \tag{114}$$

is the reproducing kernel for the Hilbert space comprising all entire functions with finite norms

$$\left\{ \frac{1}{\pi} \iint_{\mathbf{C}} |f(z) e^{\frac{z^2}{2}}|^2 e^{-|z|^2} dx dy \right\}^{\frac{1}{2}} = \left\{ \frac{1}{\pi} \iint_{\mathbf{C}} |f(z)|^2 e^{-2y^2} dx dy \right\}^{\frac{1}{2}} < \infty, \tag{115}$$

which will appear in connection with the solutions of the heat equation in Theorem 3.2.1.

15 Transforms of reproducing kernels

For the Bergman kernel

$$K_1(z, \bar{u}) = \frac{1}{(z + \bar{u})^2}$$

for the Hilbert space comprising all analytic functions $f(z)$ on R^+ with finite norms

$$\left\{ \frac{1}{\pi} \iint_{R+} |f(z)|^2 dx dy \right\}^{\frac{1}{2}} < \infty,$$

we shall consider the transform

$$z \longrightarrow z^2, u \longrightarrow u^2.$$

Then, we have the reproducing property

$$\frac{4}{\pi} \iint_{\Delta(\frac{\pi}{4})} f(z) \overline{\left(\frac{1}{(z^2 + \bar{u}^2)^2} \right)} |z|^2 dx dy = f(u)$$

79

for all analytic functions $f(z)$ on $\Delta(\frac{\pi}{4})$ with finite norms

$$\left\{ \frac{4}{\pi} \iint_{\Delta(\frac{\pi}{4})} |f(z)|^2 |z|^2 dx dy \right\}^{\frac{1}{2}} < \infty. \tag{116}$$

Hence,

$$\frac{1}{(z^2 + \overline{u}^2)^2} \tag{117}$$

is the reproducing kernel for the Hilbert space comprising all analytic functions $f(z)$ on $\Delta(\frac{\pi}{4})$ with finite norms (116).

As shown in (26) and (27),

$$K(x, y) = \frac{1}{4} e^{-|x-y|} \{1 + |x - y|\}$$

is the reproducing kernel for the Sobolev space on $(-\infty, \infty)$ comprising all absolutely continuous $f'(x)$ with finite norms

$$\left\{ \int_{-\infty}^{\infty} (|f''(x)|^2 + 2|f'(x)|^2 + |f(x)|^2) dx \right\}^{\frac{1}{2}} < \infty.$$

Meanwhile,

$$K_{11}(x, y) = \frac{\partial^2 K(x, y)}{\partial x \partial y} = \frac{1}{4} e^{-|x-y|} \{1 - |x - y|\} \tag{118}$$

is the reproducing kernel for the Hilbert space $H_{K_{11}}$ comprising all absolutely continuous functions on $(-\infty, \infty)$ with finite norms

$$\left\{ \int_{-\infty}^{\infty} \left(|f'(x)|^2 + 2|f(x)|^2 + \left| \int_{-\infty}^{x} f(t) dt \right|^2 \right) dx \right\}^{\frac{1}{2}} < \infty \tag{119}$$

as we can see directly.

This example will be formal and the result is routine. The concrete form (118) will, however, give interesting information for the RKHS $H_{K_{11}}$ from the viewpoint of the sum and the difference of reproducing kernels.

For example, for the Szegö kernel

$$K_{\frac{1}{2}}(z, \overline{u}) = \frac{1}{z + \overline{u}} \quad \text{on} \quad R^+ \times \overline{R^+},$$

we have

$$\frac{\partial^2 K_{\frac{1}{2}}(z, \overline{u})}{\partial z \partial \overline{u}} = \frac{2}{(z + \overline{u})^3} = K_{\frac{3}{2}}(z, \overline{u}).$$

80

Then, the two reproducing kernel Hilbert spaces $H_{K_{\frac{1}{2}}}$ and $H_{K_{\frac{3}{2}}}$ admit entirely different norms. See (92), (93) and (94), (95).

16 Squeezes and balloons of reproducing kernel Hilbert spaces

First we shall construct squeezes of a RKHS H_K on E, by using Theorem 3.8.

By taking a finite number of points $\{\xi_j\}_{j=1}^n$ of E as a space $\mathcal{H}$, we shall consider the space

$$\|f\|_{\mathcal{H}}^2 = \sum_{j=1}^n |f(\xi_j)|^2.$$

Then, by Theorem 3.8 the reproducing kernel $K_{\mathcal{H}}(p, q)$ for the subspace $H_{K_{\mathcal{H}}}$ of H_K equipped with the inner product

$$(f, g)_{H_{K_{\mathcal{H}}}} = (f, g)_{H_K} + \sum_{j=1}^n f(\xi_j)\overline{g(\xi_j)} \tag{120}$$

satisfies the functional equation

$$K(p, q) = K_{\mathcal{H}}(p, q) + \sum_{j=1}^n K_{\mathcal{H}}(\xi_j, q) K(p, \xi_j). \tag{121}$$

By setting $p = \xi_{j'}, j' = 1, 2, \cdots, n$, we have the equations

$$\sum_{j=1}^n (\delta_{jj'} + K(\xi_{j'}, \xi_j)) K_{\mathcal{H}}(\xi_j, q) = K(\xi_{j'}, q), \tag{122}$$

$$j' = 1, 2, \cdots, n.$$

Since the matrix

$$\|\delta_{jj'} + K(\xi_{j'}, \xi_j)\|$$

is positive definite, we denote its inverse by $\|A_{j'j}\|$. Then, from (121) and (122) we have the desired explicit representation

$$K_{\mathcal{H}}(p, q) = K(p, q) - \sum_{j,j'=1}^n K(p, \xi_j) A_{jj'} K(\xi_{j'}, q). \tag{123}$$

In particular, for one point $\xi \in E$, we have

$$K_{\mathcal{H}}(p, q) = K(p, q) - \frac{K(p, \xi) K(\xi, q)}{1 + K(\xi, \xi)}. \tag{124}$$

For the reproducing kernel $K_{\mathcal{H}}^-(p, q)$, the circumstances are more delicate for the condition (3.35). We shall consider an example.

Let D be a bounded domain on $\mathbb{C}$ and consider the Bergman space $AL_2(D)$ comprising all analytic functions $f(z)$ on D with finite norms

$$\left\{ \iint_D |f(z)|^2 \, dx\, dy \right\}^{\frac{1}{2}} < \infty.$$

We denote the Bergman reproducing kernel for $AL_2(D)$ by $K(z, \overline{u})$. Let $\{D_j\}_{j=1}^n$ be any finite number of open discs $\{|z - p_j| < r_j\}$ on D which are disjoint. Then, by the submean property of $|f(z)|^2$, we obtain the inequality

$$|f(p_j)|^2 \leqq \frac{1}{\pi r_j^2} \iint_{D_j} |f(z)|^2 dx\, dy,$$

and so we have

$$\iint_D |f(z)|^2 dx\, dy \geq \pi \sum_{j=1}^n r_j^2 |f(p_j)|^2.$$

Except for the trivial case, we can assume that for nonzero constants, the equality does not hold. Then, we can introduce the Hilbert space $H_K^-\{D_j\}$ equipped with the inner product

$$(f, g)_{H_K^-\{D_j\}} = (f, g) - \pi \sum_{j=1}^n r_j^2 f(p_j)\overline{g(p_j)}$$

$$\text{for} \quad f, g \in AL_2(D). \tag{125}$$

By Theorem 3.8, the reproducing kernel $K(z, \overline{u}; \{D_j\})$ for $H_K^-\{D_j\}$ can be determined by the functional equation

$$K(z, \overline{u}) = K(z, \overline{u}; \{D_j\}) - \pi \sum_{j=1}^n r_j^2 K(p_j, \overline{u}; \{D_j\}) K(z, \overline{p}_j). \tag{126}$$

By setting $z = p_{j'}; j' = 1, 2, \cdots, n$, we have

$$\sum_{j=1}^n \{\delta_{jj'} - \pi r_j^2 \overline{K(p_j, \overline{p}_{j'})}\} K(p_j, \overline{u}; \{D_j\}) = K(p_{j'}, \overline{u}),$$

$$j' = 1, 2, \cdots, n. \tag{127}$$

Since the matrix

$$\left\| \delta_{jj'} - \pi r_j^2 \overline{K(p_j, \overline{p}_{j'})} \right\|$$

is nonsingular, we denote its inverse by $\|B_{jj'}\|$. Then, from (126) and (127) we have the desired representation

$$K(z,\overline{u};\{D_j\}) = K(z,\overline{u}) + \pi \sum_{j,j'=1}^{n} K(z,\overline{p}_j)r_j^2 B_{jj'}K(p_{j'},\overline{u}). \qquad (128)$$

In particular, for one disc $D(p,r) = \{|z-p| < r\}$, we obtain

$$K(z,\overline{u};D(p,r)) = K(z,\overline{u}) + \frac{\pi r^2 K(z,\overline{p})K(p,\overline{u})}{1 - \pi r^2 K(p,\overline{p})}. \qquad (129)$$

17   Any inner product

For any Hilbert space $\mathcal{H}$, we shall consider the $\mathcal{H}$-valued function $\mathbf{h}(\mathbf{x})$ on $\mathcal{H}$ by the identity mapping

$$\mathbf{h} : \mathbf{x} \longrightarrow \mathbf{x} \quad \text{for} \quad \mathbf{x} \in \mathcal{H}.$$

Then, the induced linear mapping

$$x(\mathbf{y}) = (\mathbf{x},\mathbf{h}(\mathbf{y}))_{\mathcal{H}} = (\mathbf{x},\mathbf{y})_{\mathcal{H}}$$

in Theorem 1.2 transforms $\mathcal{H}$ onto the adjoint space $\mathcal{H}^*$ comprising all bounded linear functionals on $\mathcal{H}$, isometrically. Hence,

$$(\mathbf{x},\mathbf{y})_{\mathcal{H}} \quad \text{on} \quad \mathcal{H} \times \mathcal{H} \qquad (130)$$

is the reproducing kernel on $\mathcal{H}^*$ with the inner product

$$(x,y)_{\mathcal{H}^*} = (\mathbf{x},\mathbf{y})_{\mathcal{H}}. \qquad (131)$$

18   Covariance kernel for a time series

In general, for a time series $X(t)$ $(t \in T)$, the covariance kernel

$$K(t_1,t_2) = Cov[X(t_1)\overline{X(t_2)}] \qquad (132)$$

is a positive matrix on T. Hence, we can investigate the time series from the viewpoint of the theory of reproducing kernels which was discussed in detail by Parzen ([P]). See also [L1,2,3] with their references.

19   Reproducing kernels in windowed Fourier transforms

The windowed Fourier transform of f will be defined by

$$(T^{win}f)(\omega, t) = \int_{-\infty}^{\infty} f(s)g(s-t)e^{-i\omega s}ds \tag{133}$$

for some function g. In general, the function g has the properties

  (i) $g(t)$ is a nonnegative real-valued function on $(-\infty, \infty)$,
  (ii) $g(t)$ has a compact support,
  (iii) $g(t)$ is identically 1 on an interval containing the origin.

The windowed Fourier transform is a standard technique for time–frequency local-ization. See, for example, [[D]], Chapter 1. In the windowed Fourier transform (133), we can consider the corresponding reproducing kernel

$$K(\omega, t; \omega', t') = \int_{-\infty}^{\infty} g(s-t)e^{-i\omega s}\overline{g(s-t')e^{-i\omega's}}ds. \tag{134}$$

20   Reproducing kernels in wavelet transforms

Let ψ be a function of $L_2(-\infty, \infty)$ satisfying

$$\int_{-\infty}^{\infty} \psi(t)dt = 0$$

and

$$C_\psi = 2\pi \int_{-\infty}^{\infty} |\xi|^{-1}|\widehat{\psi}(\xi)|^2 d\xi < \infty$$

for the Fourier transform

$$\widehat{\psi}(\xi) = \frac{1}{\sqrt{2\pi}}\int_{-\infty}^{\infty} \psi(t)e^{-i\xi t}dt.$$

Then, we shall consider a double-indexed family of wavelets from ψ by dilating and translating

$$\psi^{a,b}(x) = |a|^{-\frac{1}{2}}\psi(\frac{x-b}{a})$$

where $a, b \in \mathbb{R}, a \neq 0$. The normalization will be done by

$$\|\psi^{a,b}\|_{L_2(-\infty, \infty)} = \|\psi\|_{L_2(-\infty, \infty)} = 1.$$

Then, the continuous wavelet transform is given by

$$(T^{wav}f)(a,b) = |a|^{-\frac{1}{2}} \int_{-\infty}^{\infty} f(x)\overline{\psi(\frac{x-b}{a})}dx. \tag{135}$$

In particular,

$$|(T^{wav}f)(a,b)| \leq \|f\|.$$

In the wavelet transform (135), a very interesting identity is known; that is, for any $f, g \in L_2(-\infty, \infty)$

$$\int_{-\infty}^{\infty}\int_{-\infty}^{\infty}(T^{wav}f)(a,b)\overline{(T^{wav}g)(a,b)}\frac{dadb}{a^2} = C_\psi \int_{-\infty}^{\infty}f(x)\overline{g(x)}dx. \tag{136}$$

Indeed, we have, directly :

$$\int_{-\infty}^{\infty}\int_{-\infty}^{\infty}(T^{wav}f)(a,b)\overline{(T^{wav}g)(a,b)}\frac{dadb}{a^2}$$

$$= \int_{-\infty}^{\infty}\int_{-\infty}^{\infty}\left[|a|^{\frac{1}{2}}\int_{-\infty}^{\infty}\hat{f}(\xi)e^{-ib\xi}\overline{\hat{\psi}(a\xi)}d\xi\right]\left[|a|^{\frac{1}{2}}\int_{-\infty}^{\infty}\overline{\hat{g}(\xi')}e^{ib\xi'}\hat{\psi}(a\xi')d\xi'\right]\frac{dadb}{a^2}$$

$$= 2\pi \int_{-\infty}^{\infty}\frac{da}{|a|}\int_{-\infty}^{\infty}\hat{f}(\xi)\overline{\hat{g}(\xi)}|\hat{\psi}(a\xi)|^2 d\xi$$

$$= C_\psi \int_{-\infty}^{\infty}f(x)\overline{g(x)}dx.$$

We thus see that for the reproducing kernel

$$K(a,b;a',b') = \int_{-\infty}^{\infty}\psi^{a',b'}(x)\overline{\psi^{a,b}(x)}dx, \tag{137}$$

the norm in the RKHS H_K is given by

$$\left\{\frac{1}{C_\psi}\int_{-\infty}^{\infty}\int_{-\infty}^{\infty}|(T^{wav}f)(a,b)|^2\frac{dadb}{a^2}\right\}^{\frac{1}{2}}. \tag{138}$$

The finiteness of the integral (138) does not, however, characterize the members in H_K; that is, H_K forms a subspace in the Hilbert space comprising all functions on $\mathbb{R} \times \mathbb{R}$ with finite norms (138). So we will be interested in the characterization of the subspace H_K. In the case of Meyer wavelets, such characterization will be examined in Chapter 3, Section 5. For wavelet theory, see, for example, [[D]] and [[Ch]].

85

21 Sampling reproducing kernels

In Theorem 1.2, for a point set $\{p_j\}_j$ of E, we assume that

$$\{\mathbf{h}(p_j)\}_j \tag{139}$$

is a complete orthonormal system of $\mathcal{H}$. Then, for any $f \in H_K$ which is expressible in the form

$$f(p) = (\mathbf{f}, \mathbf{h}(p))_{\mathcal{H}}, \quad \mathbf{f} \in \mathcal{H},$$

we obtain by the Parseval–Plancherel identity, the sampling theorem

$$f(p) = \sum_j (\mathbf{f}, \mathbf{h}(p_j))_{\mathcal{H}} \overline{(\mathbf{h}(p), \mathbf{h}(p_j))_{\mathcal{H}}}$$

$$= \sum_j f(p_j) K(p, p_j) \quad \text{on} \quad E. \tag{140}$$

Hence, by this unified method we can obtain many sampling theorems. See, for example, [J], [Y] and [Kr]. Furthermore, in connection with the condition (139), we have many complete orthonormal systems of $L_2(-\infty, \infty)$ in the form

$$\psi_{m,n}(x) = a_0^{-\frac{m}{2}} \psi(a_0^{-m} x - n b_0)$$

in the wavelet transform (135). Hence, in the wavelet transforms we can obtain the corresponding sampling theorems. For wavelet theory, see, for example, [[D]] and [[Ch]].

22 Operators on Banach spaces

Suppose that a Hilbert space H is a linear subspace of a Banach space B and the embedding

$$j : H \longrightarrow B$$

is continuous. Then, H will be called a sub-Hilbert space of B. Then, we have the following interesting theorem.

Let $R : B^* \longrightarrow B$ be a symmetric positive operator. Then, there exists a sub-Hilbert space H_R of B and the embedding

$$j_R : H_R \longrightarrow B$$

satisfies

$$R = j_R j_R^*$$

86

and, the Hilbert space H_R admits the reproducing kernel

$$K(x^*, y^*) = < Rx^*, y^* > \qquad (x^*, y^* \in B^*).$$

This theorem will mean that a symmetric positive operator $R : B^* \longrightarrow B$ can be decomposed as

$$B^* \xrightarrow{j_R^*} H_R \xrightarrow{j_R} B$$

through the reproducing kernel Hilbert space H_R.

For very interesting relationship between operators on Banach spaces and reproducing kernel Hilbert spaces, see ([[VTC]], Chapter III) for the details.

§5 Realizations of reproducing kernel Hilbert spaces

In this section, for a given positive matrix $K(p, q)$ on E, we shall discuss the methods of realizing or identifying the Hilbert space H_K admitting the reproducing kernel $K(p, q)$. These methods will be needed essentially in the applications of Theorem 1.2 to miscellaneous concrete problems.

Of course, Theorem 1.2 itself and many theorems in Sections 2 and 3 are applicable in the realization of the space H_K, as shown in Section 4.

1 Direct constructions of reproducing kernel Hilbert spaces

In all the cases, the basic method of the construction of RKHS H_K for a given positive matrix $K(p, q)$ on E is given by the fundamental Theorem 2.2. If

$$\{K(p, q), q \in E\} \tag{1}$$

is finite dimensional, then there is no problem. In this case, we can construct the RKHS H_K by Theorem 2.2. See, also Section 4.1.

If the family forms an infinite dimensional space, we take a finite point set $\mathcal{N}$ of E and we consider the subspace spanned by

$$\{K(p, q), q \in \mathcal{N}\}. \tag{2}$$

If the space spanned by (1) is separable, for a set $\mathcal{N}$ with a large cardinal number of $\mathcal{N}$, the RKHS constructed by (2) will give a good approximation of the desired RKHS H_K.

Recall, in general, the basic expression of the norm in H_K

$$\|f\|_{H_K}^2 = \sup_{\{X_p\}} \frac{|\sum_p X_p f(p)|^2}{\sum_{p,q} \overline{X_p} X_q K(p, q)}, \tag{3}$$

in the numerical sense.

In general, a RKHS H_K is formed by functions on E, but we can consider the functions as the vectors on a finite point set $\mathcal{N}$ of E, in the numerical sense. Then, the reproducing kernel $K(p, q)$ forms a positive matrix

$$\|K(q_{j'}, q_j)\| \quad (q_j \in \mathcal{N} \subset E) \tag{4}$$

and so we can construct the RKHS H_K following the method of Section 4.2.

In the case of a countable set $\mathcal{N}$ in a separable Hilbert space H_K we shall discuss the realization of the RKHS H_K in detail in the following Subsection 6.

In the sequel, we shall consider several indirect realizations of reproducing kernel Hilbert spaces.

For a positive matrix $K(p, q)$ on E, we shall use some representation in the form

$$K(p, q) = (\mathbf{h}(q), \mathbf{h}(p))_{\mathcal{H}} \quad \text{on} \quad E \times E \tag{5}$$

in terms of a Hilbert space $\mathcal{H}$-valued function $\mathbf{h}$ from E into $\mathcal{H}$. In many cases, a positive matrix $K(p, q)$ itself will be given in this form (5). Recall Theorem 1.2.

If the linear transform L induced from (5)

$$f(p) = (\mathbf{f}, \mathbf{h}(p))_{\mathcal{H}}, \quad \mathbf{f} \in \mathcal{H} \tag{6}$$

is an isometry from $\mathcal{H}$ onto H_K

$$\|f\|_{H_K} = \|\mathbf{f}\|_{\mathcal{H}} \tag{7}$$

(at this stage we know only the existence of the RKHS H_K) and if we can formulate the inversion formula for (6) in the form

$$\mathbf{f} = L^{-1}f, \tag{8}$$

then we can obtain the realization of H_K in the form

$$\|f\|_{H_K} = \|L^{-1}f\|_{\mathcal{H}}. \tag{9}$$

2 Use of a complete orthonormal system of $\mathcal{H}$

For a complete orthonormal system $\{\mathbf{v}_j\}$ of $\mathcal{H}$, we define the functions $v_j(p) \in H_K$ by

$$v_j(p) = (\mathbf{v}_j, \mathbf{h}(p))_{\mathcal{H}}. \tag{10}$$

Then, if $\{\mathbf{h}(p), p \in E\}$ is complete in $\mathcal{H}$, by Theorem 2.6 we have the representation

$$K(p,q) = \sum_j v_j(p)\overline{v_j(q)} \qquad (11)$$

which converges absolutely and $\{v_j\}_j$ forms a complete orthonormal system of H_K. Hence, we can realize the RKHS H_K in the following way.

Any member f of H_K is expressible in the form

$$f(p) = \sum_j a_j v_j(p) \quad \text{on} \quad E \qquad (12)$$

which converges absolutely on E and also in H_K, for some $\{a_j\}_j \in l^2$. The norm in H_K is given by

$$\|f\|_{H_K}^2 = \sum_j |a_j|^2. \qquad (13)$$

However, we wish to represent the coefficients $\{a_j\}$ in terms of f directly, but at this moment we do not know the inner product

$$(f, v_j)_{H_K} = a_j. \qquad (14)$$

Hence, this realization will be incomplete in this sense.

If the coefficients $\{a_j\}$ are expressible in terms of f in (12), we can, however, obtain the realization (12) with (13) in terms of f. See for example (3.4.10) and (3.4.42). Here, we need only the symmetric expansion (11) and (13), and we do not need the orthogonality of $\{\mathbf{v}_j\}_j$ in H_K.

3 Use of representations by Fourier integrals

For simplicity we shall consider the one dimensional case of

$$K(x,y) = \frac{1}{2\pi} \int_{-\infty}^{\infty} e^{ity} e^{-itx} \rho(t)dt \qquad (15)$$

for some positive continuous function ρ on $(-\infty, \infty)$. For this expression, recall the Bochner theorem: For any positive matrix $K(x-y)$ on $\mathbb{R}$, there exists a nondecreasing function $\sigma(t)$ such that

$$K(x-y) = \frac{1}{2\pi} \int_{-\infty}^{\infty} e^{it(y-x)} d\sigma(t)$$

(see, for example, [[B]], [[Don1]] and [[Wi]]). See also [D1-2] with its references. For survey articles for these positive definite functions, see [Stw] and [Jo].

From (15), the images $f(x)$ of the transform

$$f(x) = \frac{1}{2\pi} \int_{-\infty}^{\infty} F(t)e^{-itx}\rho(t)dt \tag{16}$$

for functions F satisfying

$$\frac{1}{2\pi} \int_{-\infty}^{\infty} |F(t)|^2 \rho(t)dt < \infty \tag{17}$$

belong to the RKHS H_K admitting the reproducing kernel (15) and we have the isometrical identity

$$\|f\|_{H_K}^2 = \frac{1}{2\pi} \int_{-\infty}^{\infty} |F(t)|^2 \rho(t)dt. \tag{18}$$

Meanwhile, by using the inversion formula for the Fourier transform,

$$F(t)\rho(t) = \int_{-\infty}^{\infty} f(x)e^{itx}dx \tag{19}$$

in the framework of L_2 space. We thus obtain the desired realization of the space H_K in the form

$$\|f\|_{H_K}^2 = \frac{1}{2\pi} \int_{-\infty}^{\infty} \left| \int_{-\infty}^{\infty} f(x)e^{itx}dx \right|^2 \frac{dt}{\rho(t)}. \tag{20}$$

For many ρ we can represent the norm (20) in terms of derived and integrated functions of f, by using the Parseval–Plancherel identity (Chapter 3.1.6). We applied this method many times. See, for example, [Sa8, 11, 12, 14, 17, 19, 20, 22, 23, 26].

4   Use of the Mercer expansion theorem

We shall assume that a positive matrix $K(p, q)$ is continuous on $E \times E$, where E is a closed domain in a Euclidean space. Then, by the Mercer expansion theorem, we have the expansion

$$K(p, q) = \sum_{j=1}^{\infty} (\lambda_j)^{-1} \varphi_j(p)\overline{\varphi_j(q)} \tag{21}$$

which converges absolutely and uniformly on $E \times E$. Here,

$$0 < \lambda_1 \leqq \lambda_2 \leqq \cdots , \tag{22}$$

$$\lambda_j \int_E K(p, q)\varphi_j(q)dm(q) = \varphi_j(p) \quad \text{on} \quad E, \tag{23}$$

90

and

$$\int_E \varphi_j(p)\overline{\varphi_{j'}(p)}dm(p) = \delta_{jj'}. \tag{24}$$

and dm is the usual Lebesgue measure. In particular, note that φ_j are all continuous functions on E by Theorem 2.2.8. See also, for example, ([[Yo]], pp.127-139).

From (21) we see that any member $f(p)$ of H_K is expressible in the form

$$f(p) = \sum_{j=1}^{\infty} c_j(\lambda_j)^{-1}\varphi_j(p) \quad \text{on} \quad E \tag{25}$$

for some constants $\{c_j\}$ satisfying

$$\sum_{j=1}^{\infty} |c_j|^2(\lambda_j)^{-1} < \infty \tag{26}$$

and we have

$$\|f\|_{H_K}^2 = \sum_{j=1}^{\infty} |c_j|^2(\lambda_j)^{-1}. \tag{27}$$

Meanwhile, from (25) and (24) we have

$$c_j = \lambda_j \int_E f(p)\overline{\varphi_j(p)}dm(p). \tag{28}$$

Hence we have the desired realization of H_K in the form

$$\|f\|_{H_K}^2 = \sum_{j=1}^{\infty} \lambda_j \left| \int_E f(p)\overline{\varphi_j(p)}dm(p) \right|^2. \tag{29}$$

This realization of the reproducing kernel Hilbert spaces will be of general and practical use.

However, in this realization note that reproducing kernels $K(p,q)$ are, in general, not continuous on $E \times E$ for the original closed domain E of $K(p,q)$, but they are, in general, continuous on $E_1 \times E_1$ for a subset E_1 of E, as we see from various concrete reproducing kernels in Section 4.

Furthermore, the eigenvalues and eigenfunctions are, in general, not determined explicitly in (23). Even if they can be determined explicitly, the realization of the RKHS H_K in terms of (29) is rather abstract. See, for example, (3.2.65), (3.4.42) and (3.4.48). That is one reason why we have collected many concrete reproducing kernels in Section 4.

5 Integral transforms by Green's functions

Since the definition of the Green's function is, in general, involved, we shall state the formulation abstractly and roughly. For the Green's functions, see, for example, [Ar2], [[B]], [[CJ]], [[Fo]], [[Kö]], [[Ro]], [[Sz]], [[Tr]] and [[Yo]].

On some domain D, a Green's function $G(p, q)$ will be considered as the solution of the operator L satisfying

$$L_p G(p, q) = \delta(p, q) \quad \text{on} \quad E \times E \tag{30}$$

and some boundary conditions. For many cases, $G(\cdot, q) \in L_2(D, dm)$ and the solutions $f(p)$ satisfying both

$$L_p f(p) = F(p), \quad F \in L_2(D, dm) \tag{31}$$

and the boundary conditions that $G(p, q)$ satisfies, are given by the integral transform

$$f(p) = \int_D F(q) G(p, q) dm(q). \tag{32}$$

Hence, we see that the solutions $f(p)$ in (31) form the RKHS H_K admitting the reproducing kernel

$$K(p, q) = \int_D \overline{G(q, q')} G(p, q') dm(q'). \tag{33}$$

Furthermore, (32) is, in general, one to one and so we have, following Theorem 1.2

$$\|f\|_{H_K} = \|F\|_{L_2(D, dm)}. \tag{34}$$

Therefore, we have the realization of the RKHS H_K in the form

$$\|f\|_{H_K} = \|L_p f\|_{L_2(D, dm)}. \tag{35}$$

This situation will give the basic interrelationship between the Green's function associated with the equation (30) and the reproducing kernel associated with the integral transform with the Green's function as integral kernel. For some concrete examples, see Chapter 3.1.2 and also [Sa8, 11, 12, 24].

Meanwhile, for basic relationships between Green's functions and reproducing kernels, see [Ar2].

6 Realizations by discrete points

In the three viewpoints of a realization of RKHS H_K for a positive matrix $K(p, q)$ on E, a discretization of the RKHS H_K, and the representation of the H_K functions in terms of the values on some discrete points of E, we shall discuss separable Hilbert spaces comprising functions on E based on [Kö].

The results will give a nice relationship between 'finite' and 'infinite' or 'discrete' and 'continuum' in reproducing kernel Hilbert spaces.

Let H be a separable Hilbert space comprising functions on E and let $\{e_j\}$ be a countable linearly independent basis for H. We shall define

$$K_n(p,q) = \sum_{j,j'=1}^{n} \gamma_{jj'n} e_{j'}(p)\overline{e_j(q)},$$

where $\|\gamma_{jj'n}\|_{1 \le j,j' \le n}$ is the inverse of the matrix $\|(e_j, e_{j'})_H\|_{1 \le j,j' \le n}$.

Then, we shall consider the two conditions:

(c_1) If $\{K_n(q,q)\}$ converges on E as $n \to \infty$, then any Cauchy sequence $\sum_{j=1}^{n} \alpha_{n,j} e_j$ converges everywhere on E.

(c_2) Pointwise limits of such Cauchy sequences coincide with their limits in norm.

Theorem 1   *If (c_1) and (c_2) are satisfied, then the limit $K(p,q) = \lim_{n\to\infty} K_n(p,q)$ exists on $E \times E$, and $K(p,q)$ is the reproducing kernel for H.*

   Conversely, if H admits a reproducing kernel $K(p,q)$, then (c_1) and (c_2) are satisfied and $K(p,q) = \lim_{n\to\infty} K_n(p,q)$.

Proof :   If H is finite dimensional, the theorem is apparent and so we assume that H is infinite dimensional.

Let H_n be the subspace generated by $\{e_j\}_{j=1}^{n}$. $K_n(\cdot,q)$ is, of course, an element of H_n for any $q \in E$ and $H_n \subset H_m$ for $m > n$. Then, $K_n(\cdot,q)$ is the projection of $K_m(\cdot,q)$ onto H_n. Hence, we have

$$(K_m(\cdot,q), K_n(\cdot,p))_H = K_n(p,q), \quad m > n,$$

and

$$\|K_m(\cdot,q) - K_n(\cdot,q)\|_H^2 = K_m(q,q) - K_n(q,q) > 0,$$
$$m > n. \tag{36}$$

Hence, if (c_1) holds, $\{K_n(\cdot,q)\}$ is a Cauchy sequence in H for any point q of E. Let $K(\cdot,q)$ be the limit of this sequence.

For any $f \in H$, we define the function f_n by

$$f_n(q) = (f(\cdot), K_n(\cdot,q))_H$$
$$= \sum_{j,j'=1}^{n} \beta_j \gamma_{jj'n} e_{j'}(q), \quad \beta_j = (f, e_j)_H, \tag{37}$$

which is the projection of f onto H_n. Hence, we have

$$\|f - f_n\|_H^2 = \|f\|_H^2 - \|f_n\|_H^2, \tag{38}$$

$$\|f_m - f_n\|_H^2 = \|f_m\|_H^2 - \|f_n\|_H^2, \quad m > n, \tag{39}$$

and

$$\|f_n\|_H \leqq \|f_m\|_H \leqq \|f\|_H, \quad m > n. \tag{40}$$

Hence, $\{\|f_n\|_H\}$ converges and so $\{f_n\}$ is a Cauchy sequence in H.

Suppose that

$$\hat{f}_n = \sum_{j=1}^{n} \alpha_{n,j} e_j \tag{41}$$

is a sequence converging to f. Since $\hat{f}_n \in H_n$, we have

$$(f, \hat{f}_n)_H = (f - f_n + f_n, \hat{f}_n)_H$$
$$= (f_n, \hat{f}_n)_H.$$

Then,

$$\lim_{n \to \infty} (f_n, \hat{f}_n)_H = \lim_{n \to \infty} (f, \hat{f}_n)_H$$
$$= \|f\|_H^2.$$

Hence, from (40)

$$0 \leqq \lim_{n \to \infty} \|f_n - \hat{f}_n\|_H^2$$
$$= \lim_{n \to \infty} \|f_n\|_H^2 - \|f\|_H^2 \leqq 0,$$

and so,

$$\lim_{n \to \infty} \|f_n\|_H = \|f\|_H.$$

Hence, (38) shows that $\{f_n\}$ converges to f in norm.

Further,

$$\lim_{n \to \infty} f_n(q) = \lim_{n \to \infty} (f_n, K(\cdot, q))_H$$
$$= (f, K(\cdot, q))_H$$
$$= g(q) \quad \text{(we set)}. \tag{42}$$

Hence, $\{f_n\}$ converges everywhere on E. Therefore, any Cauchy sequence of the type (41) converges also everywhere on E.

Indeed, from

$$f_n(q) - \hat{f}_n(q) = (f - \hat{f}_n, K_n(\cdot, q))_H,$$

we have

$$|f_n(q) - \hat{f}_n(q)|^2 \leqq \|f - \hat{f}_n\|_H^2 K(q, q),$$

and so,

$$\lim_{n \to \infty} |f_n(q) - \hat{f}_n(q)| = 0.$$

From the inequality

$$|g(q) - \hat{f}_n(q)| \leqq |g(q) - f_n(q)| + |f_n(q) - \hat{f}_n(q)|,$$

we see that $\{\hat{f}_n(q)\}$ converges to the same limit $g(q)$ as $\{f_n(q)\}$.

In addition, if (c_2) is satisfied, the pointwise limit and the limit in norm of Cauchy sequences of the form (41) coincide. Hence, by (42), the reproducing property

$$g(q) = f(q) = (f, K(\cdot, q))_H$$

is obtained. Also, $\{K_n(p, q)\}$ converges to $K(p, q)$ on $E \times E$. Hence, $K(p, q) = \lim_{n \to \infty} K_n(p, q)$ is the reproducing kernel for H.

Conversely, we assume that H admits a reproducing kernel $K(p, q)$. Then, (36) is valid, together with

$$\|K(\cdot, q) - K_n(\cdot, q)\|_H^2 = K(q, q) - K_n(q, q).$$

In (40), by setting $f = K(\cdot, q)$

$$K_n(q, q) < K_m(q, q) < K(q, q) \quad \text{for} \quad m > n.$$

Hence, $\{K_n(q, q)\}$ converges on E and (c_1) is satisfied. The condition (c_2) is clearly satisfied, since H admits a reproducing kernel. Hence, $\lim_{n \to \infty} K_n(p, q)$ is a reproducing kernel for H and from the uniqueness of reproducing kernels we have $K(p, q) = \lim_{n \to \infty} K_n(p, q)$.

Next, we shall examine a separable Hilbert space H_K admitting the reproducing kernel $K(p, q)$ on E. Then, the class $\{K(\cdot, q); q \in E\}$ generates H_K and so there exists a countable set S of E such that $\{K(\cdot, q_j); q_j \in S\}$ is a class of linearly independent functions forming a basis for H_K. We set $S_n = \{q_1, q_2, \cdots, q_n\} \subset S$ and $\|\Gamma_{jj'n}\|_{1 \leqq j, j' \leqq n}$ is the inverse of $\|K(q_j, q_{j'})\|_{1 \leqq j, j' \leqq n}$.

Theorem 2 *For any $f \in H_K$, the sequence of functions defined by*

$$f_n(p) = \sum_{j,j'=1}^{n} f(q_j)\Gamma_{jj'n} K(p, q_{j'}) \tag{43}$$

converges to f as $n \to \infty$ in both the senses in norm and everywhere on E.

Proof : In Theorem 1, we set

$$e_j = K(p, q_j).$$

Then,

$$K_n(p, q) = \sum_{j,j'=1}^{n} K(p, q_j)\Gamma_{jj'n} K(q_{j'}, q)$$

and (37) is identical with (43).

Note that K_n coincides with K on $S_n \times E$ and $E \times S_n$, and so, $f_n = f$ on S_n. By Theorem 1, $K_n(\cdot, q)$ converges to $K(\cdot, q)$ in both the senses in norm and everywhere on E. Hence, by Theorem 1 we see that (43) converges to f in norm and everywhere on E.

Corollary 1 *The scalar product $(f, g)_{H_K}$ is given by*

$$(f, g)_{H_K} = \lim_{n \to \infty} \sum_{j,j'=1}^{n} f(q_j)\Gamma_{jj'n} \overline{g(q_{j'})}. \tag{44}$$

Finally we obtain the desired realization and representation of the RKHS H_K functions.

Corollary 2 *For any function f defined on E satisfying*

$$\lim_{n \to \infty} \sum_{j,j'=1}^{n} f(q_j)\Gamma_{jj'n} \overline{f(q_{j'})} < \infty, \quad q_j \in S, \tag{45}$$

the sequence of functions f_n defined by

$$f_n(p) = \sum_{j,j'=1}^{n} f(q_j)\Gamma_{jj'n} K(p, q_{j'}) \tag{46}$$

is a Cauchy sequence in H_K whose limit coincides with f on E.

Conversely, any member f of H_K is obtained in this way in terms of $\{f(q_j)\}$.

Corollary 2 may be called a *generalized sampling theorem* for separable reproducing kernel Hilbert spaces.

3.   Isometrical identities and inversion formulas

In this chapter, we shall examine the typical isometrical identities and inversion formulas associated with Laplace and Fourier transforms. Further, we consider the cases of the heat equation, the wave equation, analytic and harmonic functions and of the Meyer wavelets as typical applications of fundamental theorems of linear transforms. Once these general results are established, the more specific results will directly follow. Our proofs will be derived using unified methods, in a natural way.

§1   Laplace and Fourier transforms

1   The Laplace transform

As stated in Chapter 2, Section 4.10, we have the identity, for $q \geq \frac{1}{2}$

$$K_q(z, \overline{u}) = \int_0^\infty e^{-tz} e^{-t\overline{u}} t^{2q-1} dt = \frac{\Gamma(2q)}{(z+u)^{2q}} \quad \text{on} \quad R^+ \times \overline{R^+}. \tag{1}$$

For $q > \frac{1}{2}$, $K_q(z, \overline{u})$ is the Bergman–Selberg reproducing kernel on R^+ and $K_{1/2}(z, \overline{u})$ is the Szegö reproducing kernel on R^+. See (2.4.93) and (2.4.94). Hence we obtain the following isometrical identities and inversion formulas, immediately, by Theorem 2.1.2 and Theorem 2.1.5

Theorem 1   *The Laplace transform*

$$f(z) = \int_0^\infty F(t) e^{-zt} dt \tag{2}$$

for functions F satisfying

$$\int_0^\infty |F(t)|^2 t^{1-2q} dt < 0 \tag{3}$$

for $q \geq \frac{1}{2}$, satisfies the isometrical identity

$$\|f\|_{HK_q}^2 = \int_0^\infty |F(t)|^2 t^{1-2q} dt. \tag{4}$$

Furthermore, for $q > \frac{1}{2}$ we have the inversion formula

$$F(t) = s - \lim_{N \to \infty} \frac{1}{\pi \Gamma(2q-1)} \iint_{E_N} f(z) e^{-\overline{z}t} [2Rez]^{2q-2} dx dy, \tag{5}$$

97

where $\{E_N\}_N$ is a compact exhaustion of R^+ and the limit is taken in the sense of strong convergence in $L_2((0,\infty), t^{1-2q}dt)$.

In particular, for $q = \frac{1}{2}$ we have

$$F(t) = s - \lim_{x \to +0} \frac{1}{2\pi} \int_{\partial D_x} f(x+iy) e^{-xt} e^{yti} dy, \qquad (6)$$

where $D_x = \{|\hat{z}-x| < \frac{1}{x}, Re\hat{z} > x\}$, in the sense of strong convergence in $L_2((0,\infty), dt)$.

In Theorem 2.1.2 we obtain essentially the isometrical identities in the framework of Hilbert spaces. L_p versions will be obtained in general in the form of inequalities based on the results of the isometrical identities. A basic method is provided by the following convexity theorem.

The M. Riesz–Thorin convexity theorem *(cf. [[SW]], pp.177–183) Let E_1 and E_2 be two measure spaces with measures μ_1 and μ_2, respectively. Let T be a linear operator defined for all simple functions f on E_1 and taking values in the set of complex-valued μ_2-measurable functions on E_2. Suppose that T satisfies*

$$\|Tf\|_{1/q_1} \leqq M_1 \|f\|_{1/p_1}, \|Tf\|_{1/q_2} \leqq M_2 \|f\|_{1/p_2} \qquad (7)$$

for the points (p_1, q_1) and (p_2, q_2) belonging to $0 \leqq p \leqq 1, 0 \leqq q \leqq 1$. Then, for any (p,q) satisfying

$$p = (1-t)p_1 + tp_2, q = (1-t)q_1 + tq_2 \quad (0 < t < 1)$$

we have

$$\|Tf\|_{1/q} \leqq M_1^{1-t} M_2^t \|f\|_{1/p}. \qquad (8)$$

If $p > 0$, the operator has a norm preserving extension to the whole space $L_{1/p}(E_1, d\mu_1)$.

As an example, we shall establish L_p versions for the Laplace transform based on [Sa13].

Theorem 2 *For the Laplace transform*

$$f(z) = \int_0^\infty e^{-zt} F(t) t^{2s-1} dt \qquad (9)$$

of functions F satisfying

$$\int_0^\infty |F(t)|^p t^{2s-1} dt < \infty \qquad (10)$$

98

$(1 < p \leq 2, 1/p + 1/q = 1$ *and* $s > 1/2)$, *the images* $f(z)$ *are analytic on* R^+ *and we have the inequality*

$$\left\{ \frac{1}{\pi\Gamma(2s-1)} \iint_{R^+} |f(z)|^q [2\,Re\,z]^{2s-2} dx\, dy \right\}^{\frac{1}{q}} \leq \left\{ \int_0^\infty |F(t)|^p t^{2s-1} dt \right\}^{\frac{1}{p}}. \quad (11)$$

For $s = \frac{1}{2}$, *we have*

$$\lim_{x \to +0} \left\{ \frac{1}{2\pi} \int_{-\infty}^\infty |f(x+iy)|^q dy \right\}^{\frac{1}{q}} \leq \left\{ \int_0^\infty |F(t)|^p dt \right\}^{\frac{1}{p}}. \quad (12)$$

Proof :   From (9) we have

$$|f(z)| \leqq \int_0^\infty |F(t)| t^{2s-1} dt.$$

Hence, from Theorem 1 and the M. Riesz–Thorin convexity theorem we obtain the desired inequalities, immediately.

2   The Fourier transform

We shall choose a class of Fourier transforms having weighted functions whose inversion formulas are concretely given. This convenient class will be derived from the expressions of the Green's functions of linear integro-differential equations with constant coefficients on the whole space. These examples are taken from [Sa8]. For multidimensional cases, see [Sa11].

We fix nonnegative integers m and n, and we assume that $n \geq 1$. Further we fix real constants $\{a_j\}_{j=-m}^n$ $(a_n \neq 0)$ satisfying

$$\rho(t)^{-1} = \sum_{j=-m}^n a_j t^{2j} > 0 \quad \text{on} \quad \mathbb{R}. \quad (13)$$

We shall consider the integral transform

$$f(x) = \frac{1}{2\pi} \int_{\mathbb{R}} F(t) e^{-itx} \rho(t) dt \quad (14)$$

for complex-valued functions $F \in L_2(\mathbb{R}, \rho(t)dt)$ satisfying

$$\left\{ \frac{1}{2\pi} \int_{\mathbb{R}} |F(t)|^2 \rho(t) dt \right\}^{\frac{1}{2}} < \infty. \quad (15)$$

99

Then, we shall establish the isometrical identity and inversion formula.

For integrable functions $f(t)$ on $\mathbb{R}$, we set

$$f^{(-1)}(x) = \int_{-\infty}^{x} f(t)\,dt,$$

$$f^{(-2)}(x) = \int_{-\infty}^{x} f^{(-1)}(t)\,dt$$

and so on. Then, by the Parseval–Plancherel identity, we have

$$\int_{\mathbb{R}} \left(\sum_{j=-m}^{n} a_j |f^{(j)}(x)|^2 \right) dx = \frac{1}{2\pi} \int_{\mathbb{R}} |F(t)|^2 \rho(t)\,dt. \tag{16}$$

Furthermore, we see that

$$k(x,y) = \frac{1}{2\pi} \int_{\mathbb{R}} e^{-it(x-y)} \rho(t)\,dt \tag{17}$$

is the reproducing kernel for the Hilbert space H_k comprising all functions $f(x)$ such that $f(x), f'(x), \cdots, f^{(n-1)}(x)$ are continuous and tend to zero at infinity, $f^{(n-1)}$ are absolutely continuous and $f^{(n)}(x)$ belongs to $L_2(\mathbb{R}, dx)$, and equipped with the norm in (16). Hence, from Theorem 2.1.5 we obtain

Theorem 3   *In the integral transform (14), the images $f(x)$ for $F \in L_2(\mathbb{R}, \rho(t)dt)$ form the RKHS H_k defined by (17). Furthermore, we have the isometrical identity (16) and the inversion formula*

$$F(t) = s - \lim_{N \to \infty} \int_{-N}^{N} \left(\sum_{j=-m}^{n} a_j f^{(j)}(x)(it)^j e^{itx} \right) dx \tag{18}$$

in the sense of strong convergence in $L_2(\mathbb{R}, \rho dt)$.

We note that $k(x,y)$ in (17) is the Green's function of the integro-differential equation

$$\sum_{j=-m}^{n} a_j(-1)^j Y^{(2j)}(x) = \delta(y-x) \quad \text{on} \quad \mathbb{R} \tag{19}$$

whose boundary conditions are null at infinity, that is,

$$\lim_{x \to \pm\infty} Y^{(j)}(x) = 0, \quad j = 0, 1, \cdots, n-1.$$

In order to examine the solutions for $F \in L_2(\mathbb{R}, dx)$ we shall consider the integral transform

$$f(x) = \int_{\mathbb{R}} F(t)k(x,t)dt \qquad (20)$$

whose kernel satisfies

$$\widetilde{k}(x,y) = \int_{\mathbb{R}} k(x,t)\overline{k(y,t)}dt = \frac{1}{2\pi}\int_{\mathbb{R}} e^{-it(x-y)}\rho(t)^2 dt. \qquad (21)$$

We set

$$\rho(t)^{-2} = \sum_{j=-2m}^{2n} b_j t^{2j}, \quad b_j = \sum_{\nu+\mu=j} a_\nu a_\mu.$$

Hence, the RKHS $H_{\widetilde{k}}$ can be realized as in H_k. Since the family $\{k(x,t); x \in \mathbb{R}\}$ is complete in $L_2(\mathbb{R}, dt)$, we obtain

Theorem 4 *For $F \in L_2(\mathbb{R}, dx)$, the solutions $f(x)$ for the integro-differential equation*

$$\sum_{j=-m}^{n} a_j (-1)^j Y^{(2j)}(x) = F(x) \quad on \quad \mathbb{R} \qquad (22)$$

whose boundary conditions are null at infinity, form the RKHS $H_{\widetilde{k}}$, and the integral transform (20) gives an isometrical mapping from $L_2(\mathbb{R}, dx)$ onto $H_{\widetilde{k}}$. Furthermore, we have the inversion formula

$$F(t) = s - \lim_{N \to \infty} \int_{-N}^{N} \left(\sum_{j=-2m}^{2n} b_j f^{(j)}(x) \overline{\frac{\partial^j k(x,t)}{\partial x^j}} \right) dx \qquad (23)$$

in the sense of strong convergence in $L_2(\mathbb{R}, dt)$.

We gave some concrete examples in (2.4.24), (2.4.25) and (2.4.26), (2.4.27).

3　The Paley–Wiener theorem for entire functions

We shall give an interpretation and generalization of the Paley–Wiener theorem for entire functions of exponential type in connection with the Fourier–Laplace transform on $\mathbb{R}^n$.

The important generalization of the Paley–Wiener theorem ([[PW]]) was given by Plancherel and Pólya ([PP]) (see [[Fu1]] and [[Ron]]). A further extension of Plancherel and Pólya was given by Martin ([M]) in the case of functions f analytic on the octant $\text{Im} z_k > 0$ $(k = 1, 2, \cdots, n)$. We shall discuss Martin's theorem in a general situation following our method for integral transforms. The material is taken from [Sa14].

101

For some domains $D \subset \mathbb{R}^n, \Omega \subset \mathbb{C}^n$, and for functions $F \in L_2(D, dt)$, we shall consider the Fourier–Laplace transform

$$f(z) = \left(\frac{1}{2\pi}\right)^{n/2} \int_D F(t) e^{-i(z,t)} dt \qquad (24)$$

where $(z, t) = \sum_{j=1}^n z_j t_j$. In order to examine the integral transform (24), we consider the expression

$$K(z, \overline{u}; \Omega, D) = \left(\frac{1}{2\pi}\right)^n \int_D e^{-i(z,t)} e^{i(\overline{u},t)} dt. \qquad (25)$$

Note that (25) exists on $\Omega \times \overline{\Omega}$ if and only if

$$\int_D e^{2(y,t)} dt < \infty \quad \text{on} \quad \Omega. \qquad (26)$$

The condition (26) is independent of x. Hence, we can naturally consider Ω as a tube domain of the form $T_G = \mathbb{R}^n + iG \subset \mathbb{C}^n$. Moreover, we set

$$\widehat{G}_D = \left\{ y \in \mathbb{R}^n, \int_D e^{2(y,t)} dt < \infty \right\}^\circ \quad (\circ \text{ means the interior}) \qquad (27)$$

and we can consider the maximal domain $T_{\widehat{G}_D}$ to be Ω. As we see from the inequality

$$u^\alpha v^{1-\alpha} \le \alpha u + (1 - \alpha)v \quad (u, v \ge 0, 0 < \alpha < 1),$$

$\widehat{G}_D$ is a convex domain on $\mathbb{R}^n$. Thus we consider the positive matrix $K(z, \overline{u}; T_{\widehat{G}_D}, D)$ on $T_{\widehat{G}_D} \times \overline{T_{\widehat{G}_D}}$. Then, the RKHS $H_K(T_{\widehat{G}_D}, D)$ admitting the reproducing kernel $K(z, \overline{u}; T_{\widehat{G}_D}, D)$ is composed of all holomorphic functions $f(z)$ on $T_{\widehat{G}_D}$ which are expressible in the form

$$f(z) = \left(\frac{1}{2\pi}\right)^{n/2} \int_D F(t) e^{-i(z,t)} dt \qquad (28)$$

where F is an $L_2(D, dt)$ function F. Furthermore, we have the isometrical identity

$$\|f\|^2_{H_K(T_{\widehat{G}_D}, D)} = \int_D |F(t)|^2 dt. \qquad (29)$$

On the other hand, from the Parseval–Plancherel identity we have

$$\int_{\mathbb{R}^n} |f(x)|^2 dx = \int_D |F(t)|^2 dt$$

102

and so

$$\|f\|^2_{H_K(T_{\widehat{G}_D}, D)} = \int_{\mathbb{R}^n} |f(x)|^2 dx. \qquad (30)$$

We thus see that the members $f(z)$ of $H_K(T_{\widehat{G}_D}, D)$ are analytic on $T_{\widehat{G}_D}$, $L_2(\mathbb{R}^n, dx)$ integrable, and the norms of f in $H_K(T_{\widehat{G}_D}, D)$ are given by (30). In this situation we can say that when D is a bounded interval on $\mathbb{R}$ and when D is a bounded convex domain on $\mathbb{R}^n$, the theorems of Paley–Wiener ([[PW]]) and Plancherel–Pólya ([PP]) respectively give characterizations of the members $f(z)$ of $H_K(T_{\widehat{G}_D}, D)$ in terms of the growth of $f(z)$ at infinity. Thus we can, in general, attempt, as in [Sa14], to solve these fundamental problems.

In the above situation, we may characterize the members of $H_K(T_{\widehat{G}_D}, D)$ in terms of D.

In order to give a reasonable solution for this problem, we will assume that D is convex and ∂D is a smooth hypersurface on $\mathbb{R}^n$. When $D = \Pi_{j=1}^n(-\infty, a)$ $(a > 0)$, Martin ([M]) discussed the growth of the functions in $H_K(T_{\widehat{G}_D}, D)$ at infinity, but he did not give a complete answer for the above problem in his restrictive situation.

Let O be the origin of coordinates in the t-space $\mathbb{R}^n$ and let the hyperplanes $\{\Gamma\}$ pass through it and lie parallel to the limiting positions of the tangent hyperplanes of ∂D. We consider the convex cone with vertex at the origin enveloped by these hyperplanes. The nappe of this cone lying on the same side of the hyperplanes $\{\Gamma\}$ as the domain D is called the asymptotic cone of T_D. As the asymptotic cone of a bounded domain D, we take the set $\{O\}$, that is, the origin. When V is the asymptotic cone of T_D, we shall say that the domain T_D is of type V.

We consider the conjugate cone V^* of V; that is,

$$V^* = \{t^* \in \mathbb{R}^n; \sum_{j=1}^n t_j^* t_j > 0 \quad \text{for all} \quad t \in \overline{V}(closure), t \neq 0\}.$$

If D contains a whole line, then V^* does not contain any n-dimensional sphere. Furthermore, since $\widehat{G}_D = \{\phi\}$, subsequently we assume that D does not contain any whole line.

We will consider V^* in the y-space $\mathbb{R}^n$. Then as

$$e^{(y,t)} \quad \text{is bounded on} \quad t \in D \qquad (31)$$

if and only if

$$-y \in V^* \quad \text{or} \quad y \in -V^*, \qquad (32)$$

we define the support function of the convex set $\overline{D}$ (closure) by

$$H_D(y) = \max_{t \in \overline{D}}(y, t). \qquad (33)$$

103

Then, we obtain

Theorem 5 $\widehat{G}_D = -V^*$.

Proof :   For any fixed point $y^{(0)} \in -V^*$, we set

$$(y^{(0)}, t) = |y^{(0)}||t| \cos \theta_0(t) < 0 \quad \text{on} \quad V,$$

where, of course, $\theta_0(t)$ $(|\theta_0(t)| \leq \pi)$ is the angle between the two vectors $y^{(0)}$ and t in the same space $\mathbb{R}^n$. Hence, there exists Θ such that

$$|\theta_0(t)| \geq \Theta > \frac{\pi}{2} \quad \text{on} \quad V.$$

Hence, there exist $\epsilon > 0$ and $M > 0$ such that

$$|\theta_0(t)| \geq \Theta - \epsilon > \frac{\pi}{2} \quad \text{on} \quad D \cap \{|t| \geq M\}.$$

Then, from the identity

$$\int_{|x| < N} f(|x|)dx = \frac{2\sqrt{\pi^n}}{\Gamma(n/2)} \int_0^N x^{n-1} f(x)dx \tag{34}$$

([[GR]], p.623), we have

$$\int_D e^{2(y^{(0)}, t)} dt \leq \int_{D \cap \{|t| \leq M\}} e^{2(y^{(0)}, t)} dt + \int_{D \cap \{|t| \geq M\}} e^{2|y^{(0)}||t| \cos(\Theta - \epsilon)} dt$$

$$\leq \int_{D \cap \{|t| \leq M\}} e^{2(y^{(0)}, t)} dt + \int_{\mathbb{R}^n} e^{2|y^{(0)}||t| \cos(\Theta - \epsilon)} dt$$

$$\leq \int_{D \cap \{|t| \leq M\}} e^{2(y^{(0)}, t)} dt + \lim_{N \to \infty} \frac{2\sqrt{\pi^n}}{\Gamma(n/2)} \int_0^N t^{n-1} e^{2t|y^{(0)}| \cos(\Theta - \epsilon)} dt$$

$$< \infty.$$

Hence, we have $\widehat{G}_D \supset -V^*$.

On the other hand, for any point $y^{(0)} \in (-V^*)^c$, the complement, by the definition of V^* there exists a point $t^{(0)} \in V$ such that $(y^{(0)}, t^{(0)}) > 0$. Then, there exists a narrow nondegenerate (i.e. contains an n-dimensional sphere) convex cone $\Gamma(t^{(0)})$ with vertex O, such that

$$(y^{(0)}, t) > 0 \quad \text{on} \quad \Gamma(t^{(0)}),$$

and

$$D \supset \Gamma(t^{(0)}) \cap \{|t| \geq M\} \quad \text{for some} \quad M > 0.$$

Then,

$$\int_D e^{2(y^{(0)},t)}\,dt \geq \int_{\Gamma(t^{(0)})\cap\{|t|\geq M\}} e^{2(y^{(0)},t)}\,dt$$

$$\geq \int_{\Gamma(t^{(0)})\cap\{|t|\geq M\}} dt = \infty.$$

Hence, $\widehat{G}_D \subset -V^*$ and we have completed the proof of the theorem.

Now we shall give a complete answer for the fundamental problem.

Theorem 6 $f(z)$ belongs to $H_K(T_{-V^*}, D)$ if and only if

(i) $f(z)$ is holomorphic on T_{-V^*} and $L_2(\mathbb{R}^n, dx)$ integrable, and for any $y \in -V^*$ the integral

$$\int_{\mathbb{R}^n} |f(x+iy)|^2\,dx$$

exists, and
(ii) $\frac{1}{2}\varlimsup_{\substack{\rho\to\infty \\ \rho\lambda\in -V^*}} \frac{1}{\rho}\log \int_{\mathbb{R}^n} |f(x+i\rho\lambda)|^2\,dx \leq H_D(\lambda)$.

Proof :   ⟨ Necessity ⟩ We set

$$y = \rho\lambda \quad (y_j = \rho\lambda_j, \quad \rho > 0) \quad \text{and} \quad |\lambda| = 1.$$

In (28) we have, by the Parseval–Plancherel identity

$$\int_{\mathbb{R}^n} |f(x+i\rho\lambda)|^2\,dx = \int_D |F(t)|^2 e^{2\rho(\lambda,t)}\,dt \leq e^{2\rho H_D(\lambda)}\int_D |F(t)|^2\,dt.$$

Hence, for $\rho\lambda \in -V^*$ we have

$$\frac{1}{2}\varlimsup_{\rho\to\infty} \frac{1}{\rho}\log \int_{\mathbb{R}^n} |f(x+i\rho\lambda)|^2\,dx \leq H_D(\lambda).$$

⟨ Sufficiency ⟩   We shall prove that any function $f(z)$ satisfying (i) and (ii) is an image by the Fourier–Laplace transform (28) of an $L_2(D, dt)$ function. Since $f(x) \in L_2(\mathbb{R}^n, dx)$, we can define the $L_2(\mathbb{R}^n, dx)$ function $\widetilde{F}(t)$ by

$$\widetilde{F}(t) = l.i.m._{N\to\infty} \left(\frac{1}{2\pi}\right)^{n/2} \int_{|x_j|<N} f(x)e^{i(x,t)}\,dx. \tag{35}$$

Of course,

$$f(x) = l.i.m._{N\to\infty} \left(\frac{1}{2\pi}\right)^{n/2} \int_{|t_j|<N} \widetilde{F}(t)e^{-i(x,t)}\,dt, \tag{36}$$

105

in the framework of the L_2 space.

We first assume that in addition $f(x) \in L_1(\mathbb{R}^n, dx)$. Then, (35) exists in the ordinary sense

$$\widetilde{F}(t) = \left(\frac{1}{2\pi}\right)^{n/2} \int_{\mathbb{R}^n} f(x) e^{i(x,t)} dx. \tag{37}$$

By (ii), since the integrals

$$\int_{\mathbb{R}^n} |f(x+iy)|^2 dx$$

exist for all $y \in -V^*$, the integrals

$$\left(\frac{1}{2\pi}\right)^{n/2} \int_{\mathbb{R}^n} f(x+iy) e^{i(x+iy,t)} dx \tag{38}$$

also exist for all $y \in -V^*$. Moreover, by using the Cauchy integral theorem, we see that the integrals (38) are independent of $y \in -V^*$ (see [[SW]], pp.98-101 for this argument). Hence, we set

$$\widetilde{\widetilde{F}}(t) = \left(\frac{1}{2\pi}\right)^{n/2} \int_{\mathbb{R}^n} f(x+iy) e^{i(x+iy,t)} dx.$$

Then, we see that $\widetilde{\widetilde{F}}(t)$ is continuous on $\mathbb{R}^n$ and $\widetilde{\widetilde{F}}(t) = \widetilde{F}(t)$ on $\mathbb{R}^n$. Hence,

$$\int_{\mathbb{R}^n} |f(x+iy)|^2 dx = \int_{\mathbb{R}^n} |\widetilde{F}(t)|^2 e^{2(y,t)} dt.$$

For $y = \rho\lambda \in -V^*$, we then have

$$H_D(\lambda) \geqq \frac{1}{2} \varlimsup_{\rho\to\infty} \frac{1}{\rho} \log \int_{\mathbb{R}^n} |f(x+i\rho\lambda)|^2 dx$$

$$= \frac{1}{2} \varlimsup_{\rho\to\infty} \frac{1}{\rho} \log \int_{\mathbb{R}^n} |\widetilde{F}(t)|^2 e^{2\rho(\lambda,t)} dt. \tag{39}$$

Now, for any $t_0 \in \overline{D}^c$ (complement of closure) we shall show that $\widetilde{F}(t_0) = 0$. Since $\widetilde{F}$ is continuous on $\mathbb{R}^n$, when $\widetilde{F}(t_0) \neq 0$, for some closed sphere $\overline{S(t_0)}$ with centre $t_0 (\subset \overline{D}^c)$,

$$|\widetilde{F}(t)| \geqq m > 0 \quad \text{on} \quad \overline{S(t_0)} \quad \text{for some constant} \quad m.$$

Then, since

$$(\lambda, t) > H_D(\lambda) \quad \text{on} \quad \overline{S(t_0)},$$

there exists $\epsilon > 0$ such that

$$(\lambda, t) \geqq H_D(\lambda) + \epsilon \quad \text{on} \quad \overline{S(t_0)}.$$

106

Hence, from (39) we have

$$H_D(\lambda) \geqq \frac{1}{2} \lim_{\rho \to \infty} \frac{1}{\rho} \log \int_{S(t_0)} |\widetilde{F}(t)|^2 e^{2\rho(\lambda,t)} dt$$

$$\geqq \frac{1}{2} \lim_{\rho \to \infty} \frac{1}{\rho} \log \left\{ m^2 e^{2\rho(H_D(\lambda)+\epsilon)} \int_{S(t_0)} dt \right\}$$

$$= H_D(\lambda) + \epsilon,$$

which implies a contradiction. Hence, the support of $\widetilde{F}$ is contained in $\overline{D}$. Further, since (26) is valid for any $y \in -V^*$, from (36) we obtain the desired expression

$$f(z) = \left(\frac{1}{2\pi} \right)^{n/2} \int_D \widetilde{F}(t) e^{-i(z,t)} dt.$$

When $f(x)$ is in $L_2(\mathbb{R}, dx)$ but not necessarily in $L_1(\mathbb{R}^n, dx)$ we set

$$f_\epsilon(z) = f(z) \Pi_{j=1}^n \frac{\sin \epsilon z_j}{\epsilon z_j}.$$

Then we have $f_\epsilon \in L_1(\mathbb{R}^n, dx)$. Hence, the Fourier transform $\widetilde{F}_\epsilon(t)$ of $f_\epsilon(z)$ vanishes outside the convex set $\overline{D}$. From the relation

$$\widetilde{F}_\epsilon(t) = (\frac{1}{2\epsilon})^n \int_{t_1-\epsilon}^{t_1+\epsilon} \cdots \int_{t_n-\epsilon}^{t_n+\epsilon} \widetilde{F}(\hat{t}) d\hat{t}$$

we obtain

$$\lim_{\epsilon \to 0} \widetilde{F}_\epsilon(t) = \widetilde{F}(t)$$

for almost all $t \in \mathbb{R}^n$. Hence, the support of $\widetilde{F}$ is contained in $\overline{D}$. We thus complete the proof of the theorem.

In Theorem 6, $f(x)$ are also considered as the boundary values such that

$$\lim_{\substack{y \to 0 \\ y \in -V^*}} f(x+iy) = f(x)$$

in the sense of the L_2 norm (see [[SW]], Chapter III).

4   Fourier–Laplace transforms and Bergman spaces

We shall examine Fourier–Laplace transforms on $\mathbb{R}^n$ ($n \geq 2$) whose images belong to the Bergman spaces on some domains on $\mathbb{C}^n$. The central problems are to find expressions of the Bergman kernels in terms of the Fourier–Laplace transforms and to investigate the relationship between the domains and the ranges in the expressions, based on [Sa17].

We shall first design our situation for the Fourier–Laplace transform from the viewpoint that the images belong to the Bergman spaces.

For some domains $D \subset \mathbb{R}^n, \Omega \subset \mathbb{C}^n$ and for some functions, we consider the Fourier–Laplace transform

$$f(z) = \left(\frac{1}{2\pi}\right)^{n/2} \int_D g(t)e^{i(z,t)}dt, z \in \Omega. \tag{40}$$

The Bergman space $Berg(\Omega)$ is composed of all holomorphic functions $f(z)$ on Ω equipped with the norm

$$\left\{\int_\Omega |f(z)|^2 dxdy\right\}^{\frac{1}{2}} < \infty.$$

The Bergman kernel $K(z, \overline{u}; \Omega)$ is the reproducing kernel for the space $Berg(\Omega)$. See, for example, Bochner and Martin ([[BM]]) and Fuks ([[Fu2]]).

In order to examine the case that the images by (40) belong to $Berg(\Omega)$, we shall consider the following expression. For some positive continuous function $\rho(t)$ on D,

$$K(z, \overline{u}; \Omega) = \frac{1}{(2\pi)^n} \int_D e^{i(z,t)}e^{-i(\overline{u},t)}\rho(t)dt. \tag{41}$$

Since the existence domain of the Bergman kernel $K(z, \overline{u}; \Omega)$ in the form (41) is independent of x, we see immediately that we should consider the tube domain T_G over G as Ω.

We shall determine the weight $\rho(t)$. From (40) by the Parseval–Plancherel identity we have

$$\int_{\mathbb{R}^n} |f(z)|^2 dx = \int_D e^{-2(y,t)}|g(t)|^2 dt,$$

and so, formally

$$\int_{T_G} |f(z)|^2 dxdy = \int_D |g(t)|^2 \left\{\int_G e^{-2(y,t)}dy\right\} dt. \tag{42}$$

When (42) is valid, for

$$W(t; G) = \int_G e^{-2(y,t)}dy,$$

we have the isometrical identity

$$\int_{T_G} |f(z)|^2 dxdy = \int_D |g(t)|^2 W(t; G)dt. \tag{43}$$

The above identities imply that the Bergman kernel $K(z, \overline{u}; T_G)$ has the expression

$$K(z, \overline{u}; T_G) = \frac{1}{(2\pi)^n} \int_D e^{i(z,t)} e^{-i(\overline{u},t)} W(t; G)^{-1} dt \qquad (44)$$

for some domain D. Further, the integral transform

$$f(z) = \frac{1}{(2\pi)^n} \int_D F(t) e^{i(z,t)} W(t; G)^{-1} dt \qquad (45)$$

for functions F satisfying

$$\int_D |F(t)|^2 W(t; G)^{-1} dt < \infty, \qquad (46)$$

possesses the isometrical identity

$$\int_{T_G} |f(z)|^2 dx\, dy = \frac{1}{(2\pi)^n} \int_D |F(t)|^2 W(t; G)^{-1} dt. \qquad (47)$$

Conversely, we shall determine the condition that the Bergman kernel $K(z, \overline{u}; T_G)$ is expressible in the form (44). For this problem, we obtain the fundamental theorem.

Theorem 7   *The kernel $K(z, \overline{u}; T_G)$ defined by (44) exists on $T_G \times \overline{T_G}$ (complex conjugate) and it is the Bergman kernel on T_G if and only if*

$$D = \widehat{D}_G = \{t \in \mathbb{R}^n; W(t; G) < \infty\}^\circ; \qquad (48)$$

that is, $\widehat{D}_G$ is the maximal domain satisfying $W(t; G) < \infty$.

Proof :   We first note that $\widehat{D}_G$ is a convex domain on $\mathbb{R}^n$ as in (27).

We assume that (48) is valid. Then, for any $y^{(0)} \in G$, we take a neighbourhood such that

$$U(y^{(0)}, \epsilon) = \Pi_{j=1}^n (y_j^{(0)} - \epsilon, y_j^{(0)} + \epsilon) \subset G$$

for some $\epsilon > 0$. Then, we have

$$\int_{\widehat{D}_G} e^{-2(y^{(0)},t)} W(t; G)^{-1} dt$$

$$\leqq \int_{\widehat{D}_G} e^{-2(y^{(0)},t)} \frac{dt}{\int_{U(y^{(0)},\epsilon)} e^{-2(y,t)} dy}$$

$$\leqq \int_{\mathbb{R}^n} e^{-2(y^{(0)},t)} \frac{dt}{\Pi_{j=1}^n (1/2t_j)\{\exp[-2(y_j^{(0)} - \epsilon)t_j] - \exp[-2(y_j^{(0)} + \epsilon)t_j]\}}$$

$$\leqq 2^n \Pi_{j=1}^n \int_{-\infty}^{\infty} \exp(-2y_j^{(0)} t_j) \frac{t_j}{\exp[-2(y_j^{(0)} - \epsilon)t_j] - \exp[-2(y_j^{(0)} + \epsilon)t_j]} dt_j$$

$$< \infty.$$

109

Hence, the kernel $K(z, \overline{u}; T_G)$ defined by (44) on $\widehat{D}_G$ exists on $T_G \times \overline{T_G}$.

Let $H(T_G, \widehat{D}_G)$ be the RKHS admitting the reproducing kernel $K(z, \overline{u}; T_G)$. The space $H(T_G, \widehat{D}_G)$ is composed of all holomorphic functions $f(z)$ on T_G which are expressible in the form

$$f(z) = \frac{1}{(2\pi)^n} \int_{\widehat{D}_G} F(t)e^{i(z,t)}W(t;G)^{-1}dt \qquad (49)$$

for functions $F \in L_2(\widehat{D}_G, W(t;G)^{-1}dt)$. Further, we have the isometrical identity

$$\|f\|^2_{H(T_G, \widehat{G}_D)} = \frac{1}{(2\pi)^n} \int_{\widehat{G}_D} |F(t)|^2 W(t;G)^{-1}dt. \qquad (50)$$

Meanwhile, from (49) and the Parseval–Plancherel identity, we have

$$\int_{T_G} |f(z)|^2 dxdy = \frac{1}{(2\pi)^n} \int_{\widehat{D}_G} |F(t)|^2 W(t;G)^{-1}dt. \qquad (51)$$

This implies that the Hilbert space $H(T_G, \widehat{D}_G)$ is a subspace of the Bergman space $Berg(T_G)$.

On the other hand, for any $f \in Berg(T_G)$, we set

$$\tilde{g}(t) = \left(\frac{1}{2\pi}\right)^{n/2} \int_{\mathbb{R}^n} f(z)e^{-i(z,t)}dt. \qquad (52)$$

Of course, this integral exists and, further, $\tilde{g}$ is independent of $y \in G$ as we see from the Cauchy integral. In addition, $\tilde{g}(t)$ is continuous. Further, by the Parseval–Plancherel identity we have

$$\int_{T_G} |f(z)|^2 dxdy = \int_G \int_{\mathbb{R}^n} e^{-2(y,t)} |\tilde{g}(t)|^2 dtdy. \qquad (53)$$

Hence, in particular, a.e. on the support of $\tilde{g}$, $W(t;G) < \infty$, and so, on the support of $\tilde{g}$, $W(t;G) < \infty$. We thus see that any $f \in Berg(T_G)$ is expressible in the form (49) and

$$\tilde{g}(t) = \begin{cases} \frac{1}{(2\pi)^{n/2}}F(t)W(t;G)^{-1} & \text{on } \widehat{G}_D \\ 0 & \text{on } \mathbb{R}^n \setminus \widehat{G}_D. \end{cases} \qquad (54)$$

Hence, we have the desired result $H(T_G, \widehat{G}_D) = Berg(T_G)$.

From Theorem 7 and Theorem 2.1.5, we have

Theorem 8   *For the domain $\widehat{D}_G$ satisfying (48) and for the integral transform (45) we have the isometrical identity*

$$\|f\|^2_{Berg(T_G)} = \frac{1}{(2\pi)^n} \int_{\widehat{D}_G} |F(t)|^2 W(t;G)^{-1} dt \tag{55}$$

between the whole spaces $L_2(\widehat{D}_G, W(t;G)^{-1} dt)$ and $Berg(T_G)$. Further, we have the inversion formula

$$F(t)W(t;G) = s - \lim_{N \to \infty} \int_{E_N} f(z)e^{-i(\bar{z},t)} dx dy \tag{56}$$

in the sense of strong convergence in $L_2(\widehat{D}_G, W(t;G)^{-1} dt)$. Here, $\{E_N\}$ is any compact exhaustion of T_G.

For a given $\widehat{D}_G$, we take the maximal domain in the sense that

$$\widehat{G} = \left\{ y \in \mathbb{R}^n; \int_{\widehat{D}_G} e^{-2(y,t)} W(t;G)^{-1} dt < \infty \right\}^{\circ} \tag{57}$$

or

$$\widehat{G} = \{ y \in \mathbb{R}^n; K(z,\bar{z};T_G) < \infty \}^{\circ}. \tag{58}$$

Then, we see that $\widehat{G}$ is also convex. This fact corresponds to the classical fact that every holomorphic function on T_G has an analytic continuation to a holomorphic function on $T_{G'}$, where G' is the convex hull of G. See [[BM]] and [[Fu2]]. In our case, we see that any member f of $Berg(T_G)$ can be continued analytically onto $T_{\widehat{G}}$ and its continuation belongs to $Berg(T_{\widehat{G}})$.

When G or $\widehat{G}$ contains a whole line, then we see that the corresponding domain $\widehat{D}_G$ or $\widehat{D}_{\widehat{G}}$ is void. This fact implies that a necessary and sufficient condition for $Berg(T_G)$ or $Berg(T_{\widehat{G}})$ to contain a function that is not identically zero is that G or $\widehat{G}$ does not contain a whole line, respectively. See also, [[SW]], p. 94.

We now obtain

Theorem 9   *(i) If $\widehat{G}$ is a bounded domain on $\mathbb{R}^n$, then we have $\widehat{D}_{\widehat{G}} = \mathbb{R}^n$.   (ii) If $\widehat{G}$ is a convex cone Γ with vertex at the origin, then we have $\widehat{D}_{\widehat{G}} = \Gamma^*$. (iii)   If $T_{\widehat{G}}$ is of type V, then we have $\widehat{D}_{\widehat{G}} = V^*$.*

Proof :   (i) is clear and we see (ii) from the following proof of (iii).
For any fixed $t^{(0)} \in V^*$, we set

$$(y, t^{(0)}) = |y||t^{(0)}| \cos \theta_0(y) > 0 \quad \text{on} \quad V,$$

$$(|\theta_0(y)| < \frac{\pi}{2}).$$

111

Hence, there exists Θ such that

$$|\theta_0(y)| < \Theta < \frac{\pi}{2} \quad \text{on} \quad V.$$

Hence, there exist $\epsilon > 0$ and $M > 0$ such that

$$|\theta_0(y)| < \Theta - \epsilon < \frac{\pi}{2} \quad \text{on} \quad \widehat{G} \cap \{|y| \geqq M\}.$$

Then, from (34) we have

$$\int_{\widehat{G}} e^{-2(y,t^{(0)})} dy \leqq \int_{\widehat{G} \cap \{|y| \leqq M\}} \exp[-2(y, t^{(0)})] dy$$

$$+ \int_{\widehat{G} \cap \{|y| \geqq M\}} \exp[-2|y||t^{(0)}| \cos(\Theta - \epsilon)] dy$$

$$\leqq \int_{\widehat{G} \cap \{|y| \leqq M\}} \exp[-2(y, t^{(0)})] dy$$

$$+ \int_{\mathbf{R}^n} \exp[-2|y||t^{(0)}| \cos(\Theta - \epsilon)] dy$$

$$\leqq \int_{\widehat{G} \cap \{|y| \leqq M\}} \exp[-2(y, t^{(0)})] dy$$

$$+ \lim_{N \to \infty} \frac{2\sqrt{\pi^n}}{\Gamma(n/2)} \int_0^N y^{n-1} \exp[-y(2|t^{(0)}| \cos(\Theta - \epsilon))] dy$$

$$< \infty.$$

Hence, we have $V^* \subset \widehat{D}_{\widehat{G}}$.

On the other hand, for any $t^{(0)} \in \mathbf{R}^n \setminus V^*$, by definition of V^*, there exists a point $y^{(0)} \in V$ such that $(y^{(0)}, t^{(0)}) < 0$. Then, there exists a narrow, nondegenerate, convex cone $\Gamma(y^{(0)})$ with vertex at the origin such that $(y, t^{(0)}) < 0$ on $\Gamma(y^{(0)})$, and

$$\widehat{G} \supset \Gamma(y^{(0)}) \cap \{|y| \geqq M\} \quad \text{for some constant} \quad M > 0.$$

Then, we have

$$\int_{\widehat{G}} e^{-2(y,t^{(0)})} dy \geqq \int_{\Gamma(y^{(0)}) \cap \{|y| \geqq M\}} e^{-2(y,t^{(0)})} dy \geqq \int_{\Gamma(y^{(0)}) \cap \{|y| \geqq M\}} dy = \infty.$$

Hence, $V^* \supset \widehat{D}_{\widehat{G}}$ and so we have the desired result $V^* = \widehat{D}_{\widehat{G}}$.

It is, in general, difficult to compute the Bergman kernel $K(z, \overline{u}; T_G)$ concretely, but for the concrete expressions of the Bergman kernels of the classical domains, see [[Hu]] and [Gi1], [Gi2].

112

We will be able to discuss the weighted Bergman kernel on T_G and inequalities of Hausdorff–Young type for general $L_p(D, W(t; G)^{-1}dt)(p \geqq 1)$ functions as in [Sa13]. But, the arguments are similar and formal as in Theorem 2, and so we do not refer to them here.

5 Hilbert spaces of Szegö type and Fourier–Laplace transforms

Let G be a domain on $\mathbb{R}^n$ $(n \geq 2)$ and $T_G = \mathbb{R}^n + iG \subset \mathbb{C}^n$ be the tube domain over G. Then, we shall examine the Fourier–Laplace transforms on some subset of $\mathbb{R}^n$ whose images belong to the Hilbert spaces of Szegö type on T_G, in a general situation, based on [Sa23]. From the viewpoints of the 'Szegö space' on $\mathbb{C}^n$ and the Fourier–Laplace transform on $\mathbb{R}^n$, we shall define a kernel form $K(z, \overline{w})$ on $\mathbb{C}^n$ in a general situation and we shall examine the Hilbert space admitting the reproducing kernel $K(z, \overline{w})$. Some important and typical cases were investigated by Gindikin ([Gi1-2]) (see [[Fu1]]). The situation of Gindikin is perfect in a reasonable sense, but the situation will be restrictive in two senses. The first sense is on the assumption for the boundary values of the members of the Szegö space. Indeed, he assumes that the members of the Szegö space can be extended continuously up to the boundary. Therefore, his space does not form a Hilbert space. In order to define his space as a Hilbert space, we must investigate the properties of boundary values of the members of the space. The second sense is on the assumption for the existence domain of the Szegö space. His existence domain is natural in the sense of holomorphic function theory and perfect in this sense. At the same time, his reproducing kernels will be strongly restrictive for this fact. Compare our explicit examples of reproducing kernels with Gindikin's examples. We will first give a kernel form in a relaxed form and examine the Hilbert space admitting the reproducing kernel. We will state our situation clearly in the next paragraph.

In order to find a reasonable and general definition of Hilbert spaces of Szegö type, we first design our situation for the Fourier–Laplace transforms from the viewpoint that the images belong to the Hilbert spaces of Szegö type.

For some domain $D \subset \mathbb{R}^n$ and for some functions g on D, we consider the Fourier–Laplace transform (40).

We will discuss the case where the images $f(z)$ belong to a Hilbert space of Szegö type. Here, a Hilbert space of Szegö type on some domain Ω on $\mathbb{C}^n$ will mean that:

(i) $\partial\Omega$ is composed of a finite number of piecewise smooth hypersurfaces on $\mathbb{C}^n$.

(ii) We consider the class $H_2(\Omega)$ of all holomorphic functions $f(z)$ on Ω such that

$$\overline{\lim_{\varepsilon \to 0}} \int_{\partial\Omega_\varepsilon} |f(z)|^2 d\mu_\varepsilon < \infty$$

for $\Omega_\epsilon = \{z \in \Omega | g(z, w) > \epsilon\}$ for the Green's function $g(z, w)$ of Ω with pole at any fixed $w \in \Omega$ and for the surface measure $d\mu_\epsilon$ of $\partial\Omega_\epsilon$. Then, for any measurable set $S \subset \partial\Omega$, $f(z)$ have nontangential boundary values $f(\xi)$ a.e. on S and

$$\left\{ \int_S |f(\xi)|^2 \omega(\xi) d\mu(\xi) \right\}^{\frac{1}{2}} < \infty$$

for the surface measure $d\mu$ of S and for any positive continuous function ω on S.

(iii) $H_2(\Omega; S, \omega)$ denotes a Hilbert space equipped with the inner product

$$(f, g)_{H_2(\Omega;S,\omega)} = \int_S f(\xi)\overline{g(\xi)}\omega(\xi) d\mu(\xi).$$

(iv) $H_2(\Omega; S, \omega)$ admits a reproducing kernel $K(z, \overline{w}; \Omega, S, \omega)$ such that for any fixed $w \in \Omega, K(z, \overline{w}; \Omega, S, \omega) \in H_2(\Omega; S, \omega)$ and, for any $f \in H_2(\Omega; S, \omega)$ and for any $w \in \Omega$,

$$f(w) = (f(\cdot), K(\cdot, \overline{w}; S, \omega))_{H_2(\Omega;S,\omega)}.$$

The reproducing kernel $K(z, \overline{w}; \Omega, S, \omega)$ and the Hilbert space $H_2(\Omega; S, \omega)$ are called the Szegö kernel and the Szegö space on Ω, respectively, for both cases of $S = \partial\Omega$ and a subset S of $\partial\Omega$. See, for example, [[SW]], [[Fu1]] and [[AY]].

In order to examine the case where the images $f(z)$ by (40) belong to $H_2(\Omega; S, \omega)$, we consider the following expression, for a positive and continuous function $\rho(t)$ on D

$$K(z, \overline{w}; \Omega, S, \omega) = \frac{1}{(2\pi)^n} \int_D e^{i(z,t)} e^{-i(\overline{w},t)} \rho(t) dt. \tag{59}$$

Since the existence domain of the kernel (59) is independent of the real parts of z and w, we see immediately that we should take the tube domain T_G as Ω. Since now S is a subset of ∂T_G or ∂T_G itself, we will assume that S is a tube of the form $T_{S_0} = \mathbb{R}^n + iS_0 \ (S_0 \subset \partial G)$.

We will determine formally the weight $\rho(t)$. From (40), we have by Plancherel's theorem

$$\int_{\mathbb{R}^n} |f(z)|^2 dx = \int_D e^{-2(y,t)} |g(t)|^2 dt$$

and so, we have formally, for the surface measure $d\mu$ of S_0

$$\int_{T_{S_0}} |f(z)|^2 \omega(y) dx d\mu(y) = \int_D |g(t)|^2 \left\{ \int_{S_0} e^{-2(y,t)} \omega(y) d\mu(y) \right\} dt. \tag{60}$$

Here, we exchanged the function $\omega(y)$ and the measure $d\mu$ as follows. The function $\omega(y)$ is a positive and continuous function on S_0 which is independent of x, and $d\mu$ is the surface measure of S_0 which is not the measure on ∂T_{S_0}. Since S_0 is a subset

114

of ∂G, we must give an interpretation for the 'boundary' values $f(z)$ on $\mathbb{R}^n \times S_0$ in (60). When the equation (60) is valid, for

$$W(t; S_0, \omega) = \int_{S_0} e^{-2(y,t)} \omega(y) d\mu(y) \quad \text{on} \quad D,$$

we obtain the isometrical identity

$$\int_{TS_0} |f(z)|^2 \omega(y) dx d\mu(y) = \int_D |g(t)|^2 W(t; S_0, \omega) dt. \tag{61}$$

The above arguments will mean that the 'Szegö' kernel has the expression

$$K(z, \overline{w}; T_G, S_0, \omega) = \frac{1}{(2\pi)^n} \int_D e^{i(z,t)} e^{-i(\overline{w},t)} W(t; S_0, \omega)^{-1} dt \tag{62}$$

for some domains T_G and D, under the condition that the 'Szegö' kernel is expressible in terms of the Fourier–Laplace transform (40).

Here, we shall consider the kernel form (62), in a general and natural situation and we will examine the Hilbert space admitting the reproducing kernel (62). Then, we shall be able to obtain the Hilbert spaces of Szegö type in a general situation. Moreover, the reproducing kernels are explicit and simple, but unusual Hilbert spaces of Szegö type will occur in our investigations. For these, we shall give several explicit examples of reproducing kernels which are not covered by Gindikin.

For a set S_0 of a finite number of piecewise smooth surfaces on $\mathbb{R}^n$ of topological dimension $k < n$ and for a positive and continuous function ω on S_0, we set

$$W(t; S_0, \omega) = \int_{S_0} e^{-2(y,t)} \omega(y) d\mu(y)$$

for the surface measure $d\mu$ of S_0.

Note that $W(t; S_0, \omega)$ is a convex function in t and so it is continuous in t on the interior of a set where $W(t; S_0, \omega)$ is bounded. Let us take any domain $D(S_0, \omega)$ on $\mathbb{R}^n$ contained in the set

$$\{t \in \mathbb{R}^n; 0 < W(t; S_0, \omega) < \infty\}.$$

This mild condition for the arbitrariness of $D(S_0, \omega)$ will enable us to consider a wide class of reproducing kernels containing Gindikin's reproducing kernels.

Next, we consider the following maximal domain

$$G[D(S_0, \omega)] = \left\{ y \varepsilon \mathbb{R}^n; \int_{D(S_0, \omega)} e^{-2(y,t)} W(t; S_0, \omega)^{-1} dt < \infty \right\}^{\circ}.$$

Then, we see that $G[D(S_0, \omega)]$ is a convex domain on $\mathbb{R}^n$. Then, the function

$$K(z, \overline{w}; T_{G[D(S_0,\omega)]}) = \frac{1}{(2\pi)^n} \int_{D(S_0,\omega)} e^{i(z,t)} e^{-i(\overline{w},t)} W(t; S_0, \omega)^{-1} dt \qquad (63)$$

exists on $T_{G[D(S_0,\omega)]} \times \overline{T_{G[D(S_0,\omega)]}}$ (complex conjugate). Further, note that it is analytic on $T_{G[D(S_0,\omega)]} \times \overline{T_{G[D(S_0,\omega)]}}$ and is a positive matrix on $T_{G[D(S_0,\omega)]}$. Hence, there exists a uniquely determined Hilbert space $H(T_{G[D(S_0,\omega)]})$ comprised of holomorphic functions on $T_{G[D(S_0,\omega)]}$ and admitting the reproducing kernel $K(z, \overline{w}; T_{G[D(S_0,\omega)]})$.

On the other hand, as we see from the expression (63), this Hilbert space $H(T_{G[D(S_0,\omega)]})$ is composed of all holomorphic functions $f(z)$ on $T_{G[D(S_0,\omega)]}$ which are expressible in the form

$$f(z) = \frac{1}{(2\pi)^n} \int_{D(S_0,\omega)} F(t) e^{i(z,t)} W(t; S_0, \omega)^{-1} dt \qquad (64)$$

for functions $F \in L_2(D(S_0, \omega), W(t; S_0, \omega)^{-1} dt)$ satisfying

$$\left\{ \frac{1}{(2\pi)^n} \int_{D(S_0,\omega)} |F(t)|^2 W(t; S_0, \omega)^{-1} dt \right\}^{\frac{1}{2}} < \infty. \qquad (65)$$

Furthermore, we have the isometrical identity

$$\|f\|^2_{H(T_{G[D(S_0,\omega)]})} = \frac{1}{(2\pi)^n} \int_{D(S_0,\omega)} |F(t)|^2 W(t; S_0, \omega)^{-1} dt. \qquad (66)$$

From the identity

$$\int_{S_0} \left\{ \int_{D(S_0,\omega)} |F(t)|^2 e^{-2(y,t)} W(t; S_0, \omega)^{-2} dt \right\} \omega(y) d\mu(y)$$

$$= \int_{D(S_0,\omega)} |F(t)|^2 W(t; S_0, \omega)^{-1} dt < \infty,$$

we see that for almost all y of S_0 with respect to $d\mu$

$$\int_{D(S_0,\omega)} |F(t)|^2 e^{-2(y,t)} W(t; S_0, \omega)^{-2} dt < \infty. \qquad (67)$$

We denote the convex hull of S_0 by $\hat{S}_0$. Then, for almost all y of $\hat{S}_0$, (67) is valid. Hence, for almost all y of $\hat{S}_0$, the function $f(x + iy)$ defined by (64) can be extended

116

onto $\mathbb{R}^n$ by means of the Fourier integral in the framework of L_2 space and belongs to $L_2(\mathbb{R}^n, d\boldsymbol{x})$. In this sense, we obtain the isometrical identity

$$\int_{T_{S_0}} |f(z)|^2 \omega(y) d\boldsymbol{x} d\mu(y) = \frac{1}{(2\pi)^n} \int_{D(S_0,\omega)} |F(t)|^2 W(t; S_0, \omega)^{-1} dt \qquad (68)$$

and so we obtain a 'realization' of the norm in $H(T_{G[D(S_0,\omega)]})$

$$\|f\|^2_{H(T_{G[D(S_0,\omega)]})} = \int_{T_{S_0}} |f(z)|^2 \omega(y) d\boldsymbol{x} d\mu(y), \qquad (69)$$

by means of the Fourier integral. As to the relationship of S_0 to $G[D(S_0, \omega)]$, we obtain

Theorem 10   *The interior $(\hat{S}_0)^\circ$ of $\hat{S}_0$, satisfies the inclusion*

$$(\hat{S}_0)^\circ \subset G[D(S_0, \omega)].$$

Proof :   For any $F \in L_2(D(S_0, \omega), W(t; S_0, \omega)^{-1} dt)$, we have (67) for a.e. y of $\hat{S}_0$. For any $y \in (\hat{S}_0)^\circ$, we can easily take a line through the point and connecting two points y_1 and y_2 on $\hat{S}_0$ such that (67) is valid for y_1 and y_2. Hence, we see that for the point y, (67) is valid. Then, for all $F \in L_2(D(S_0, \omega), W(t; S_0, \omega)^{-1} dt)$,

$$|f(\boldsymbol{x} + iy)| = \frac{1}{(2\pi)^n} \left| \int_{D(S_0,\omega)} F(t) e^{i(\boldsymbol{x},t)} e^{-(y,t)} W(t; S_0, \omega)^{-1} dt \right| < \infty.$$

Hence, by Landau's theorem for the converse of Hölder's inequality, we have the desired result

$$\int_{D(S_0,\omega)} e^{-2(y,t)} W(t; S_0, \omega)^{-1} dt < \infty.$$

We thus obtain Theorem 10.

By Theorem 10, if $(\hat{S}_0)^\circ \neq \phi$, a point of S_0 is contained in $G[D(S_0, \omega)]$ or is a boundary point of $G[D(S_0, \omega)]$. When a point y_0 of S_0 is a boundary point of $G[D(S_0, \omega)]$, we want to define $f(z_0)$ $(z_0 = \boldsymbol{x} + iy_0)$ as a boundary value of the analytic function $f(z)$ on $T_{G[D(S_0,\omega)]}$ in (69). Now S_0 is composed of a finite number of piecewise smooth surfaces on $\mathbb{R}^n$ of topological dimension $k < n$. Then, for y_0 when we can take an open polyhedron P_{y_0} (i.e. the interior of the convex hull of a finite number of points) contained in $G[D(S_0, \omega)]$ and having y_0 as a boundary point such that for some constant A

$$\int_{\mathbb{R}^n} |f(z)|^2 d\boldsymbol{x} = \frac{1}{(2\pi)^n} \int_{D(S_0,\omega)} |F(t)|^2$$

$$\times e^{-2(y,t)} W(t; S_0, \omega)^{-2} dt \le A < \infty \quad \text{on} \quad P_{y_0}, \qquad (70)$$

117

$f(x + iy)$ tends to $f(x + iy_0)$ in the L_2 norm as y tends to y_0 within P_{y_0}. Here, $f(x + iy_0)$ is defined by the Fourier transform (64). See [[SW]], Chapter III, Section 2. The inequality (70) is valid when

(i) $(\hat{S}_0)^\circ \neq \phi$ and there exists an open polyhedron $P_{y_0}{}'$ contained in $(\hat{S}_0)^\circ$ and having y_0 as a boundary point such that for $y \in P_{y_0}{}'$,

$$\int_{\mathbb{R}^n} |f(x + iy)|^2 dx = \int_{D(S_0,\omega)} |F(t)|^2 e^{-2(y_0,t)} W(t; S_0, \omega)^{-2} dt < \infty$$

and

$$P_{y_0} \subset P_{y_0}{}',$$

or

(ii) $e^{-2(y,t)} W(t; S_0, \omega)^{-1}$ is bounded for $t \in D(S_0, \omega)$ and $y \in P_{y_0}$.

When (70) is valid a.e. on $S_0 \cap \partial G[D(S_0, \omega)]$, we can define the *restricted* boundary values $f(x + iy_0)$ in the above sense. Then, of course, (69) is valid. Hence, we can regard (69) as a ' Szegö' norm. So we shall call this space $H(T_{G[D(S_0,\omega)]})$ and this kernel $K(z, \overline{w}; T_{G[D(S_0,\omega)]})$ a Hilbert space and a reproducing kernel of Szegö type, respectively, in our sense.

In order to investigate the interrelationship between $D(S_0, \omega)$ and $G[D(S_0, \omega)]$, we assume that $D(S_0, \omega)$ is a convex domain on $\mathbb{R}^n$, does not contain whole lines, and $\partial D(S_0, \omega)$ is a piecewise smooth hypersurface.

Then, we obtain

Theorem 11   *We assume that* $T_{D(S_0,\omega)}$ *is of type V.*

(I) If for any constant $L_1 > 0$,

$$W(t; S_0, \omega)^{-1} dt \quad \text{is integrable on} \quad D(S_0, \omega) \cap \{|t| \leq L_1\}$$

and there exist constants m_0 *and* L_2 *such that*

$$W(t; S_0, \omega) \geq m_0 > 0 \quad on \quad D(S_0, \omega) \cap \{|t| \geq L_2\},$$

then we have

$$G[D(S_0, \omega)] \supset V^*.$$

(II) If there exist constants M_0 *and* L *such that*

$$W(t; S_0, \omega) \leq M_0 \quad on \quad D(S_0, \omega) \cap \{|t| \geq L\},$$

then we have

$$G[D(S_0, \omega)] \subset V^*.$$

Proof :   For any $y_0 \in V^*$, we have

$$(y_0, t) = |y_0||t| \cos \theta_0(t) > 0 \quad \text{on} \quad V,$$

where, of course, $\theta_0(t)$ $(0 \leq \theta_0(t) < \frac{\pi}{2})$ is the angle between the two vectors y_0 and t in the same space $\mathbb{R}^n$. Hence, there exists a constant θ such that

$$\theta_0(t) < \theta < \frac{\pi}{2} \quad \text{on} \quad V.$$

Hence, there exist constants $\varepsilon > 0$ and $L_3 > 0$ such that

$$\theta_0(t) < \theta - \varepsilon < \frac{\pi}{2} \quad \text{on} \quad D(S_0, \omega) \cap \{|t| \geq L_3\}.$$

Then, we have:

$$\int_{D(S_0,\omega)} e^{-2(y_0,t)} W(t; S_0, \omega)^{-1} dt$$

$$\leq \int_{D(S_0,\omega) \cap \{|t| \leq \max(L_2,L_3)\}} e^{-2(y_0,t)} W(t; S_0, \omega)^{-1} dt$$

$$+ \frac{1}{m_0} \int_{D(S_0,\omega) \cap \{|t| \geq \max(L_2,L_3)\}} e^{-2|y_0||t| \cos(\theta-\varepsilon)} dt$$

$$\leq \int_{D(S_0,\omega) \cap \{|t| \leq \max(L_2,L_3)\}} e^{-2(y_0,t)} W(t; S_0, \omega)^{-1} dt$$

$$+ \lim_{N \to \infty} \frac{2\sqrt{\pi^n}}{m_0 \Gamma(\frac{n}{2})} \int_0^N t^{n-1} \exp[-t\{2|y_0| \cos(\theta - \varepsilon)\}] dt$$

$$< \infty.$$

Hence, we have the desired result (I).

Next, for any $y_0 \in \mathbb{R}^n - V^*$, by definition of V^*, there exists a point $t_0 \in V$ such that $(y_0, t_0) < 0$. Then, there exists a narrow nondegenerate (i.e. contains an n-dimensional sphere) convex cone $\Gamma(t_0)$ with vertex at the origin such that

$$(y_0, t) < 0 \quad \text{on} \quad \Gamma(t_0)$$

and

$$D(S_0, \omega) \supset \Gamma(t_0) \cap \{|t| \geq M\} \quad \text{for some constant} \quad M > 0.$$

Then, we have

$$\int_{D(S_0,\omega)} e^{-2(y_0,t)} W(t; S_0, \omega)^{-1} dt \geq \frac{1}{M_0} \int_{\Gamma(t_0) \cap \{|t| \geq \max(L,M)\}} e^{-2(y_0,t)} dt = \infty.$$

119

Hence, we see that $y_0 \notin G[D(S_0, \omega)]$ and so we have the desired result (II).

The members $f(z)$ of $H(T_{G[D(S_0,\omega)]})$ are analytic on $T_{G[D(S_0,\omega)]}$ and have finite norms (69). Conversely, in order to give a characterization of the members of $H(T_{G[D(S_0,\omega)]})$, note that the members $f(z)$ are characterized by the expression (64) with (65).

Theorem 12 *If (70) is valid; that is, $f(x + iy)$ are $L_2(\mathbb{R}^n, dx)$ integrable and*

$$\int_{\mathbb{R}^n} |f(z)|^2 dx \quad \text{are bounded on some} \quad P_{y_0} \tag{71}$$

for a.e. y_0 of $S_0 \cap \partial G[D(S_0, \omega)]$ and $D(S_0, \omega)$ is the maximal domain in the sense of

$$D(S_0, \omega) = D^m(S_0, \omega) = \{t \in \mathbb{R}^n; W(t; S_0, \omega) < \infty\}^\circ,$$

then the analytic functions $f(z)$ on $T_{G[D(S_0,\omega)]}$ having finite norms (69) in the sense of the restricted boundary values $f(z)$ belong to the space $H(T_{G[D(S_0,\omega)]})$.

Proof : For any analytic function $f(z)$ on $T_{G[D(S_0,\omega)]}$ having a finite norm (69), we set

$$g_y(t) = \frac{1}{(2\pi)^{\frac{n}{2}}} \int_{\mathbb{R}^n} f(z) e^{-i(z,t)} dx. \tag{72}$$

From our assumption (71), we see that for almost all y of S_0 and on the corresponding polyhedron P_y, $f(x + iy) \in L_2(\mathbb{R}^n)$. Hence, the integrals in (72) can be defined by means of the strong convergence and the functions $g_y(t)$ belong to $L_2(\mathbb{R}^n, dt)$. By Theorem 10, $(\hat{S}_0)^\circ \subset G[D(S_0, \omega)]$, and so $g_y(t)$ are independent of $y \in G[D(S_0, \omega)]$ as we see from the Cauchy integral formula. Hence, we set $g_y(t) = g(t)$. Then, we have by Plancherel's theorem

$$\int_{T_{S_0}} |f(z)|^2 \omega(y) dx d\mu(y) = \int_{\mathbb{R}^n} |g(t)|^2 \left\{ \int_{S_0} e^{-2(y,t)} \omega(y) d\mu(y) \right\} dt < \infty.$$

Hence, we see that a.e. on $\mathbb{R}^n - D^m(S_0, \omega)$, $g(t) = 0$. Note that the maximal domain $D^m(S_0, \omega)$ is convex. Therefore, we see that $f(z)$ is expressible in the form (64) with (65) for

$$g(t) = \begin{cases} \frac{1}{(2\pi)^{\frac{n}{2}}} F(t) W(t; S_0, \omega)^{-1} & \text{on} \quad D^m(S_0, \omega) \\ 0 & \text{on} \quad \mathbb{R}^n - D^m(S_0, \omega). \end{cases}$$

We thus have the desired result that $f(z) \in H(T_{G[D(S_0,\omega)]})$.

In Theorem 12, when the domain $D(S_0, \omega)$ is not the maximal domain $D^m(S_0, \omega)$, in order to characterize the members of $H(T_{G[D(S_0,\omega)]})$, we need more additional conditions for analytic functions $f(z)$ on $T_{G[D(S_0,\omega)]}$ with finite norms (69). See example (1).

As a typical case, we shall determine the case where the solutions of Schrödinger's equation belong to the Hilbert space $H(T_{G[D(S_0,\omega)]})$ of our Szegö type.

The fundamental solution of the Schrödinger equation

$$\frac{\partial u(x,t)}{\partial t} - ib\Delta u(x,t) = 0 \quad (b > 0)$$

is given by

$$E(x,t) = Y(t)\left(\frac{1-i}{2\sqrt{2\pi bt}}\right)^n \exp\frac{ix^2}{4bt}$$

(cf. [[Sz]], pp. 371–374). So we shall examine the integral transform

$$u(z,t) = \left(\frac{1-i}{2\sqrt{2\pi bt}}\right)^n \int_D \exp\{\frac{i}{4bt}\sum_{j=1}^n (z_j - \xi_j)^2\} g(\xi) d\xi \tag{73}$$

for some functions $g(\xi)$ and for some domain $D \subset \mathbb{R}^n$. For this purpose, let us consider the kernel form

$$K_S(z, \overline{w}; \rho) = \int_D \left(\frac{1-i}{2\sqrt{2\pi bt}}\right)^n \exp\{\frac{i}{4bt}\sum_{j=1}^n (z_j - \xi_j)^2\}$$
$$\cdot \left(\frac{1-i}{2\sqrt{2\pi bt}}\right)^n \exp\{\frac{i}{4bt}\sum_{j=1}^n (w_j - \xi_j)^2\} \rho(\xi) d\xi \tag{74}$$

for a positive and continuous function ρ on D. Then, we have

$$K_S(z, \overline{w}; \rho) = \left(\frac{1}{4\pi bt}\right)^n \exp\{\frac{i}{4bt}\sum_{j=1}^n z_j^2\} \exp\{\frac{-i}{4bt}\sum_{j=1}^n \overline{w}_j^2\}$$
$$\cdot \int_D \exp\{\frac{-i}{2bt}(z, \xi)\} \exp\{\frac{i}{2bt}(\overline{w}, \xi)\} \rho(\xi) d\xi$$
$$= \exp\{\frac{i}{4bt}\sum_{j=1}^n z_j^2\} \exp\{\frac{-i}{4bt}\sum_{j=1}^n \overline{w}_j^2\}$$
$$\cdot \frac{1}{(2\pi)^n} \int_{-\frac{1}{2bt}D} e^{i(z,\xi)} e^{-i(\overline{w},\xi)} \rho(-2bt\xi) d\xi$$
$$= \exp\{\frac{i}{4bt}\sum_{j=1}^n z_j^2\} \exp\{\frac{-i}{4bt}\sum_{j=1}^n \overline{w}_j^2\}$$
$$\cdot K\left(z, \overline{w}; T_{G[-\frac{1}{2\pi t}D(S_0,\omega)]}\right), \tag{75}$$

121

for

$$\rho(\xi) = W(-\frac{\xi}{2bt}; S_0, \omega)^{-1}.\qquad(76)$$

Here, $-\frac{1}{2bt}D$ denotes the domain $\{-\frac{\xi}{2bt}; \xi \in D\}$. The above arguments imply that in the integral transform

$$v(z,t) = \left(\frac{1-i}{2\sqrt{2\pi bt}}\right)^n \int_{-\frac{1}{2bt}D} \exp\{\frac{i}{4bt}\sum_{j=1}^{n}(z_j - \xi_j)^2\}F(\xi)\rho(\xi)d\xi \qquad(77)$$

for functions $F(\xi)$ satisfying

$$\int_{-\frac{1}{2bt}D} |F(\xi)|^2 \rho(\xi)d\xi < \infty,$$

we obtain the isometrical identity

$$\|v(z,t)\exp\{\frac{-i}{4bt}\sum_{j=1}^{n}z_j^2\}\|^2_{H(T_{G[D(S_0,\omega)]})} = \int_{-\frac{1}{2bt}D} |F(\xi)|^2 \rho(\xi)d\xi. \qquad(78)$$

Hence, by setting $g(\xi) = F(\xi)\rho(\xi)$ we obtain

Theorem 13   *We construct S_0, ω, a domain $D(S_0, \omega)$ on $\mathbb{R}^n$ contained in the set*

$$\{t \in R^n; 0 < W(t; S_0, \omega) < \infty\},$$

and the maximal domain $G[D(S_0, \omega)]$. We assume that

$$0 \in G[D(S_0, \omega)].$$

Then, the solution $u(x, t)$ for the Schrödinger equation satisfying

$$u(x, 0) = g(x) \quad on \quad \mathbb{R}^n$$

with

$$\int_{-\frac{1}{2bt}D} |g(\xi)|^2 W(-\frac{\xi}{2bt}; S_0, \omega)d\xi < \infty$$

and

$$g(\xi) = 0 \quad on \quad \mathbb{R}^n \backslash (-\frac{1}{2bt}D),$$

can be extended analytically onto the tube domain $T_{G[D(S_0,\omega)]}$ and the extension $u(z, t)$ satisfies the identity

$$\|u(z,t)\exp\{\frac{-i}{4bt}\sum_{j=1}^{n}z_j^2\}\|^2_{H(T_{G[D(S_0,\omega)]})} = \int_{-\frac{1}{2bt}D} |g(\xi)|^2 W(-\frac{\xi}{2bt}; S_0, \omega)d\xi.$$

Several explicit examples for reproducing kernels of Szegö type were given by [Gi1-2] (see [[Fu2]]). We shall give several explicit examples not covered by Gindikin and which illustrate Theorems 10, 11 and 12.

(1)
$$S_0 = \{(y_1, 0); y_1 > 0\}.$$
$$\omega(y) = 1 \quad \text{and} \quad D(S_0, \omega) = \{(t_1, t_2); t_1, t_2 > 0\}.$$

Then,
$$W(t; S_0, \omega) = \frac{1}{2t_1}.$$

Furthermore,
$$G[D(S_0, \omega)] = \{(y_1, y_2); y_1, y_2 > 0\}$$

and
$$K(z, \overline{z}; T_{G[D(S_0, \omega)]}) = \frac{1}{16\pi^2 y_1^2 y_2}.$$

Hence, we have
$$K(z, \overline{w}; T_{G[D(S_0, \omega)]}) = \frac{-1}{\pi(z_1 - \overline{w}_1)^2} \frac{i}{2\pi(z_2 - \overline{w}_2)}.$$

For the set S_0, for all $y_0 \in S_0$, we see that (70) is valid for the sake of property (ii). Meanwhile, the maximal domain is $D^m(S_0, \omega) = \{(t_1, t_2); t_1 > 0\}$ and then $G[D^m(S_0, \omega)] = \phi$. Then, for any fixed $a > 0$ and for the analytic function $f(z_1, z_2)$ on $\{\text{Im} z_1 > 0\} \times \mathbb{C}$ such that

$$f(z_1, z_2) = \frac{1}{(2\pi)^2} \int_{D^m(S_0, \omega)} (e^{-at_1})^{\frac{1}{2}} (e^{-at_2^2})^{\frac{1}{2}} e^{iz_1 t_1} e^{iz_2 t_2} (2t_1) dt_1 dt_2,$$

we have

$$\int_{\mathbb{R}^2} \int_{S_0} |f(z_1, x_2)|^2 dx_1 dx_2 dy_1 = \frac{1}{(2\pi)^2} \int_{D^m(S_0, \omega)} e^{-at_1} e^{-at_2^2} (2t_1) dt_1 dt_2 < \infty,$$

but

$$f(z_1, z_2) \notin H(T_{G[D(S_0, \omega)]}).$$

For this latter fact, recall the characterization of the members of $H(T_{G[D(S_0, \omega)]})$ by expression (64) with (65).

123

This example shows that $f(z_1, z_2)$ is analytic on $T_{G[D(S_0,\omega)]}$ and has a finite norm

$$\left\{\int_{\mathbb{R}^2}\int_{S_0}|f(z_1, x_2)|^2 dx_1 dx_2 dy_1\right\}^{\frac{1}{2}} < \infty \qquad ,$$

in the sense of the restricted boundary values $f(z_1, x_2)$ on T_{S_0} which was stated below (70). However,

$$f(z_1, z_2) \notin H(T_{G[D(S_0,\omega)]}).$$

We will find a characterization of the members $f(z_1, z_2) \in H(T_{G[D(S_0,\omega)]})$. For any $\varepsilon > 0$, on

$$G_\varepsilon = \{(y_1, y_2); 0 < \arg(y_1 + iy_2) < \frac{\pi}{2} - \varepsilon\},$$

there exists a constant M_ε such that

$$|2t_1 e^{-2y_1 t_1} e^{-2y_2 t_2}| \le M_\varepsilon \quad \text{on} \quad D(S_0, \omega).$$

Hence, for $f(z) \in H(T_{G[D(S_0,\omega)]})$ in the expression (64) with (65), we have

$$\int_{\mathbb{R}^2}|f(z)|^2 dx = \frac{1}{4\pi^2}\int_{D(S_0,\omega)}|F(t_1, t_2)|^2$$
$$\times e^{-2y_1 t_1} e^{-2y_2 t_2}(2t_1)^2 dt_1 dt_2$$
$$\le \frac{M_\varepsilon}{4\pi^2}\int_{D(S_0,\omega)}|F(t_1, t_2)|^2(2t_1)dt_1 dt_2 < \infty.$$

Hence, the functions $f(z) \in H(T_{G[D(S_0,\omega)]})$ belong to the analytic Hardy class H^2 on T_{G_ε}. In this example, the functions $f(z) \in H(T_{G[D(S_0,\omega)]})$ are conversely characterized by the conditions that $f(z)$ belong to the analytic Hardy class H^2 on T_{G_ε} for any $\varepsilon > 0$ and have finite norms (69). See [[SW]], Chapter III, Section 3.

Indeed, by Theorem 3.1 in [[SW]], pp. 101–102, if $f(z)$ belongs to the analytic Hardy class H^2 on T_{G_ε} for any $\varepsilon > 0$, then $f(z)$ is expressible in the form

$$f(z) = \int_{G_\varepsilon^*} e^{i(z,t)}g(t)dt$$

for a function $g(t)$ satisfying

$$\int_{G_\varepsilon^*}|g(t)|^2 dt < \infty.$$

Hence, from the finiteness of the integral (69), we see that $f(z)$ is expressible in the form (64) with (65). Hence, $f \in H(T_{G[D(S_0,\omega)]})$.

124

(2)
$$S_0 = \{(y_1, y_2); y_2 = y_1, y_1 > 0\},$$
$$\omega(y) = 1 \quad \text{and} \quad D(S_0, \omega) = \{(t_1, t_2); t_1, t_2 > 0\}.$$

Then,
$$W(t; S_0, \omega) = \frac{1}{\sqrt{2}(t_1 + t_2)}.$$

Furthermore,
$$G[D(S_0, \omega)] = \{(y_1, y_2); y_1, y_2 > 0\}$$

and
$$K(z, \overline{z}; T_{G[D(S_0,\omega)]}) = \frac{\sqrt{2}}{32\pi^2} \frac{y_1 + y_2}{y_1^2 y_2^2}.$$

Hence,
$$K(z, \overline{w}; T_{G[D(S_0,\omega)]}) = \frac{\sqrt{2}}{4\pi^2 i} \frac{(z_1 + z_2) - \overline{(w_1 + w_2)}}{(z_1 - \overline{w_1})^2 (z_2 - \overline{w_2})^2}$$
$$= \frac{1}{\sqrt{2}} \left\{ \frac{i}{2\pi(z_1 - \overline{w_1})} \frac{-1}{\pi(z_2 - \overline{w_2})^2} + \frac{-1}{\pi(z_1 - \overline{w_1})^2} \frac{i}{2\pi(z_2 - \overline{w_2})} \right\}.$$

The kernels $\frac{i}{2\pi(z_1 - \overline{w_1})}$ and $\frac{-1}{\pi(z_1 - \overline{w_1})^2}$ are the Szegö kernel and the Bergman kernel on the upper half plane of z_1, respectively. Hence, Examples (1) and (2) give the interrelationships among reproducing kernels. Hence, we can know the interrelationships among the associated reproducing kernel Hilbert spaces. The reproducing kernels in (1) and (2) have simple forms and they are the reproducing kernels of Szegö type in our sense. However, these reproducing kernels do not appear in the framework of Gindikin's Szegö kernels.

In the following examples, we will see the interrelationship among $S_0, D(S_0, \omega)$ and $G[D(S_0, \omega)]$.

(3) $S_0 = \{(y_1, y_2); y_2 = -y_1 + a, y_1 > 0\}$ $(a > 0)$ and $\omega(y) = 1$.

Then, we have
$$W(t; S_0, \omega) = \frac{e^{-2at_2}}{\sqrt{2}(t_1 - t_2)} \quad \text{for} \quad t_2 < t_1.$$

(i) $D(S_0, \omega) = \{(t_1, t_2); 0 < t_2 < t_1\}$.
Then,
$$G[D(S_0, \omega)] = \{(y_1, y_2); y_2 > -y_1 + a, y_1 > 0\}$$

and

$$K(z, \overline{z}; T_{G[D(S_0, \omega)]}) = \frac{\sqrt{2}}{32\pi^2} \frac{1}{y_1^2(y_1 + y_2 - a)}.$$

Hence, we have

$$K(z, \overline{w}; T_{G[D(S_0, \omega)]}) = \frac{\sqrt{2}}{4\pi^2 i(z_1 - \overline{w}_1)^2 \{(z_1 + z_2) - \overline{(w_1 + w_2)} - 2ai\}}.$$

(ii) $D(S_0, \omega) = \{(t_1, t_2); t_2 < 0, t_1 > 0\}$.
Then,

$$G[D(S_0, \omega)] = \{(y_1, y_2); y_2 < a, y_1 > 0\}$$

and

$$K(z, \overline{z}; T_{G[D(S_0, \omega)]}) = \frac{\sqrt{2}}{32\pi^2} \frac{a + y_1 - y_2}{y_1^2(a - y_2)^2}.$$

Hence, we have

$$K(z, \overline{w}; T_{G[D(S_0, \omega)]}) = \frac{\sqrt{2}[2ai + \{(z_1 - z_2) - \overline{(w_1 - w_2)}\}]}{4\pi^2 i(z_1 - \overline{w}_1)^2(2ai - (z_2 - \overline{w}_2))^2}.$$

(iii) $D(S_0, \omega) = \{(t_1, t_2); t_2 < t_1 < 0\}$.
Then,

$$G[D(S_0, \omega)] = \{(y_1, y_2); y_2 < -y_1 + a, y_2 < a\}$$

and

$$K(z, \overline{z}; T_{G[D(S_0, \omega)]}) = \frac{\sqrt{2}}{32\pi^2} \frac{1}{(a - y_1 - y_2)(a - y_2)^2}.$$

Hence, we have

$$K(z, \overline{w}; T_{G[D(S_0, \omega)]}) = \frac{\sqrt{2}}{4\pi^2 i} \frac{1}{\{2ai - (z_1 + z_2) + \overline{(w_1 + w_2)}\}\{2ai - (z_2 - \overline{w}_2)\}^2}.$$

Next, we will examine a typical case of $(\hat{S}_0)^\circ \neq \phi$, and we will determine the maximal domain $D^m(S_0, \omega)$ and $G[D^m(S_0, \omega)]$.

(4) $S_0 = \{(0, y_2); y_2 > 0\} \cup \{(y_1, 0); y_1 > 0\}$ and $\omega(y) = 1$.

Then,

$$W(t; S_0, \omega) = \frac{1}{2}(\frac{1}{t_1} + \frac{1}{t_2}) \quad \text{for} \quad t_1, t_2 > 0.$$

Hence,

$$D^m(S_0, \omega) = \{(t_1, t_2); t_1, t_2 > 0\},$$

126

and so,

$$G[D^m(S_0, \omega)] = \{(y_1, y_2); y_1, y_2 > 0\}.$$

Then,

$$
\begin{aligned}
&K(z, \bar{z}; T_{G[D^m(S_0,\omega)]}) \\
&= \frac{1}{(2\pi)^2} \int_0^\infty \int_0^\infty e^{-2y_1 t_1} e^{-2y_2 t_2} \frac{2t_1 t_2}{t_1 + t_2} dt_1 dt_2 \quad (t_1 - t_2 = y, t_1 + t_2 = x) \\
&= \frac{1}{16\pi^2} \iint_{\substack{-x<y<x \\ x>0}} e^{-(y_1+y_2)x} e^{-(y_1-y_2)y} \frac{x^2 - y^2}{x} dx dy \\
&= \begin{cases} \frac{1}{16\pi^2(y_1-y_2)^2} \left\{ \frac{1}{y_1} + \frac{1}{y_2} - \frac{2\log(y_1 \div y_2)}{y_1 - y_2} \right\}, & y_1 \neq y_2 \\ \frac{1}{48\pi^2 y^3}, & y_1 = y_2 = y. \end{cases}
\end{aligned}
$$

Hence, we have

$$
K(z, \bar{w}; T_{G[D^m(S_0,\omega)]}) = \frac{-1}{4\pi^2\{(z_1 - z_2) - \overline{(w_1 - w_2)}\}^2}
$$
$$
\cdot \left\{ \frac{2i}{z_1 - \overline{w_1}} + \frac{2i}{z_2 - \overline{w_2}} - 4i\frac{\log\{(z_1 - \overline{w_1}) \div (z_2 - \overline{w_2})\}}{(z_1 - z_2) - \overline{(w_1 - w_2)}} \right\}.
$$

By taking a point mass as $\omega(y)d\mu(y)$, we can deal with the Cauchy–Szegö kernels for radial tube domains. For the Cauchy–Szegö kernels, see [[SW]], Chapter III.

(5)   For $S_0 = \{y^{(0)}\}$ and $n = 2$,   we set $W(t; S_0, \omega) = e^{-2(y^{(0)}, t)}$

and

$$D(S_0, \omega) = \{(t_1, t_2); t_1, t_2 > 0\}.$$

Then,

$$G[D(S_0, \omega)] = \{(y_1, y_2); y_1 > y_1^{(0)}, y_2 > y_2^{(0)}\}$$

and

$$K(z, \bar{z}; T_{G[D(S_0,\omega)]}) = \frac{1}{4\pi^2} \frac{1}{2(y_1 - y_1^{(0)})2(y_2 - y_2^{(0)})}.$$

Hence, we have

$$K(z, \bar{w}; T_{G[D(S_0,\omega)]}) = \frac{1}{16\pi^2(\frac{z_1 - \overline{w_1}}{2i} - y_1^{(0)})(\frac{z_2 - \overline{w_2}}{2i} - y_2^{(0)})}.$$

(6)   In the case (3) (iii), Theorem 13 can be stated explicitly as follows.

The solutions $u(x,t)$ of the Schrödinger equation

$$\frac{\partial u(x,t)}{\partial t} - ib\Delta u(x,t) = 0 \qquad (b > 0, n = 2)$$

satisfying the condition

$$u(x,0) = g(x)$$

with

$$\int_{\xi_2 > \xi_1 > 0} |g(\xi)|^2 \frac{\exp \frac{a\xi_2}{bt}}{\xi_2 - \xi_1} d\xi < \infty$$

and

$$g(\xi) = 0 \quad \text{on} \quad \mathbb{R}^2 \backslash \{\xi_2 > \xi_1 > 0\}$$

can be extended analytically onto $T_{\{y_2 < -y_1 + a, y_2 < a\}}$, and the analytic extensions $u(z,t)$ satisfy the identity

$$\|u(z,t)\exp\{\frac{-i}{4bt}(z_1^2 + z_2^2)\}\|^2_{H(T_{\{y_2 < -y_1 + a, y_2 < a\}})}$$

$$= \int_{-\infty}^{\infty} \int_{-\infty}^{\infty} \int_0^{\infty} |u(x_1 + iy_1, x_2 + i(-y_1 + a), t)|^2$$

$$\cdot \exp\{\frac{x_1 y_1 - x_2 y_1 + x_2 a}{bt}\} dx_1 dx_2 dy_1$$

$$= \frac{2bt}{\sqrt{2}} \int_{\xi_2 > \xi_1 > 0} |g(\xi)|^2 \frac{\exp \frac{a\xi_2}{bt}}{\xi_2 - \xi_1} d\xi.$$

For similar representations of Bergman kernels and Szegö kernels for certain tube domains, see also [Has1], [K], [KS] and [N]. For some applications of the representations and related recent topics, see [CG], [Has2] and [K1-3].

§2   The heat equation

We shall examine several isometrical identities and inversion formulas in the heat equation

$$u_t(x,t) = u_{xx}(x,t) \quad \text{on} \quad \mathbb{R} \times T_+ \quad (T_+ = \{t > 0\}) \tag{1}$$

satisfying the initial condition

$$u(x,0) = F(x) \quad \text{on} \quad \mathbb{R} \tag{2}$$

for functions $F \in L_2(\mathbb{R})$. Using the Fourier transform we formally obtain a representation of the solution $u(x,t)$

$$u(x,t) = \frac{1}{\sqrt{4\pi t}} \int_{\mathbb{R}} F(\xi) \exp\left[-\frac{(x - \xi)^2}{4t}\right] d\xi. \tag{3}$$

128

Then, for $F \in L_2(\mathbb{R})$ we have

$$\lim_{t \to 0} \|u(x,t) - F(x)\|_{L_2(\mathbb{R})} = 0.$$

For $F \in C^1(\mathbb{R})$ we have the classical solution $u(x,t)$ satisfying

$$\lim_{t \to 0} u(x,t) = F(x) \quad \text{on} \quad \mathbb{R}$$

and this convergence is uniform.

Meanwhile, for the solution $u(x,t)$ of the heat equation

$$u_t(x,t) = u_{xx}(x,t) \quad \text{on} \quad \mathbb{R}_+ \times T_+ \tag{4}$$

satisfying the conditions

$$\lim_{x \to +0} u(x,t) = F(t) \quad \text{on} \quad (0,t), \tag{5}$$

$$u(x,t)\big|_{t=0} = 0 \quad \text{on} \quad \mathbb{R}_+, \tag{6}$$

with $F \in L_2(0,t)$, we may obtain a formal integral representation, by using Fourier and Laplace transforms

$$u(x,t) = \int_0^t F(\xi) \frac{x \exp\left\{\frac{-x^2}{4(t-\xi)}\right\}}{2\sqrt{\pi}(t-\xi)^{\frac{3}{2}}} d\xi. \tag{7}$$

See, for example, [[CJ]], Chapter 4 and [[Kö]] for the details. Apart from the classical solutions $u(x,t)$ we shall establish typical isometrical identities and inversion formulas for the integral transforms (3) and (7); see [Sa7], [Sa20] and [Sa21]. For some general partial differential equations of parabolic type, see [Sa12] and for the case of Schrödinger's equation, see [Sa19].

1 Space dependent

For any fixed $t > 0$, we first examine the integral transform (3). From

$$k(x,t) = \frac{1}{\sqrt{4\pi t}} \exp\left[-\frac{x^2}{4t}\right],$$

we form the kernel

$$K(x,x';t) = \int_{\mathbb{R}} k(x-\xi,t)k(x'-\xi,t)d\xi$$

$$= \frac{1}{2\sqrt{2\pi t}} \exp\left[\frac{-x^2}{8t}\right] \exp\left[\frac{-x'^2}{8t}\right] \exp\left[\frac{xx'}{4t}\right]. \tag{8}$$

129

Note here that the kernel $K(x, x'; t)$ is extended analytically onto $\mathbb{C} \times \overline{\mathbb{C}}$ in the form

$$K(z, \overline{u}; t) = \frac{1}{2\sqrt{2\pi t}} \exp\left[\frac{-z^2}{8t}\right] \exp\left[\frac{-\overline{u}^2}{8t}\right] \exp\left[\frac{z\overline{u}}{4t}\right]. \tag{9}$$

By Corollary 2.3.3 and (2.4.39) or (2.4.114) and (2.4.115), we see that the RKHS H_K admitting the reproducing kernel (9) is composed of all entire functions $f(z)$ with finite norms

$$\left\{\frac{1}{\sqrt{2\pi t}} \iint_{\mathbb{C}} |f(z)|^2 \exp[\frac{-y^2}{2t}] dx dy\right\}^{\frac{1}{2}} < \infty. \tag{10}$$

This integral is also expressible in terms of the trace $f(x)$ of $f(z)$ to the real line.

By using the identity

$$k(x - \xi, t) = \frac{1}{2\pi} \int_{\mathbb{R}} \exp\{-p^2 t + ip(x - \xi)\} dp, \tag{11}$$

we have

$$K(x, x'; t) = \frac{1}{2\pi} \int_{\mathbb{R}} \exp\{-2p^2 t + ip(x - x')\} dp. \tag{12}$$

This implies that any member $f(x)$ of H_K is expressible in the form

$$f(x) = \frac{1}{2\pi} \int_{\mathbb{R}} g(p) e^{ipx} e^{-2p^2 t} dp \tag{13}$$

for a function g satisfying

$$\int_{\mathbb{R}} |g(p)|^2 e^{-2p^2 t} dp < \infty$$

and we have the isometrical identity

$$\|f\|_{H_K}^2 = \frac{1}{2\pi} \int_{\mathbb{R}} |g(p)|^2 e^{-2p^2 t} dp. \tag{14}$$

By the Fourier transform, from (13) we have

$$g(p) = \sqrt{2\pi} \hat{f}(p) e^{2p^2 t}$$

in the framework of the L_2 space. Hence, from (14) we obtain, using the Parseval–Plancherel identity

$$\|f\|_{H_K}^2 = \sum_{j=0}^{\infty} \frac{(2t)^j}{j!} \int_{\mathbb{R}} |f^{(j)}(x)|^2 dx. \tag{15}$$

Clearly, for any fixed $t > 0$

$$\{k(x - \xi, t); \xi \in \mathbb{R}\}$$

is complete in $L_2(\mathbb{R})$. Hence, we have, by Theorem 2.1.2

Theorem 1 *In the integral transform (3) of $L_2(\mathbb{R})$ functions F, the images $u(x, t)$ are extended analytically onto $\mathbb{C}$ in the form $u(z, t)$ and we obtain the isometrical identities*

$$\int_{\mathbb{R}} |F(\xi)|^2 d\xi = \frac{1}{\sqrt{2\pi t}} \iint_{\mathbb{C}} |u(z, t)|^2 \exp[\frac{-y^2}{2t}] dx dy$$

$$= \sum_{j=0}^{\infty} \frac{(2t)^j}{j!} \int_{\mathbb{R}} |\partial_x^j u(x, t)|^2 dx \qquad (16)$$

for any fixed $t > 0$.

Corollary 1 *If a C^{∞} function $f(x)$ has a finite integral on the right hand side in (15), then the function is extended analytically onto $\mathbb{C}$ and we have the identity*

$$\sum_{j=0}^{\infty} \frac{(2t)^j}{j!} \int_{\mathbb{R}} |\partial_x^j f(x)|^2 dx = \frac{1}{\sqrt{2\pi t}} \iint_{\mathbb{C}} |f(z)|^2 \exp[\frac{-y^2}{2t}] dx dy. \qquad (17)$$

In the inversion formula in Theorem 2.1.5, if the norm in H_K is given by the series (15), then as the integrals on E_N we can take, for example, the $N - sum$

$$\sum_{j=0}^{N} \frac{(2t)^j}{j!} \int_{\mathbb{R}} |f^{(j)}(x)|^2 dx. \qquad (18)$$

Hence we have the inversion formulas

Theorem 2 *In Theorem 1, for the integral transform (3) and for any fixed $t > 0$ we have*

$$F(\xi) = s - \lim_{N \to \infty} \frac{1}{\sqrt{2\pi t}} \iint_{E_N} u(z, t) \overline{k(z - \xi, t)} \exp[\frac{-y^2}{2t}] dx dy$$

$$= s - \lim_{N \to \infty} \sum_{j=0}^{N} \frac{(2t)^j}{j!} \int_{\mathbb{R}} \partial_x^j u(x, t) \partial_x^j k(x - \xi, t) dx$$

$$= s - \lim_{N \to \infty} \sum_{j=0}^{N} \frac{(-2t)^j}{j!} \int_{\mathbb{R}} u(x, t) \partial_x^{2j} k(x - \xi, t) dx,$$

where $\{E_N\}$ is any exhaustion of $\mathbb{C}$ by compact sets and the limits mean strong convergence in $L_2(\mathbb{R})$.

2   Time dependence

In the integral transform (3) we form the corresponding reproducing kernel

$$K(t, t'; x) = \int_{\mathbb{R}} k(x - \xi, t) k(x - \xi, t') d\xi = \frac{1}{2\sqrt{\pi}} \frac{1}{\sqrt{t + t'}} \quad \text{for} \quad t, t' > 0. \tag{19}$$

This implies that the kernel is extended analytically on the right half plane $R^+(\tau = t + i\hat{t}, t > 0)$ in the form

$$K(\tau, \overline{\tau'}; x) = \frac{1}{2\sqrt{\pi}\sqrt{\tau + \overline{\tau'}}} \quad \text{on} \quad R^+ \times \overline{R^+}. \tag{20}$$

Clearly, for any fixed x, the family

$$\{k(x - \xi, \tau), \tau \in R^+\}$$

is not complete in $L_2(\mathbb{R})$. Hence, at this moment we see that for any fixed x, the images $u(x, t)$ are extended analytically onto the right half plance R^+ in the form $u(x, \tau)$ and we have the inequality

$$\|u(x, \tau)\|_{H_K}^2 \leq \int_{\mathbb{R}} |F(\xi)|^2 d\xi \tag{21}$$

for the RKHS H_K admitting the reproducing kernel (20). For the kernel (20) note that

$$K(\tau, \overline{\tau'}; x)^2 = \frac{1}{4\pi(\tau + \overline{\tau'})} \tag{22}$$

and

$$\frac{1}{\tau + \overline{\tau'}}$$

is the Szegö reproducing kernel on R^+ with finite norms

$$\|f\|^2 = \frac{1}{2\pi} \sup_{t>0} \int_{-\infty}^{\infty} |f(t + i\hat{t})|^2 d\hat{t}$$

$$= \frac{1}{2\pi} \int_{-\infty}^{\infty} |f(i\hat{t})|^2 d\hat{t}$$

stated in (2.4.94). By Theorem 2.3.3, for any $f \in H_K$ we shall consider, in particular, the product $f(\tau)^2$. Then $f(\tau)^2$ belongs to the Szegö space and we have the inequality

$$2 \int_{-\infty}^{\infty} |f(i\hat{t})|^2 d\hat{t} \leq \|f\|_{H_K}^4 \tag{23}$$

132

(recall also Corollary 2.3.3). Hence, from (21) and (23) we obtain

Theorem 3   *For any fixed $x \in \mathbb{R}$, the images $u(x,t)$ of (3) for $L_2(\mathbb{R})$ functions are extended analytically onto the right half plane R^+ in the form $u(x,\tau)$ ($\tau = t + i\hat{t}$). $u(x,\tau)$ have the nontangential boundary values on the imaginary axis from R^+ belonging to $L_2(i\mathbb{R}, d\hat{t})$ and we have the inequality*

$$2 \int_{\mathbb{R}} |u(x, i\hat{t})|^4 d\hat{t} \leq \left\{ \int_{\mathbb{R}} |F(\xi)|^2 d\xi \right\}^2.$$

We shall examine a more general integral transform in order to examine the property of $D^\alpha u(x,t)$ for

$$D^\alpha = \frac{\partial^\alpha}{\partial x^\alpha} \quad (\alpha \geq 1)$$

in the integral transform

$$D^\alpha u(x,t) = \frac{1}{\sqrt{4\pi t}} \int_{\mathbb{R}} F(\xi) D^\alpha \exp\left[-\frac{(x-\xi)^2}{4t} \right] d\xi. \tag{24}$$

By using the identity (12) we calculate the reproducing kernel

$$
\begin{aligned}
K_{D_\alpha}(\tau, \overline{\tau'}; x) &= \int_{\mathbb{R}} \frac{1}{\sqrt{4\pi\tau}} D^\alpha \exp\left[-\frac{(x-\xi)^2}{4\tau} \right] \\
&\quad \cdot \frac{1}{\sqrt{4\pi\tau'}} D^\alpha \exp\left[-\frac{(x-\xi)^2}{4\overline{\tau'}} \right] d\xi \\
&= \frac{\Gamma(\alpha + \frac{1}{2})}{2\pi} \frac{1}{(\tau + \overline{\tau'})^{\alpha + \frac{1}{2}}} \quad \text{on} \quad R^+ \times \overline{R^+}.
\end{aligned}
\tag{25}
$$

Since $2\pi K_{D_\alpha}(\tau, \overline{\tau'}; x)$ is the Bergman–Selberg reproducing kernel on R^+ for $q = \frac{\alpha}{2} + \frac{1}{4}$ in (2.4.92), we obtain

Theorem 4   *In the integral transform (24) for $L_2(\mathbb{R})$ functions F, for any fixed x, the images $D^\alpha u(x,t)$ are extended analytically onto the right half plane R^+ in the form $D^\alpha u(x,\tau)$ and we have the inequality*

$$\frac{2^{\alpha - \frac{1}{2}}(\alpha - 1)}{\Gamma(\alpha + \frac{1}{2})} \iint_{R^+} |D^\alpha u(x,\tau)|^2 t^{\alpha - \frac{3}{2}} dt d\hat{t} \leq \int_{\mathbb{R}} |F(\xi)|^2 d\xi.$$

Corollary 2   *For any fixed x, for the integral transform (3) for $L_2(\mathbb{R})$ functions F the images $\partial_t u(x,t)$ are extended analytically onto R^+ in the form $\partial_\tau u(x,\tau)$ and we have the inequality*

$$\frac{4\sqrt{2}}{\sqrt{\pi}} \iint_{R^+} |\partial_\tau u(x,\tau)|^2 \sqrt{t} \, dt d\hat{t} \leq \int_{\mathbb{R}} |F(\xi)|^2 d\xi.$$

133

As regards the time dependence of $u(x,t)$ for $L_2(\mathbb{R}, e^{ax^2}dx)$ initial functions F, we have in general,

Theorem 5 *For any fixed x, the images $u(x,t)$ of the integral transform (3) for $L_2(\mathbb{R}, e^{ax^2}dx)$ functions F are extended analytically in the form $u(x,\tau)$ to the following domain D_a of the τ plane:*

(i)

$$for \quad a > 0, D_a = \left\{ \left| \tau + \frac{1}{4a} \right| > \frac{1}{4a} \right\} \setminus \left\{ t \leq -\frac{1}{2a}, \hat{t} = 0 \right\},$$

(ii)

$$for \quad a = 0, \qquad D_a = R^+,$$

and

(iii)

$$for \quad a < 0, \qquad D_a = \left\{ \left| \tau + \frac{1}{4a} \right| < \frac{1}{4|a|} \right\}.$$

In particular, we have the inequality

$$\frac{1}{2\pi} \int_0^\infty \left| \int_{\mathbb{R}} u\left(x, \frac{1}{t + i\hat{t}} \right) \frac{e^{i\xi\hat{t}}}{\sqrt{t + i\hat{t}}} d\hat{t} \right|^2 \frac{\sqrt{\xi} e^{2\xi(t+2a)}}{\cosh(4ax\sqrt{\xi})} d\xi \leq \int_{\mathbb{R}} |F(\xi)|^2 e^{a\xi^2} d\xi$$

for any $t > -2a$ and for any point x.

Proof : We consider the kernel

$$K_a(\tau, \overline{\tau'}; x) = \int_{\mathbb{R}} k(x - \xi, \tau) k(x - \xi, \overline{\tau'}) e^{-a\xi^2} d\xi$$

$$= \frac{1}{2\sqrt{\pi}(\tau + \overline{\tau'} + 4a\tau\overline{\tau'})^{\frac{1}{2}}} \exp\left[\frac{-ax^2(\tau + \overline{\tau'})}{\tau + \overline{\tau'} + 4a\tau\overline{\tau'}} \right]$$

on the domain $D_a \times \overline{D_a}$. From this identity we see that any member $g(\tau)$ in H_{K_a} is expressible in the form

$$g(\tau) = \int_{\mathbb{R}} G(\xi) k(x - \xi, \tau) e^{-a\xi^2} d\xi$$

for some function G satisfying

$$\int_{\mathbb{R}} |G(\xi)|^2 e^{-a\xi^2} d\xi < \infty$$

134

and we have the identity

$$\|g\|_{H_{K_a}}^2 = \int_{\mathbb{R}} |G(\xi)|^2 e^{-a\xi^2} d\xi.$$

Hence, by setting $F(\xi) = G(\xi)e^{-a\xi^2}$ we have the desired analyticity of $u(x, \tau)$.
Next, we consider the kernel

$$K_a^*(\tau, \overline{\tau'}; x) = K_a(\frac{1}{\tau}, \frac{1}{\overline{\tau'}}; x)$$

$$= \frac{\tau^{\frac{1}{2}} \tau'^{\frac{1}{2}}}{2\pi} \int_0^\infty e^{-\tau\xi} e^{-\overline{\tau'}\xi} \frac{\cosh(4ax\sqrt{\xi})}{\sqrt{\xi}} e^{-4a\xi} d\xi \qquad (26)$$

for $\mathrm{Re}\tau, \mathrm{Re}\tau' > -2a$. In order to avoid the multiple-valuedness of the function $\tau^{\frac{1}{2}}$,
we shall consider the kernel on the half plane $R_{-2a}^+ = \{\mathrm{Re}\tau > -2a\} \backslash \{-2a \leq \mathrm{Re}\tau \leq 0\}$ cut by the slit $[-2a, 0]$ for $a > 0$. The expression (26) implies that any members
$\varphi(\tau)$ in $H_{K_a^*}$ are analytic on R_{-2a}^+ and are expressible in the form

$$\varphi(\tau) = \frac{\tau^{\frac{1}{2}}}{2\pi} \int_0^\infty \psi(\xi)e^{-\tau\xi} \frac{\cosh(4ax\sqrt{\xi})}{\sqrt{\xi}} e^{-4a\xi} d\xi \qquad (27)$$

for some functions ψ satisfying

$$\int_0^\infty |\psi(\xi)|^2 \frac{\cosh(4ax\sqrt{\xi})}{\sqrt{\xi}} e^{-4a\xi} d\xi < \infty.$$

Furthermore, we have the identity

$$\|\varphi\|_{H_{K_a^*}}^2 = \frac{1}{2\pi} \int_0^\infty |\psi(\xi)|^2 \frac{\cosh(4ax\sqrt{\xi})}{\sqrt{\xi}} e^{-4a\xi} d\xi. \qquad (28)$$

By using the inversion formula for the Fourier transform in the framework of the L_2
space we have from (28)

$$\|\varphi\|_{H_{K_a^*}}^2 = \frac{1}{2\pi} \int_0^\infty \left| \int_{\mathbb{R}} \frac{\varphi(t + i\hat{t})}{\sqrt{t + i\hat{t}}} e^{i\xi\hat{t}} d\hat{t} \right|^2$$

$$\times \frac{\sqrt{\xi}}{\cosh(4ax\sqrt{\xi})} e^{2t(\xi+2a)} d\xi, \qquad (29)$$

for any $t > -2a$. Hence, from the expression

$$u(x, \frac{1}{t}) = \frac{\sqrt{t}}{\sqrt{4\pi}} \int_{\mathbb{R}} \exp\left[-\frac{(x-\xi)^2 t}{4}\right] G(\xi)e^{-a\xi^2} d\xi$$

135

for $L_2(\mathbb{R}, e^{-a\xi^2} d\xi)$ functions G and, from (26) and (29), we have the desired result by setting $F(\xi) = G(\xi)e^{-a\xi^2}$.

For the solutions $u(x, t)$ of the heat equation

$$\partial_t u = \Delta u \quad \text{on} \quad \mathbb{R}_+ \times \mathbb{R}^n (n \geq 2)$$

satisfying the initial condition

$$u(x, 0) = F(x) \quad \text{on} \quad \mathbb{R}^n$$

and for the miscellaneous derived functions, we can obtain the corresponding results, similarly. See, for example, [Sa21].

3 The case with variable boundary values

Next, we shall examine the integral transform (7) for $L_2((0, t), d\xi)$ functions F and for any fixed $t > 0$. In the complex form, we calculate the kernel form

$$
\begin{aligned}
K(z, \overline{u}; t) &= \int_0^t \frac{z \exp[\frac{-z^2}{4(t-\xi)}] \, \overline{u} \exp[\frac{-\overline{u}^2}{4(t-\xi)}]}{2\sqrt{\pi}(t-\xi)^{\frac{3}{2}} \, 2\sqrt{\pi}(t-\xi)^{\frac{3}{2}}} d\xi \\
&= \frac{z\overline{u}}{4\pi} \int_{\frac{1}{t}}^{\infty} \eta \exp\left[\frac{-(z^2 + \overline{u}^2)}{4}\eta\right] d\eta \\
&= z \exp\left[\frac{-z^2}{4t}\right] \overline{u} \exp\left[\frac{-\overline{u}^2}{4t}\right] \left\{ \frac{4}{\pi(z^2 + \overline{u}^2)^2} + \frac{1}{t\pi(z^2 + \overline{u}^2)} \right\}.
\end{aligned}
\tag{30}
$$

We see that the kernel $K(z, \overline{u}; t)$ is analytic on $\Delta(\frac{\pi}{4}) \times \overline{\Delta(\frac{\pi}{4})}$ for $\Delta(\frac{\pi}{4}) = \{|\arg z| < \frac{\pi}{4}\}$. Here, note that

$$K_1(z, \overline{u}) = \frac{4}{\pi(z^2 + \overline{u}^2)^2}$$

is the reproducing kernel for the Hilbert space H_{K_1} comprising all analytic functions $f(z)$ on $\Delta(\frac{\pi}{4})$ with finite norms

$$\left\{ \iint_{\Delta(\frac{\pi}{4})} |f(z)|^2 |z|^2 dx dy \right\}^{\frac{1}{2}} < \infty \tag{31}$$

as stated in (2.4.116) and (2.4.117). Similarly,

$$K_2(z, \overline{u}) = \frac{1}{\pi(z^2 + \overline{u}^2)}$$

136

is the reproducing kernel for the Hilbert space H_{K_2} comprising all analytic functions $f(z)$ on $\Delta(\frac{\pi}{4})$ with finite norms

$$\left\{ \int_{\partial\Delta(\frac{\pi}{4})} |f(z)|^2 |z||dz| \right\}^{\frac{1}{2}} < \infty. \tag{32}$$

Here, the members $f(z)$ of H_{K_2} are obtained by the functions such that $f(z^2)$ belong to the Szegö space on the right half plane. See (2.4.94) and (2.4.95). Hence, the properties of the images $u(x,t)$ of the integral transform (7) can be interpreted by Theorem 2.3.2 and Corollary 2.3.3. In order to avoid the sum of reproducing kernels, note the decomposition of the kernel $K(z,\overline{u};t)$

$$\begin{aligned}
K(z,\overline{u};t) &= \int_0^t \frac{z\exp[\frac{-z^2}{4(t-\xi)}]}{2\sqrt{\pi}(t-\xi)^{\frac{3}{2}}} \frac{\overline{u}\exp[\frac{-\overline{u}^2}{4(t-\xi)}]}{2\sqrt{\pi}(t-\xi)^{\frac{3}{2}}} \cdot \left\{ \frac{\xi}{t} + \frac{t-\xi}{t} \right\} d\xi \\
&= z\exp\left[\frac{-z^2}{4t}\right] \overline{u}\exp\left[\frac{-\overline{u}^2}{4t}\right] \cdot \left\{ K_1(z,\overline{u};t) + \frac{1}{t}K_2(z,\overline{u};t) \right\}.
\end{aligned} \tag{33}$$

We thus obtain

Theorem 6   *In the integral transform*

$$\varphi(x,t) = \frac{1}{t}\int_0^t F(\xi)\frac{x\exp\left[\frac{-x^2}{4(t-\xi)}\right]}{2\sqrt{\pi}(t-\xi)^{\frac{3}{2}}}\xi d\xi \tag{34}$$

for functions F satisfying

$$\int_0^t |F(\xi)|^2 \xi d\xi < \infty, \tag{35}$$

the images $\varphi(x,t)$ can be extended analytically onto $\Delta(\frac{\pi}{4})$ and we have the isometrical identity

$$\frac{1}{t}\int_0^t |F(\xi)|^2 \xi d\xi = \iint_{\Delta(\frac{\pi}{4})} |\varphi(z,t)|^2 \exp\left\{ \frac{x^2-y^2}{2t} \right\} dxdy. \tag{36}$$

Moreover, we have the inversion formula

$$F(\xi) = \frac{1}{2\sqrt{\pi}(t-\xi)^{\frac{3}{2}}}s - \lim_{N\to\infty} \iint_{E_N} \varphi(z,t)\overline{z}\exp\left[\frac{-\overline{z}^2}{4(t-\xi)}\right]$$

$$\cdot \exp\left\{ \frac{x^2-y^2}{2t} \right\} dxdy \tag{37}$$

137

in the sense of strong convergence in the space satisfying (35) for any compact exhaustion $\{E_N\}$ of $\Delta(\frac{\pi}{4})$.

In Theorem 6, note that $u(x,t) = t\varphi(x,t)$ are the solutions for the heat equation (4) satisfying

$$u(0,t) = tF(t) \quad \text{on} \quad T_+, \tag{38}$$

and

$$u(x,0) = 0 \quad \text{on} \quad \mathbb{R}_+. \tag{39}$$

In particular, we have

Corollary 3 *For the images $\phi(x,t)$ of (34) for (35), and for any fixed $\xi > 0$ and $j \geq 0$, we have*

$$\lim_{t \to +0} t \left| \partial_\xi^j [\phi(\xi,t) \exp \frac{\xi^2}{4t}] \right|^2 = 0. \tag{40}$$

Proof : By Theorem 6 and the reproducing kernel on $\Delta(\frac{\pi}{4})$, we have

$$\phi(\xi,t) \exp \frac{\xi^2}{4t} = \iint_{\Delta(\frac{\pi}{4})} \phi(z,t) \exp \frac{z^2}{4t} \cdot \overline{\left(\frac{4z\xi}{\pi(z^2 + \xi^2)^2} \right)} \, dx dy.$$

Hence, by Theorem 6 we have the desired result

$$\lim_{t \to +0} t \left| \partial_\xi^j \left(\phi(\xi,t) \exp \frac{\xi^2}{4t} \right) \right|^2$$

$$\leq \lim_{t \to +0} t \iint_{\Delta(\frac{\pi}{4})} \left| \phi(z,t) \exp \frac{z^2}{4t} \right|^2 \, dx dy$$

$$\times \iint_{\Delta(\frac{\pi}{4})} \left| \frac{\partial^j}{\partial \xi^j} \left(\frac{4z\xi}{\pi(z^2 + \xi^2)^2} \right) \right|^2 \, dx dy$$

$$= \lim_{t \to +0} t \frac{1}{t} \int_0^t |F(\xi)|^2 \xi d\xi \cdot \left[\frac{\partial^j}{\partial z^j} \frac{\partial^j}{\partial \xi^j} \frac{4z\xi}{\pi(z^2 + \xi^2)^2} \right]_{z=\xi}$$

$$= 0.$$

For the members of the Bergman space $H_B(\Delta(\frac{\pi}{4}))$ on $\Delta(\frac{\pi}{4})$, in general, we have

Corollary 4 *For any fixed $j \geq 0$ and for $f \in H_B(\Delta(\frac{\pi}{4}))$,*

$$\lim_{x \to \infty} x^{j+1} \partial_x^j f(x) = 0 \tag{41}$$

138

and

$$\lim_{x \to +0} x^{j+1} \partial_x^j f(x) = 0. \tag{42}$$

Proof :   By the Cauchy integral formula, for $f \in H_B(\Delta(\frac{\pi}{4}))$

$$\partial_\xi^j f(\xi) = \frac{j!}{2\pi i} \int_{|z-\xi|=r, r < \frac{\xi}{\sqrt{2}}} \frac{f(z)}{(z-\xi)^{j+1}} dz$$

we obtain by the Schwarz inequality

$$\xi^{j+1} |\partial_\xi^j f(\xi)| \leq \frac{j!(j+2)2^{\frac{1}{2}(j+1)}}{2\pi} \left\{ \iint_{|z-\xi| < \frac{\xi}{\sqrt{2}}} |f(z)|^2 dx dy \right\}^{\frac{1}{2}},$$

which implies the desired results.

Theorem 7   *For the images $\varphi(x,t)$ of the integral transform*

$$\varphi(x,t) = \frac{1}{t} \int_0^t F(\xi) \frac{x \exp\left[\frac{-x^2}{4(t-\xi)}\right]}{2\sqrt{\pi}(t-\xi)^{\frac{3}{2}}} (t-\xi) d\xi \tag{43}$$

for functions F satisfying

$$\int_0^t |F(\xi)|^2 (t-\xi) d\xi < \infty, \tag{44}$$

$\varphi(x,t)$ *can be extended analytically onto $\Delta(\frac{\pi}{4})$ in the form $\varphi(z,t)$ and*

$$\varphi(z^2, t) \frac{1}{z^2} e^{\frac{z^4}{4t}} \tag{45}$$

belong to the Szegö space on the right half plane. Furthermore, we have the isometrical identity

$$\frac{1}{t} \int_0^t |F(\xi)|^2 (t-\xi) d\xi = \int_{\partial \Delta(\frac{\pi}{4})} |\varphi(z,t)|^2 \frac{|dz|}{|z|}. \tag{46}$$

Moreover, we have the inversion formula

$$F(\xi) = \frac{1}{2\sqrt{\pi}(t-\xi)^{\frac{3}{2}}} s - \lim_{N \to \infty} \int_{\partial E_N} \varphi(z,t) \overline{z} \exp\left(\frac{-\overline{z}^2}{4(t-\xi)}\right) \cdot \exp\frac{x^2 - y^2}{2t} \frac{|dz|}{|z|} \tag{47}$$

139

in the sense of strong convergence in the space of functions F satisfying (44) and for any regular exhaustion $\{E_N\}$ of $\Delta(\frac{\pi}{4})$ satisfying

(i)   *∂E_N are analytic Jordan curves on $\Delta(\frac{\pi}{4})$,*
(ii)   *$\overline{E_N}$ (closure) are compact on $\Delta(\frac{\pi}{4})$,*
(iii)   *$E_N \subset E_{N+1}$,*
and
(iv)   *$\Delta(\frac{\pi}{4}) = \bigcup_N E_N$.*

In Theorem 7, note that

$$u(x,t) = \int_0^t F(\xi) \frac{1}{\sqrt{\pi(t-\xi)}} \exp\left[\frac{-x^2}{4(t-\xi)}\right] d\xi \qquad (48)$$

can be related to the solutions of the heat equation (4) satisfying

$$u_x(0,t) = -F(t) \quad \text{on} \quad T_+ \qquad (49)$$

and

$$u(x,0) = 0 \quad \text{on} \quad \mathbb{R}_+ \qquad (50)$$

(cf. [[CJ]], Chapter 4). Hence, $-\frac{2t}{x}\varphi(x,t)$ can be related to the solutions of the heat equation (4) satisfying (49) and (50).

The material of this subsection is taken from [Sa20]. The more delicate properties of the solutions of the heat equation in Theorem 6 containing the identity Chapter 1, (22) are given by [AHS1]. The n-dimensional versions are given by [Sa26] whose results are quite different from the one-dimensional case. For the corresponding results for the Schrödinger equation, see [Sa19].

4   Inverse formulas on bounded domains

We shall give the simplest inverse formula representing the initial temperature distribution in terms of the present time temperature distribution on a bounded domain $D \subset \mathbb{R}^n$ subject to the n-dimensional heat equation

$$\nabla^2 u = \frac{\partial u}{\partial t}. \qquad (51)$$

The boundary condition imposed on the solutions u will be taken to be

$$u(p,t) = 0 \quad \text{on} \quad \partial D \quad \text{for all} \quad t. \qquad (52)$$

Further, for initial temperature distributions f we assume that

$$u(p,0) = f(p) \quad \text{on} \quad D, \qquad (53)$$

and
$$\int_D f(p)^2 dm(p) < \infty. \tag{54}$$

Then, the solutions $u(p, t)$ satisfying (51)–(54) are expressible in the form

$$u(p, t) = \int_D f(q) G(q, p; t) dm(q). \tag{55}$$

Here, the Green's function $G(q, p; t)$ for equations (51) and (52) is expanded in terms of the eigenvalues $\{\lambda_n\}_{n=1}^\infty$ determined by the problem

$$\nabla^2 v(p) + \lambda v(p) = 0 \quad \text{on} \quad D; \quad v(p) = 0 \quad \text{on} \quad \partial D, \tag{56}$$

and the corresponding orthonormal eigenfunctions $\{u_n\}_{n=1}^\infty$ in the form

$$G(p, q; t) = \sum_{n=1}^\infty e^{-\lambda_n t} u_n(p) u_n(q), \quad t > 0. \tag{57}$$

(see, for example [[Ro]], Chapter 9, Section 7).

We shall give a complete solution of (55) for $f \in L_2^\circ(D, dm)$ for any $t > 0$, where $L_2^\circ(D, dm)$ is the Hilbert space spanned by the family of the eigenfunctions $\{u_n\}_{n=1}^\infty$.

For the identification of the images $u(p, t)$ of (55), we form the reproducing kernel

$$K(p, q; t) = \int_D G(r, q; t) G(r, p; t) dm(r)$$
$$= \sum_{n=1}^\infty e^{-2\lambda_n t} u_n(p) u_n(q), \tag{58}$$

which converges absolutely on $D \times D$ for any fixed $t > 0$. Hence, $u(p, t)$ are expressible in the form

$$u(p, t) = \sum_{n=1}^\infty c_n e^{-2\lambda_n t} u_n(p) \tag{59}$$

for uniquely determined l^2 elements $\{c_n\}$. Furthermore, we have the isometrical identity for the RKHS H_K

$$\|u\|_{H_K}^2 = \sum_{n=1}^\infty c_n^2 e^{-2\lambda_n t}$$
$$= \sum_{n=1}^\infty \left(\int_D u(p, t) u_n(p) dm(p) \right)^2 e^{2\lambda_n t}$$
$$= \int_D f(p)^2 dm(p). \tag{60}$$

141

In order to solve (55), we shall represent $f \in L_2^\circ(D, dm)$ in the form

$$f(p) = \sum_{n=1}^{\infty} A_n u_n(p). \tag{61}$$

Then, of course,

$$A_n = (f, u_n)_{L_2^\circ(D, dm)}. \tag{62}$$

Since the integral transform (55) gives an isometrical mapping L from $L_2^\circ(D, dm)$ onto H_K, we have $L^{-1} = L^*$, for the adjoint operator L^* of L. Hence we have

$$\begin{aligned}
A_n &= \left(L^{-1} u(\cdot, t), u_n\right)_{L_2^\circ(D, dm)} \\
&= \left(L^* u(\cdot, t), u_n\right)_{L_2^\circ(D, dm)} \\
&= \left(u(\cdot, t), L u_n\right)_{H_K} \\
&= \left(u(\cdot, t), e^{-\lambda_n t} u_n\right)_{H_K} \\
&= \left(\int_D u(r, t) u_n(r) dm(r)\right) e^{\lambda_n t}.
\end{aligned} \tag{63}$$

Hence, we have the desired inverse formula

$$f(p) = \sum_{n=1}^{\infty} e^{\lambda_n t} \left(\int_D u(r, t) u_n(r) dm(r)\right) u_n(p), \tag{64}$$

for any fixed $t > 0$.

For the simplest case of

$$\begin{aligned}
u_t &= u_{xx}, \quad 0 < x < \pi; \\
u(0, t) &= u(\pi, t) = 0, \quad u(x, 0) = f(x),
\end{aligned} \tag{65}$$

we have the solution

$$f(x) = \frac{2}{\pi} \sum_{n=1}^{\infty} e^{n^2 t} \left(\int_0^\pi u(\xi, t) \sin n\xi d\xi\right) \sin nx, \tag{66}$$

(cf. [[Gr]], 15–16).

As we see from the above arguments, we will be able to obtain miscellaneous inverse formulas for various situations for the heat equations, similarly.

The result (64) is, of course, obtained formally and directly, by using the properties of eigenvalues and eigenfunctions in (56). We, however, derived the formula naturally by our unified principle using the theory of reproducing kernels and we can apply the theory to the solutions $u(p, t)$ to derive their properties. We can obtain miscellaneous concrete reproducing kernels $K(p, q; t)$ by (58), and from the concrete forms we will be able to derive the corresponding detailed properties of the solutions $u(p, t)$.

§3   The wave equation

We shall consider the simplest example of a second-order wave equation $u_{tt}(x,t) = c^2 u_{xx}(x,t)$ on the real line $\mathbb{R}$ satisfying typical initial and boundary conditions in partial differential equations of hyperbolic type. In each case, we shall establish a natural isometrical identity and inversion formula between the source function and the response function. The material is taken from [Sa10].

We consider the solutions of the wave equation

$$\frac{\partial^2 u(x,t)}{\partial t^2} = c^2 \frac{\partial^2 u(x,t)}{\partial x^2} \qquad (c : constant, > 0) \tag{1}$$

satisfying the conditions

$$u(0,t) = F(t), \quad u(x,0) = 0 \quad \text{on} \quad \mathbb{R}, \tag{2}$$
$$u_t(x,t)\big|_{t=0} = F(x), \quad u(x,0) = 0 \quad \text{on} \quad \mathbb{R}, \tag{3}$$

and

$$u(x,0) = F(x), \quad u_t(x,t)\big|_{t=0} = 0 \quad \text{on} \quad \mathbb{R}, \tag{4}$$

respectively.

By du Hamel's method we have a formal integral representation of the solution $u(x,t)$ of (1) which satisfies (2)

$$u(x,t) = \frac{\partial}{\partial t} \int_0^t F(\xi) U(x, t - \xi) d\xi. \tag{5}$$

Meanwhile, by d'Alembert's formula the solutions of (1) which satisfy (3) and (4) are

$$u(x,t) = \frac{1}{2c} \int_{x-ct}^{x+ct} F(\xi) d\xi$$
$$= \frac{1}{2c} \int_{\mathbb{R}} F(\xi)\theta(ct - |x - \xi|) d\xi \tag{6}$$

and

$$u(x,t) = \frac{1}{2}(F(x + ct) - F(x - ct))$$
$$= \frac{1}{2c} \frac{\partial}{\partial t} \int_{\mathbb{R}} F(\xi)\theta(ct - |x - \xi|) d\xi, \tag{7}$$

respectively. Here,

$$U(x,t) = \begin{cases} 1 & \text{for} \quad |x| < ct \\ 0 & \text{for} \quad |x| > ct \end{cases}$$

143

and

$$\theta(x) = \begin{cases} 1 & x > 0 \\ 0 & x < 0. \end{cases}$$

See, for example, [[Fo]], [[Kö]], [[Ro]], [[Sz]] and [[Tr]] for the basic properties for the wave equation (1).

For (5) and (7), we have formally, for any fixed $T > 0$

$$f_T(x) = \int_0^T u(x,t)dt = \int_0^T F(\xi)U(x,T-\xi)d\xi \qquad (8)$$

and

$$\int_0^T u(x,t)dt = \frac{1}{2c}\int_{\mathbb{R}} F(\xi)\theta(cT - |x - \xi|)d\xi. \qquad (9)$$

In these integral transforms (6), (8) and (9) we shall establish natural isometrical identities and inversion formulas apart from the solutions of the wave equation, in general, for $F \in L_2(\mathbb{R})$ or $L_2(0,\infty)$.

For example, in d'Alembert's formulas (6) and (7), note that for $F \in C^1(\mathbb{R})$ and $F \in C^2(\mathbb{R})$, $u(x,t)$ in (6) and (7) give the classical solutions of (1) satisfying

$$u \in C^2(T_+ \times \mathbb{R}),$$
$$u, u_t, u_x \in C^1(T_+ \times \mathbb{R}),$$

and

$$(3) \quad \text{and} \quad (4),$$

respectively.

1 The case with variable boundary values

In order to examine the integral transform (8), we form the reproducing kernel

$$K_1(x_1, x_2) = \int_0^T U(x_1, T - \xi)U(x_2, T - \xi)d\xi$$
$$= \min\left(T - \frac{x_1}{c}, T - \frac{x_2}{c}\right), \quad \text{on} \quad [0, cT] \times [0, cT]. \qquad (10)$$

Note that $K_1(x_1, x_2)$ is the reproducing kernel for the Hilbert space H_{K_1} comprising all functions $f(x)$ on $[0, cT]$ satisfying the condition that $f(x)$ are absolutely continuous on $[0, cT]$, $f'(x) \in L_2(0, cT)$, $f(cT) = 0$, and equipped with the inner product

$$(f_1, f_2)_{H_{K_1}} = c\int_0^{cT} f_1'(x)f_2'(x)dx.$$

144

Note that the family
$$\{U(x, T - \xi); x \in [0, cT]\} \tag{11}$$
is complete in $L_2(0, T)$. Hence, we have

Theorem 1 *For any fixed $T > 0$, the images $f_T(x)$ by the integral transform (8) for $F \in L_2(0, T)$ form the RKHS H_{K_1} and we have the isometrical identity*
$$\int_0^T F(\xi)^2 d\xi = c \int_0^{cT} f_T'(x)^2 dx.$$

Corollary 1 *For the classical solutions $u(x, t)$ of (1) satisfying (2), we have the isometrical identity*
$$\int_0^T F(\xi)^2 d\xi = c \int_0^{cT} \left(\frac{\partial}{\partial x} \int_0^T u(x, t) dt \right)^2 dx.$$

We shall consider the inversion formula of (8). Note that for any $\xi \in (0, T)$, $U(x, T - \xi)$ is not absolutely continuous on $[0, cT]$. Hence, we cannot apply Theorem 2.1.5 to (8), directly. Hence, for a reasonable subfamily of H_{K_1}, we shall establish the inversion formula.

We assume that $f_T \in H_{K_1}$ and further

$$f_T'(x) \quad \text{is absolutely continuous on} \quad [0, cT], \tag{12}$$
$$f_T''(x) \in L_1(0, cT), \tag{13}$$

and

$$f_T'(0) = 0. \tag{14}$$

Then, we obtain, by using the reproducing property of $K_1(x, x_2)$ for H_{K_1}

$$f_T(x_2) = (f_T(x), K_1(x, x_2))_{H_{K_1}}$$

$$= c \int_0^{cT} \left\{ f_T'(x) \frac{\partial}{\partial x} \left(\int_0^T U(x, T - \xi) U(x_2, T - \xi) d\xi \right) \right\} dx$$

$$= c \left[f_T'(x) \int_0^T U(x, T - \xi) U(x_2, T - \xi) d\xi \right]_{x=0}^{cT}$$

$$- c \int_0^{cT} \left(f_T''(x) \int_0^T U(x, T - \xi) U(x_2, T - \xi) d\xi \right) dx$$

$$= \int_0^T \left(-c \int_0^{cT} f_T''(x) U(x, T - \xi) \right) U(x_2, T - \xi) d\xi.$$

145

Hence, from (8) and (11) we obtain

Theorem 2 · *For $f_T \in H_{K_1}$ satisfying (12)-(14) we have the inversion formula of (8)*

$$F(\xi) = -c \int_0^{cT} f_T''(x)U(x, T - \xi)dx.$$

2   The case with initial boundary values

We examine the integral transform (6). We form the reproducing kernel for $x_1 \neq x_2$

$$K_2(x_1, x_2; t_1, t_2) = \frac{1}{2c} \int_{\mathbb{R}} \theta(ct_1 - |x_1 - \xi|)\theta(ct_2 - |x_2 - \xi|)d\xi$$

$$= \begin{cases} \frac{1}{2c}(c(t_1 + t_2) - |x_1 - x_2|) & \text{for} \quad |x_1 - x_2| \leqq c(t_1 + t_2) \\ 0 & \text{for} \quad |x_1 - x_2| \geqq c(t_1 + t_2). \end{cases} \tag{15}$$

Meanwhile, for $x_1 = x_2 = x$ we have

$$K_2(x_1, x_2; t_1, t_2) = K_{2,1}(t_1, t_2; x)$$

$$= \frac{1}{2c} \int_{\mathbb{R}} \theta(ct_1 - |x - \xi|)\theta(ct_2 - |x - \xi|)d\xi$$

$$= \min(t_1, t_2) \quad \text{for} \quad t_1, t_2 > 0. \tag{16}$$

Note that $K_{2,1}(t_1, t_2; x)$ is the reproducing kernel for the Hilbert space $H_{K_{2,1}}$ comprising all functions $f(t)$ on $[0, \infty)$ such that $f(t)$ are absolutely continuous on $[0, \infty)$, $f(0) = 0$, $f'(t) \in L_2(0, \infty)$ and equipped with the inner product

$$\int_0^\infty f_1'(t)f_2'(t)dt,$$

as stated in (2.4.11) and (2.4.12). The family

$$\{\theta(ct - |x - \xi|); t \in [0, \infty), \quad \text{for fixed} \quad x\}$$

is not complete in $L_2(\mathbb{R})$. Indeed,

$$\int_{\mathbb{R}} F(\xi)\theta(ct - |x - \xi|)d\xi = 0 \quad \text{for all} \quad t > 0;$$

that is,

$$\int_{x-ct}^{x+ct} F(\xi)d\xi = 0 \quad \text{for all} \quad t > 0$$

146

implies that

$$F(x + ct) = -F(x - ct) \quad \text{for all} \quad t > 0. \tag{17}$$

In general, $F \in L_2(\mathbb{R})$ is decomposed uniquely in the form

$$F(\xi) = F_{e(x)}(\xi) + F_{0(x)}(\xi); \; F_{e(x)}, F_{0(x)} \in L_2(\mathbb{R}),$$

where

$$F_{e(x)}(\xi) = \frac{1}{2}(F(\xi) + F(2x - \xi)) \quad \text{and} \quad F_{0(x)}(\xi) = \frac{1}{2}(F(\xi) - F(2x - \xi)).$$

The odd part of F satisfies (17) and so

$$\int_{\mathbb{R}} F(\xi)\theta(ct - |x - \xi|)d\xi = \int_{\mathbb{R}} F_{e(x)}(\xi)\theta(ct - |x - \xi|)d\xi.$$

In addition,

$$\int_{\mathbb{R}} F_{e(x)}(\xi) F_{0(x)}(\xi)d\xi = 0.$$

We thus have

Theorem 3   *For any fixed x, the images $f(t, x)$ by the integral transform (6) for functions $F \in L_2(\mathbb{R})$ form the RKHS $H_{K_{2,1}}$. Further, we obtain the isometrical identity*

$$\int_0^\infty \left(\frac{\partial f(t, x)}{\partial t}\right)^2 dt = \min \frac{1}{2c} \int_{\mathbb{R}} F(\xi)^2 d\xi,$$

where the minimum is taken over all functions F satisfying

$$f(t, x) = \frac{1}{2c} \int_{\mathbb{R}} F(\xi)\theta(ct - |x - \xi|)d\xi, \quad F \in L_2(\mathbb{R}). \tag{18}$$

Moreover, the minimum is attained by F^* if and only if F^* is the even part $F_{e(x)}$ of any F satisfying (18).

It seems that Theorem 3 itself has an interesting physical sense for the wave. Note that in the classical solutions $u(x, t)$ of (1) satisfying (3), the integral

$$\frac{1}{2} \int_{\mathbb{R}} F(\xi)^2 d\xi$$

can be understood as the energy of the wave.

147

For $f(t, x) \in H_{K_{2,1}}$, we take F^* such that

$$\int_0^\infty f_t(t, x)^2 dt = \frac{1}{2c} \int_{\mathbb{R}} F^*(\xi)^2 d\xi$$

in Theorem 3. Then, we obtain the inversion formula, as in Theorem 2.

Theorem 4 *For any $f(t, x) \in H_{K_{2,1}}$ such that*

$$f_t(t, x) \quad \text{is absolutely continuous and} \quad f_{tt}(t, x) \in L_1(0, \infty),$$

we have the inversion formula for (6)

$$F^*(\xi) = -\int_0^\infty \frac{\partial^2 f(t, x)}{\partial t^2} \theta(ct - |x - \xi|) dt.$$

Next, for any fixed $T > 0$, we consider the integral transform (6). Then, we set $K_2(x_1, x_2; T, T) = K_{2,2}(x_1, x_2; T)$. It seems that RKHS $H_{K_{2,2}}$ has essentially a much more complicated structure than H_{K_1} and $H_{K_{2,1}}$. In order to realize the norm in $H_{K_{2,2}}$, recall the identity

$$\frac{1}{2\pi} \int_{\mathbb{R}} \left(\frac{\sin(\xi/2)}{\xi/2} \right)^2 e^{-ix\xi} d\xi = \begin{cases} 1 - |x| & \text{for} \quad |x| \leqq 1 \\ 0 & \text{for} \quad |x| \geqq 1 \end{cases}$$

(see, for example [[BN]]). We thus have the expression

$$K_{2,2}(x_1, x_2; T) = \frac{T}{2\pi} \int_{\mathbb{R}} \exp\left(\frac{-ix_1\xi}{2cT} \right) \overline{\exp\left(\frac{-ix_2\xi}{2cT} \right)} W_1(\xi) d\xi, \qquad (19)$$

where

$$W_1(\xi) = \left(\frac{\sin(\xi/2)}{\xi/2} \right)^2.$$

Note that the family

$$\left\{ \exp\left(\frac{-ix\xi}{2cT} \right); x \in \mathbb{R} \right\}$$

is complete in the space $L_2(W_1(\xi)d\xi)$. Hence, any member f of $H_{K_{2,2}}$ is expressible in the form

$$f(x) = \frac{T}{2\pi} \int_{\mathbb{R}} g(\xi) \exp\left(\frac{-ix\xi}{2cT} \right) W_1(\xi) d\xi \qquad (20)$$

148

for a uniquely determined $g \in L_2(W_1(\xi)d\xi)$ and we have

$$\|f\|^2_{H_{K_{2,2}}} = \frac{T}{2\pi} \int_{\mathbb{R}} |g(\xi)|^2 W_1(\xi)d\xi.$$

By using the inversion formula for Fourier integrals in the framework of the L_2 space to (20), the norm in $H_{K_{2,2}}$ can be realized as follows :

$$\|f\|^2_{H_{K_{2,2}}} = \frac{2\pi}{T} \int_{\mathbb{R}} \left| \underset{L\to\infty}{\mathrm{l.\,i.\,m.}} \int_{-L}^{L} f(x) \exp\left(\frac{ix\xi}{2cT}\right) dx \right|^2 \frac{d\xi}{W_1(\xi)}. \tag{21}$$

Since the family

$$\{\theta(cT - |x - \xi|); x \in \mathbb{R}\}$$

is complete in $L_2(\mathbb{R})$, we have, as stated in Chapter 1, (14)

Theorem 5 *For any fixed $T > 0$, the images $f_T(x)$ by the integral transform (6) for functions $F \in L_2(\mathbb{R})$ form the RKHS $H_{K_{2,2}}$ and we have the isometrical identity*

$$\frac{1}{2} \int_{\mathbb{R}} F(\xi)^2 d\xi = \frac{2\pi c}{T} \int_{\mathbb{R}} \left| \underset{L\to\infty}{\mathrm{l.\,i.\,m.}} \int_{-L}^{L} f_T(x) \exp\left(\frac{ix\xi}{2cT}\right) dx \right|^2 \frac{d\xi}{W_1(\xi)}.$$

Next, we consider the inversion formula for (6). From the reproducing property of $K_{2,2}(x, x_2; T)$ for $H_{K_{2,2}}$, we have, for any $f_T \in H_{K_{2,2}}$

$$\begin{aligned}
f_T(x_2) &= (f_T(x), K_{2,2}(x, x_2; T))_{H_{K_{2,2}}} \\
&= \frac{2\pi}{T} \int_{\mathbb{R}} \left\{ \underset{L\to\infty}{\mathrm{l.\,i.\,m.}} \int_{-L}^{L} f_T(\xi_1) \exp\left(\frac{ix\xi_1}{2cT}\right) d\xi_1 \right\} \\
&\quad \cdot \left\{ \underset{L\to\infty}{\mathrm{l.\,i.\,m.}} \int_{-L}^{L} \left(\frac{1}{2c} \int_{\mathbb{R}} \theta(cT - |\xi_2 - \xi_3|)\theta(cT - |x_2 - \xi_3|)d\xi_3 \right) \right. \\
&\quad \left. \cdot \exp\left(\frac{ix\xi_2}{2cT}\right) d\xi_2 \right\} \frac{dx}{W_1(x)}.
\end{aligned}$$

Since

$$\underset{L\to\infty}{\mathrm{l.\,i.\,m.}} \int_{-L}^{L} \theta(cT - |\xi_2 - \xi_3|) \exp\left(\frac{ix\xi_2}{2cT}\right) d\xi_2 = 2cTW_1(x) \exp\left(\frac{-ix\xi_3}{2cT}\right),$$

149

we have

$$
f_T(x_2) = 2\pi \int_{\mathbb{R}} \left\{ \underset{L \to \infty}{\text{l.i.m.}} \int_{-L}^{L} f_T(\xi_1) \exp\left(\frac{ix\xi_1}{2cT} \right) d\xi_1 \right\}
$$

$$
\cdot \left\{ \int_{\mathbb{R}} \theta(cT - |x_2 - \xi_3|) \exp\left(\frac{-ix\xi_3}{2cT} \right) d\xi_3 \right\} dx
$$

$$
= \frac{1}{2c} \lim_{M \to \infty} \int_{\mathbb{R}} \left[4\pi c \int_{-M}^{M} \left\{ \underset{L \to \infty}{\text{l.i.m.}} \int_{-L}^{L} f_T(\xi_1) \exp\left(\frac{ix\xi_1}{2cT} \right) d\xi_1 \right\} \right.
$$

$$
\left. \cdot \exp\left(\frac{-ix\xi_3}{2cT} \right) dx \right] \theta(cT - |x_2 - \xi_3|) d\xi_3.
$$

Hence, we have

Theorem 6   *For any $f_T \in H_{K_{2,2}}$, we have the inversion formula for (6)*

$$
F(\xi) = s - \lim_{M \to \infty} 4\pi c \int_{-M}^{M} \left\{ \underset{L \to \infty}{\text{l.i.m.}} \int_{-L}^{L} f_T(\xi_1) \exp\left(\frac{ix\xi_1}{2cT} \right) d\xi_1 \right\} \exp\left(\frac{-ix\xi}{2cT} \right) dx,
$$

in the space $L_2(\mathbb{R})$.

3   The inhomogeneous case

We consider the inhomogeneous equation

$$
\left(\frac{\partial^2}{\partial x^2} - \frac{1}{c^2} \frac{\partial^2}{\partial t^2} \right) u(x, t) = -\rho(x, t)
$$

with the conditions

$$
u(x, 0) = u_t(x, t)\big|_{t=0} = 0.
$$

Then, for a suitable ρ, we obtain the integral representation of the solutions $u(x, t)$

$$
u(x, t) = \frac{c}{2} \int_0^t \int_{\mathbb{R}} \rho(\xi, \hat{t}) \theta(c(t - \hat{t}) - |x - \xi|) d\xi d\hat{t}. \tag{22}
$$

So, we shall examine the integral transform (22).

First, for any fixed $T > 0$ and for functions F satisfying

$$
\int_0^T \int_{\mathbb{R}} F(\xi, t)^2 d\xi dt < \infty, \tag{23}
$$

we consider the integral transform

$$f_T(x) = \int_0^T \int_{\mathbb{R}} F(\xi, t)\theta(c(T - t) - |x - \xi|)d\xi dt. \tag{24}$$

We form the reproducing kernel

$$K_{3,1}(x_1, x_2; T) = \int_0^T \int_{\mathbb{R}} \theta(c(T - t) - |x_1 - \xi|)\theta(c(T - t) - |x_2 - \xi|)d\xi dt$$

$$= \begin{cases} cT^2 \left(1 - \frac{|x_1 - x_2|}{2cT}\right)^2 & \text{for} \quad |x_1 - x_2| \leq 2cT \\ 0 & \text{for} \quad |x_1 - x_2| \geq 2cT. \end{cases}$$

Then, for

$$W_2(\xi) = \int_{\mathbb{R}} W_1(x)W_1(\xi - x)dx$$

we have

$$\frac{1}{2\pi} \int_{\mathbb{R}} e^{-ix\xi} W_2(\xi)d\xi = \begin{cases} (1 - |x|)^2 & \text{for} \quad |x| \leq 1 \\ 0 & \text{for} \quad |x| \geq 1. \end{cases}$$

Hence, we obtain the expression

$$K_{3,1}(x_1, x_2; T) = cT^2 \int_{\mathbb{R}} \exp\left(\frac{-ix_1\xi}{2cT}\right) \overline{\exp\left(\frac{-ix_2\xi}{2cT}\right)} W_2(\xi)d\xi.$$

Hence, we obtain, as in Theorem 3

Theorem 7   *For any fixed $T > 0$, the images $f_T(x)$ by the integral transform (24) for functions F satisfying (23) form the RKHS $H_{K_{3,1}}$ comprising all functions f_T on $\mathbb{R}$ with finite norms*

$$\|f_T\|_{H_{K_{3,1}}}^2 = \frac{1}{2\pi cT^2} \int_{\mathbb{R}} \left| \operatorname*{l.i.m.}_{L\to\infty} \int_{-L}^L f_T(x) \exp\left(\frac{ix\xi}{2cT}\right) dx \right|^2 \frac{d\xi}{W_2(\xi)}.$$

Furthermore, we have the isometrical identity

$$\|f_T\|_{H_{K_{3,1}}}^2 = \int_0^T \int_{\mathbb{R}} F(\xi, t)^2 d\xi dt.$$

As in Theorem 6, we obtain

151

Theorem 8 *For any fixed $T > 0$ and $f_T \in H_{K_{3,1}}$ we have the inversion formula for (24)*

$$F(\xi, t) = \frac{2}{\pi T} s - \lim_{M \to \infty} \int_{-M}^{M} \left\{ \frac{1}{x} \sin \frac{x(T-t)}{2T} \exp\left(\frac{-ix\xi}{2cT}\right) \right\}$$

$$\cdot \left\{ \operatorname*{l.i.m.}_{L \to \infty} \int_{-L}^{L} f_T(\xi_1) \exp\left(\frac{i\xi_1 x}{2cT}\right) d\xi_1 \right\} \frac{dx}{W_2(x)}.$$

Next, for any fixed x, we consider the integral transform (22) for functions $\rho(\xi, t)$ satisfying

$$\int_0^\infty \int_{\mathbb{R}} \rho(\xi, t)^2 d\xi dt < \infty. \tag{25}$$

We form the reproducing kernel

$$K_{3,2}(t_1, t_2; x) = \frac{1}{2} \int_0^\infty \int_{\mathbb{R}} \theta(c(t_1 - \hat{t}) - |x - \xi|)\chi(\hat{t}; (0, t_1))$$

$$\cdot \theta(c(t_2 - \hat{t}) - |x - \xi|)\chi(\hat{t}; (0, t_2)) d\xi d\hat{t}, \tag{26}$$

where

$$\chi(\hat{t}; (0, t)) = \begin{cases} 1 & \text{for } 0 < \hat{t} < t \\ 0 & \text{for } \hat{t} \le 0, t \le \hat{t}. \end{cases}$$

Then, we have

$$K_{3,2}(t_1, t_2; x) = \frac{c^2}{2} \min\{t_1^2, t_2^2\} \quad \text{for } t_1, t_2 \ge 0. \tag{27}$$

Note that $K_{3,2}(t_1, t_2; x)$ is the reproducing kernel for the Hilbert space $H_{K_{3,2}}$ comprising all functions $f(t)$ such that $f(t)$ are absolutely continuous on $[0, \infty)$ and $f(0) = 0$, and equipped with finite norms

$$\left\{ \frac{1}{c^2} \int_0^\infty f'(t)^2 \frac{dt}{t} \right\}^{\frac{1}{2}} < \infty, \tag{28}$$

as stated in (2.4.16). Since, the family

$$\{\theta(c(t - \hat{t}) - |x - \xi|)\chi(\hat{t}; (0, t)); t \in (0, \infty)\}$$

is not complete in $L_2(\mathbb{R} \times (0, \infty))$, we obtain

152

Theorem 9   *For any fixed x, the images $u(x,t)$ by the integral transform (22) for functions ρ satisfying (25) form the RKHS $H_{K_{3,2}}$ and we obtain the isometrical identity*

$$\frac{1}{c^2} \int_0^\infty \left(\frac{\partial u(x,t)}{\partial t} \right)^2 \frac{dt}{t} = \min \frac{c}{2} \int_0^\infty \int_{\mathbb{R}} \rho(\xi, \hat{t})^2 d\xi d\hat{t}$$

$$= \frac{c}{2} \int_0^\infty \int_{\mathbb{R}} \rho^*(\xi, \hat{t})^2 d\xi d\hat{t},$$

where the minimum is taken over all ρ satisfying (22) and (25).

Further, we have, as in Theorem 2

Theorem 10   *For any $u(x,t) \in H_{K_{3,2}}$ satisfying*

$$\int_0^\infty \left| \frac{\partial^2 u(x,t)}{\partial t^2} \right| \frac{dt}{t} < \infty,$$

we obtain the inversion formula for (22) and (25)

$$\rho^*(\xi, \hat{t}) = -\frac{1}{c^2} \int_{\hat{t}}^\infty \frac{\partial}{\partial t} \left(\frac{1}{t} \frac{\partial u(x,t)}{\partial t} \right) \theta(c(t - \hat{t}) - |x - \xi|) \frac{dt}{t}.$$

The Green's functions $G_2(x, y; x', y'; t)$ and $G_3(x, y, z; x', y', z'; t)$ for $\mathbb{R}^2 = \{(x, y)\}$ and $\mathbb{R}^3 = \{(x, y, z)\}$ for the corresponding wave equations are given by

$$G_2(x, y; x', y'; t) = \frac{c\theta(ct - \sqrt{(x - x')^2 + (y - y')^2})}{2\pi \sqrt{c^2 t^2 - [(x - x')^2 + (y - y')^2]}}$$

and

$$G_3(x, y, z; x', y', z'; t) = \frac{c\delta(ct - \sqrt{(x - x')^2 + (y - y')^2 + (z - z')^2})}{4\pi \sqrt{(x - x')^2 + (y - y')^2 + (z - z')^2}},$$

respectively.

Hence, the arguments in this subsection are not valid directly on the spaces $\mathbb{R}^2$ and $\mathbb{R}^3$ for the singularities of the Green's functions. We can, however, deduce the corresponding results by applying the method Chapter 1, (7), (c) to the integral kernels.

Let $u(f)$ be the solution to the hyperbolic equation in a bounded domain $\Omega \subset \mathbb{R}^n$:

$$\partial_t^2 u(x,t) = \Delta u(x,t) + \sigma(t)f(x)(x \in \Omega, 0 < t < T),$$
$$u(x,0) = \partial_t u(x,0) = 0 \quad (x \in \Omega),$$
$$u(x,t) = 0 \quad (x \in \partial\Omega, 0 < t < T).$$

We assume that $\sigma \in C^1[0,T]$ is a known function, $\sigma(0) \neq 0$, and $f \in L_2(\Omega)$ is unknown, and $\Gamma \subset \partial\Omega$ is given. Then, Yamamoto [Ya2] examines an inverse problem of determining $f(x)$ $(x \in \Omega)$ from $[\partial u(f)/\partial n](x,t)$ $(x \in \Gamma, 0 < t < T)$. Further, he discusses the stability estimates of $\|f\|_{L_2(\Omega)}$ by $\|\partial u(f)/\partial n\|_{H^1(0,T;L_2(\Omega))}$, a reconstruction of f from $\partial u(f)/\partial n$ and a Tikhonov regularization. For heat source problems in a similar situation, see also [Ya1].

§4   Analytic and harmonic functions of class L_2

Let D be a regular region in the finite complex plane whose boundary ∂D consists of a finite number of analytic Jordan curves. Let E be a compact subset of D of positive planar (or linear) measure. Let $AL_2(\partial D)$ and $HL_2(\partial D)$ denote the analytic and harmonic Hardy classes on D with index 2, respectively. For the basic properties of the Hardy classes, see for example [[Du]]. Of course, the restrictions $f\big|_E$ and $u\big|_E$ for $f \in AL_2(\partial D)$ and $u \in HL_2(\partial D)$ determine f and u on the whole region D, respectively. We shall give a general method which determines all the values of f and u on D from $f\big|_E$ and $u\big|_E$, respectively, based on [Sa9]. Further, in some cases our method will give a necessary and sufficient condition such that a given analytic function f_1 on E is extensible analytically onto D and the extension of f_1 belongs to $AL_2(\partial D)$. Moreover, we will be able to see an interesting property from the Cauchy integral formula. With regard to representing analytic functions in terms of their boundary values, the Cauchy integral formula is not natural, in the sense of computing its inversion.

We use the Mercer expansion theorem for a positive definite kernel to realize the reproducing kernel Hilbert spaces and our method works in miscellaneous families of functions and regions.

1   The Cauchy transform

We first recall the Cauchy integral formula

$$f(z) = \frac{1}{2\pi i}\int_{\partial D}\frac{f(\xi)}{\xi - z}d\xi \quad \text{for} \quad z \in E \quad \text{and} \quad f \in AL_2(\partial D), \tag{1}$$

where $f(\xi)$ means Fatou's nontangential boundary value of f at $\xi \in \partial D$ belonging to $L_2(\partial D)$.

154

We shall consider the integral transform (1) from the Hilbert space $L_2(\partial D)$ equipped with the norm

$$\left\{\frac{1}{2\pi}\int_{\partial D}|F(z)|^2|dz|\right\}^{\frac{1}{2}}$$

onto the uniquely determined Hilbert space $A(E)$ comprising analytic functions on E admitting the reproducing kernel

$$K_A(z,\overline{u}) = \frac{1}{2\pi}\int_{\partial D}\frac{d\xi}{i(\xi-z)|d\xi|}\overline{\left(\frac{d\xi}{i(\xi-u)|d\xi|}\right)}|d\xi|$$

$$= \frac{1}{2\pi}\int_{\partial D}\frac{|d\xi|}{(\xi-z)\overline{(\xi-u)}} \quad (z,u \in E). \tag{2}$$

Since $K_A(z,\overline{u})$ is continuous on $E \times \overline{E}$ and a positive definite kernel on E, by the Mercer expansion theorem, we have the expansion

$$K_A(z,\overline{u}) = \sum_{j=1}^{\infty}(\lambda_j)^{-1}\varphi_j(z)\overline{\varphi_j(u)}, \tag{3}$$

which converges absolutely and uniformly on $E \times E$. Here, of course,

$$0 < \lambda_1 \leqq \lambda_2 \leqq \cdots, \tag{4}$$

$$\lambda_j\int_E K_A(z,\overline{u})\varphi_j(u)dm(u) = \varphi_j(z) \quad \text{on} \quad E, \tag{5}$$

and

$$\int_E \varphi_j(z)\overline{\varphi_{j'}(z)}dm(z) = \delta_{jj'}. \tag{6}$$

Here, $dm(z)$ denotes the Lebesgue plane (or linear) measure. In particular, note that any φ_j is extensible analytically onto D (Theorem 2.2.9 and Theorem 2.2.10). Hence, we see that any member $f_1 \in A(E)$ is expressible in the form

$$f_1(z) = \sum_{j=1}^{\infty}c_j(\lambda_j)^{-1}\varphi_j(z) \tag{7}$$

for some constants c_j satisfying

$$\sum_{j=1}^{\infty}|c_j|^2(\lambda_j)^{-1} < \infty. \tag{8}$$

Here, (7) converges absolutely and uniformly on E. Note that the constants c_j are uniquely determined, as we see from (6); that is,

$$c_j = \lambda_j \int_E f_1(z)\overline{\varphi_j(z)}dm(z). \tag{9}$$

Hence, the norm of f_1 in $A(E)$ is given by

$$\|f_1\|^2_{A(E)} = \sum_{j=1}^{\infty} \lambda_j \left| \int_E f_1(z)\overline{\varphi_j(z)}dm(z) \right|^2. \tag{10}$$

We now consider the inversion of (1) for $L_2(\partial D)$ functions; that is,

$$f_E(z) = \frac{1}{2\pi i} \int_{\partial D} \frac{F(\xi)}{\xi - z}d\xi \quad \text{for} \quad z \in E \quad \text{and} \quad F \in L_2(\partial D), \tag{11}$$

and for f_E we let F^* be the function such that

$$f_E(z) = \frac{1}{2\pi i} \int_{\partial D} \frac{F^*(\xi)}{\xi - z}d\xi \quad \text{for} \quad z \in E \quad \text{and} \quad F^* \in L_2(\partial D), \tag{12}$$

and

$$\int_{\partial D} |F^*(z)|^2|dz| = \min \int_{\partial D} |F(z)|^2|dz|, \tag{13}$$

where the minimum is taken over all F satisfying (11). Then, we regard F^* as the inverse of f_E in the integral transform (11). Let $L_2^o(\partial D)$ denote the subspace of $L_2(\partial D)$ comprising all functions F satisfying

$$\int_{\partial D} \frac{F(\xi)}{\xi - z}d\xi = 0 \quad \text{on} \quad E \quad \text{and so on} \quad D. \tag{14}$$

Let $(L_2^o(\partial D))^{\perp}$ denote the orthogonal complementary subspace of $L_2^o(\partial D)$ in $L_2(\partial D)$. Then, of course, F^* is characterized by (12) and

$$F^* \in (L_2^o(\partial D))^{\perp}. \tag{15}$$

Furthermore, we have the isometrical identity

$$\|f_E\|^2_{A(E)} = \frac{1}{2\pi} \int_{\partial D} |F^*(z)|^2|dz|. \tag{16}$$

Note that the Cauchy integral formula is one to one between $AL_2(\partial D)$ and $A(E)$, but, in general, not an isometry, as we shall see in the following subsection.

156

Now, from the reproducing property of $K_A(z, \overline{u})$ for $A(E)$, we have for $f \in A(E)$

$$f(z) = (f(\cdot), K_A(\cdot, \overline{z}))_{A(E)}$$

$$= \sum_{j=1}^{\infty} \lambda_j \int_E f(\eta)\overline{\varphi_j(\eta)}dm(\eta) \overline{\int_E K_A(\xi', \overline{z})\overline{\varphi_j(\xi')}dm(\xi')}$$

$$= \sum_{j=1}^{\infty} \lambda_j \int_E f(\eta)\overline{\varphi_j(\eta)}dm(\eta)$$

$$\cdot \int_E \left\{ \frac{1}{2\pi} \int_{\partial D} \frac{|d\xi|}{(\xi - \xi')(\xi - z)} \varphi_j(\xi') \right\} dm(\xi')$$

$$= \lim_{N \to \infty} \frac{1}{2\pi i} \int_{\partial D} \frac{F_N(\xi)}{(\xi - z)} d\xi; \qquad (17)$$

where

$$F_N(\xi) = \frac{1}{i} \frac{|d\xi|}{d\xi} \sum_{j=1}^{N} \lambda_j \int_E f(\eta)\overline{\varphi_j(\eta)}dm(\eta) \int_E \frac{\varphi_j(\xi')}{(\xi' - \xi)}dm(\xi'). \qquad (18)$$

Hence, we obtain

$$F^*(\xi) = s - \lim_{N \to \infty} F_N(\xi) \qquad (19)$$

in the sense of strong convergence in $L_2(\partial D)$. By using F^* we can obtain the value of f at any point $z \in D$ in term of $f|_E$

$$f(z) = \frac{1}{2\pi i} \int_{\partial D} \frac{F^*(\xi)}{\xi - z} d\xi. \qquad (20)$$

We thus obtain

Theorem 1   *For any integrable function f_1 on E, there exists an analytic function $f \in AL_2(\partial D)$ satisfying*

$$\int_E f_1(z)\overline{\varphi_j(z)}dm(z) = \int_E f(z)\overline{\varphi_j(z)}dm(z) \quad for \quad all \quad j \qquad (21)$$

if and only if

$$\sum_{j=1}^{\infty} \lambda_j \left| \int_E f_1(z)\overline{\varphi_j(z)}dm(z) \right|^2 < \infty.$$

If $f(z)$ exists, it is uniquely determined and $f(z)$ is expressible in the form (20) with (18) and (19). Moreover, we have the isometrical identity

$$\frac{1}{2\pi} \int_{\partial D} |F^*(z)|^2 |dz| = \sum_{j=1}^{\infty} \lambda_j \left| \int_E f_1(z)\overline{\varphi_j(z)}dm(z) \right|^2. \qquad (22)$$

157

In the expansion (3) of $K_A(z, \overline{u})$, note that the orthogonality (6) is not essential, but it is essential that c_j are expressible in terms of f_1. See the following subsection 4 for examples.

For a smooth function f_1 on E, from (21) when we can deduce that

$$f_1(z) = f(z) \quad \text{on} \quad E, \tag{23}$$

we can obtain a complete condition such that f_1 is extensible analytically onto D and its extension belongs to $AL_2(\partial D)$.

2 Inverses for the Cauchy transform

For the inverse F^* for the Cauchy transform (20), we shall examine here when the identity

$$F^*(\xi) = f(\xi) \quad \text{on} \quad \partial D \tag{24}$$

is valid following [Sa16]. For this problem, we obtain

Theorem 2 *For $f \in AL_2(\partial D)$, the boundary value function $f(\xi)$ coincides with the inverse F^* of $f(z)$ for the Cauchy integral in the sense of the minimum norm, if and only if*

$$\int_{\partial D} f(z) \frac{|dz|}{\overline{z} - \overline{u}} = 0 \quad for \quad all \quad u \in D. \tag{25}$$

Proof : We need the elementary properties of the classical Szegö reproducing kernel $\widehat{K}(z, \overline{u})$ on D for the Hilbert space $AL_2(\partial D)$ equipped with the inner product

$$\int_{\partial D} f(z)\overline{g(z)}|dz|.$$

Let $\widehat{L}(z, u)$ be the adjoint L kernel for $\widehat{K}(z, \overline{u})$; that is, it is a uniquely determined meromorphic function on $D \cup \partial D$ with one simple pole at u such that

$$\widehat{L}(z, u) = \frac{1}{2\pi(z - u)} + \text{regular terms, around} \quad z = u,$$

which satisfies the boundary relation

$$\overline{\widehat{K}(z, \overline{u})}|dz| = \frac{1}{i}\widehat{L}(z, u)dz \quad \text{along} \quad \partial D. \tag{26}$$

158

Then, we see that

$$\int_{\partial D} f(z)\overline{\hat{L}(z,u)}|dz| = 0 \quad \text{for all} \quad f \in AL_2(\partial D). \tag{27}$$

Further, the function

$$\hat{l}(z,u) = \hat{L}(z,u) - \frac{1}{2\pi(z-u)}$$

is analytic on $\overline{D} \times \overline{D}$ (closure) and we have the identity

$$\frac{1}{4\pi^2}\int_{\partial D}\frac{|d\xi|}{(\xi-z)\overline{(\xi-u)}} = \hat{K}(z,\overline{u}) + \int_{\partial D}\hat{l}(\xi,z)\overline{\hat{l}(\xi,u)}|d\xi|. \tag{28}$$

Let H_c denote the Hilbert space admitting the reproducing kernel

$$\frac{1}{4\pi^2}\int_{\partial D}\frac{|d\xi|}{(\xi-z)\overline{(\xi-u)}}.$$

Then, any member $f(z) \in H_c$ is expressible in the form

$$f(z) = \frac{1}{4\pi^2}\int_{\partial D}\frac{F(\xi)}{\xi-z}|d\xi| \tag{29}$$

for some $L_2(\partial D)$ functions F, and its norm in H_c is given by

$$\|f\|_{H_c}^2 = \min\frac{1}{4\pi^2}\int_{\partial D}|F(\xi)|^2|d\xi|. \tag{30}$$

Here the minimum is taken over all functions $F \in L_2(\partial D)$ satisfying (29). From the expression

$$f(z) = \frac{1}{4\pi^2}\int_{\partial D}\frac{1}{\xi-z}\frac{2\pi f(\xi)d\xi}{i|d\xi|}|d\xi|,$$

in general, we have the inequality

$$\|f\|_{H_c}^2 \leqq \int_{\partial D}|f(\xi)|^2|d\xi|. \tag{31}$$

Meanwhile, from (28) we see that $f \in H_c$ is expressible in the form

$$f(z) = f_1(z) + \int_{\partial D}\hat{l}(\xi,z)G_f(\xi)|d\xi| \quad \text{on} \quad D, \tag{32}$$

159

for some $f_1 \in AL_2(\partial D)$ and for some $G_f(\xi) \in L_2(\partial D)$. Furthermore, the norm $\|f\|_{H_c}$ is given by

$$\|f\|_{H_c}^2 = \min \left\{ \int_{\partial D} |f_1(\xi)|^2 |d\xi| + \int_{\partial D} |G_f(\xi)|^2 |d\xi| \right\}, \tag{33}$$

where the minimum is taken over all expressions satisfying (32). In particular, note that for any $G \in L_2(\partial D)$,

$$\int_{\partial D} \widehat{l}(\xi, z) G(\xi) |d\xi| \in AL_2(\partial D),$$

and so we have the expression, as in (32)

$$f(z) = \left(f(z) - \int_{\partial D} \widehat{l}(\xi, z) G(\xi) |d\xi| \right) + \int_{\partial D} \widehat{l}(\xi, z) G(\xi) |d\xi|. \tag{34}$$

Hence, for f satisfying the equality in (31), we get from (34) and (33), the inequality

$$\int_{\partial D} |f(\xi)|^2 |d\xi| \leqq \int_{\partial D} \left| f(z) - \int_{\partial D} \widehat{l}(\xi, z) G(\xi) |d\xi| \right|^2 |dz| + \int_{\partial D} |G(\xi)|^2 |d\xi|,$$

that is,

$$0 \leqq \int_{\partial D} -2 \, \mathrm{Re} \left\{ f(z) \int_{\partial D} \overline{\widehat{l}(\xi, z) G(\xi)} |d\xi| \right\} |dz|$$

$$+ \int_{\partial D} \left| \int_{\partial D} \widehat{l}(\xi, z) G(\xi) |d\xi| \right|^2 |dz| + \int_{\partial D} |G(\xi)|^2 |d\xi|,$$

$$\text{for all} \quad G \in L_2(\partial D).$$

Since G was arbitrary, we have

$$0 = \int_{\partial D} \left\{ f(z) \int_{\partial D} \overline{\widehat{l}(\xi, z) G(\xi)} |d\xi| \right\} |dz| = \int_{\partial D} \left\{ \int_{\partial D} f(z) \overline{\widehat{l}(\xi, z)} |dz| \right\} \overline{G(\xi)} |d\xi|;$$

that is,

$$\int_{\partial D} f(z) \overline{\widehat{l}(\xi, z)} |dz| = 0 \quad \text{for all} \quad \xi \in \partial D.$$

Since the function $\widehat{l}(\xi, z)$ is analytic on $\overline{D} \times \overline{D}$ (closures),

$$\int_{\partial D} f(z) \overline{\widehat{l}(\xi, z)} |dz| = 0 \quad \text{for all} \quad \xi \in D.$$

160

From (26), $\widehat{L}(z, u) = -\widehat{L}(u, z)$ on $\overline{D} \times \overline{D}$, and so $\widehat{l}(\xi, z) = -\widehat{l}(z, \xi)$ on $\overline{D} \times \overline{D}$. Hence, from the orthogonality (27) of $\widehat{L}(z, \overline{u})$ we have the desired condition (25).

From the above argument, the sufficiency of (25) is clear.

Note that $\widehat{l}(\xi, z) \equiv 0$ if and only if D is a disc ([Oz]). Hence, when D is a disc, for all the functions $f \in AL_2(\partial D)$, their boundary value functions $f(\xi)$ coincide with the inverses of f for the Cauchy integral in the sense of the minimum norm.

3   The case of harmonic functions

We start with the Green's formula

$$u(z) = \frac{1}{2\pi} \int_{\partial D} F(\xi) \frac{\partial G(\xi, z)}{\partial \nu_\xi} |d\xi| \tag{35}$$

for $F \in L_2(\partial D)$. Here, $G(\xi, z)$ is the Green's function of D for the Laplace equation with pole at z and $\partial/\partial\nu$ denotes differentiation along the inward normal with respect to D.

For a compact set E of D with a positive plane measure, we consider the reproducing kernel

$$K_H(z, w) = \frac{1}{2\pi} \int_{\partial D} \frac{\partial G(\xi, z)}{\partial \nu_\xi} \frac{\partial G(\xi, w)}{\partial \nu_\xi} |d\xi| \tag{36}$$

$$\text{on} \quad E \times E.$$

Since $K_H(z, w)$ is harmonic on $E \times E$ and a positive definite kernel on E, we have the expansion as in (3) by using the eigenfunctions

$$0 < \mu_1 \leqq \mu_2 \leqq \cdots,$$

$$\mu_j \int_E K_H(z, w)\psi_j(w)dm(w) = \psi_j(z) \quad \text{on} \quad E,$$

$$\int_E \psi_j(z)\psi_{j'}(z)dm(z) = \delta_{jj'},$$

and

$$K_H(z, w) = \sum_{j=1}^{\infty}(\mu_j)^{-1}\psi_j(z)\psi_j(w)$$

which converges absolutely and uniformly on $E \times E$. Then, we have, as in Theorem 2

Theorem 3   *For any integrable function u_1 on E, there exists a harmonic function $u \in HL_2(\partial D)$ (harmonic Hardy class with index 2) satisfying*

$$\int_E u_1(z)\psi_j(z)dm(z) = \int_E u(z)\psi_j(z)dm(z) \tag{37}$$

161

if and only if

$$\sum_{j=1}^{\infty} \mu_j \left| \int_E u_1(z)\psi_j(z)dm(z) \right|^2 < \infty. \tag{38}$$

When u exists, it is uniquely determined and u is expressible in the form

$$u(z) = \lim_{N \to \infty} \frac{1}{2\pi} \int_{\partial D} \left\{ \sum_{j=1}^{N} \mu_j \int_E u_1(z)\psi_j(z)dm(z) \right.$$

$$\left. \cdot \int_E \psi_j(\eta) \frac{\partial G(\eta, \xi)}{\partial \nu_\eta} dm(\eta) \right\} \frac{\partial G(\xi, z)}{\partial \nu_\xi} |d\xi|$$

in terms of (37). Moreover, we have the isometrical identity

$$\frac{1}{2\pi} \int_{\partial D} |u(z)|^2 |dz| = \sum_{j=1}^{\infty} \mu_j \left| \int_E u_1(z)\psi_j(z)dm(z) \right|^2.$$

Note that $H_K(z, w)$ itself is the reproducing kernel for the Hilbert space $H L_2(\partial D)$ equipped with the inner product

$$\frac{1}{2\pi} \int_{\partial D} u_1(z)\overline{u_2(z)}|dz|.$$

4   Examples

(I)   We consider a region D surrounded by $N + 1$ circles $c_0 = \{|z| = 1\}, c_k = \{|z - a_k| = r_k; r_k > 0\}$ $(k = 1, 2, \cdots, N)$ and, for any k $(1 \leq k \leq N), c_k \subset \{|z| < 1\}$ and $\{|z - a_k| < r_k\} \cap \{|z - a_{k'}| < r_{k'}\} = \phi$ for $k \neq k'$. Then, we have the expansion of (2)

$$K_A(z, \overline{u}) = \frac{1}{2\pi} \int_{\partial D} \frac{|d\xi|}{(\xi - z)\overline{(\xi - u)}}$$

$$= \sum_{n=0}^{\infty} z^n \overline{u}^n + \sum_{k=1}^{N} \sum_{n=0}^{\infty} \frac{r_k^{2n+1}}{\overline{(u - a_k)}^{n+1}(z - a_k)^{n+1}} \tag{39}$$

$$\text{on} \quad D \times \overline{D}.$$

Hence any member $f(z)$ of H_{K_A} is expressible in the form

$$f(z) = \sum_{n=0}^{\infty} B_n z^n + \sum_{k=1}^{N} \sum_{n=0}^{\infty} \frac{B_n^{(k)} r_k^{2n+1}}{(z - a_k)^{n+1}} \quad \text{on} \quad D, \tag{40}$$

162

for some constants B_n and $B_n^{(k)}$ satisfying

$$\sum_{n=0}^{\infty} |B_n|^2 + \sum_{k=1}^{N}\sum_{n=0}^{\infty} |B_n^{(k)}|^2 r_k^{2n+1} < \infty. \tag{41}$$

In order to realize the norm $\|f\|_{HK_A}$ in a reasonable way we take the set $\{\tilde{c}_0, \tilde{c}_1, \cdots, \tilde{c}_N\}$ of smooth Jordan curves as E such that $\tilde{c}_k \subset D$ and each $\tilde{c}_k$ is homologous to c_k. Then we have the expression of the constants in terms of the values of f on E

$$B_n = \frac{1}{2\pi i} \int_{\tilde{c}_0} \frac{f(\xi)}{\xi^{n+1}} d\xi$$

and

$$B_n^{(k)} = \frac{-1}{2\pi i r_k^{2n+1}} \int_{\tilde{c}_k} f(\xi)(\xi - a_k)^n d\xi.$$

Hence, we have the isometrical identity

$$\|f\|_{HK_A}^2 = \frac{1}{4\pi^2} \sum_{n=0}^{\infty} \left| \int_{\tilde{c}_0} \frac{f(\xi)}{\xi^{n+1}} d\xi \right|^2$$

$$+ \sum_{k=1}^{N}\sum_{n=0}^{\infty} \frac{1}{4\pi^2 r_k^{2n+1}} \left| \int_{\tilde{c}_k} f(\xi)(\xi - a_k)^n d\xi \right|^2$$

$$= \frac{1}{2\pi} \int_{\partial D} |F^*(\xi)|^2 |d\xi|, \tag{42}$$

in Theorem 1. Of course, we can determine $f(z)$ for any $z \in D$ by the values of f on E as in Theorem 1.

(II) Let $R(r)$ be the annulus region $r < |z| < 1$ and take an analytic Jordan curve $c \subset R(r)$ as E which separates $\{|z| = r\}$ and $\{|z| = 1\}$. Then, for the Szegö kernel $K_{R(r)}(z, \overline{u})$ on $R(r)$, we consider the integral transform

$$f(z) = \int_{\partial R(r)} F(\xi) \overline{K_{R(r)}(\xi, \overline{z})} |d\xi| \quad \text{for} \quad z \in c \tag{43}$$

for $F \in L_2(\partial R(r))$, and the identity

$$K_{R(r)}(z, \overline{u}) = \int_{\partial R(r)} K_{R(r)}(\xi, \overline{u}) \overline{K_{R(r)}(\xi, \overline{z})} |d\xi| \tag{44}$$

163

which correspond to (1) and (2), respectively. For $K_{R(r)}(z, \overline{u})$, we have

$$K_{R(r)}(z, \overline{u}) = \frac{1}{2\pi} \sum_{j=-\infty}^{\infty} \frac{(z\overline{u})^j}{1 + r^{2j+1}}. \tag{45}$$

See, for example [[Ne]], p.391. Note that for any integrable function f_1 on c,

$$\int_c f_1(z)\overline{z}^j |dz| = 0 \quad \text{for all} \quad j \tag{46}$$

implies that $f_1(z) = 0$ a.e. on c. Hence, for any integrable function f_1 on c, there exists an analytic function $f \in AL_2(\partial R(r))$ satisfying

$$f_1(z) = f(z) \quad a.e. \quad \text{on} \quad c \tag{47}$$

if and only if

$$\sum_{j=-\infty}^{\infty} (1 + r^{2j+1}) \left| \int_c \frac{f_1(z)dz}{z^{j+1}} \right|^2 < \infty. \tag{48}$$

(Ⅲ) On the unit disc $|z| < 1$, we consider the Poisson integral, for $z = re^{i\theta}$

$$u(r, \theta) = \frac{1}{2\pi} \int_{-\pi}^{\pi} f(\tilde{\theta}) \frac{1 - r^2}{1 - 2r\cos(\theta - \tilde{\theta}) + r^2} d\tilde{\theta} \tag{49}$$

for real-valued functions $f \in L_2(-\pi, \pi)$. We form the reproducing kernel (36)

$$K_H(r, \theta; r', \theta') = 1 + 2\sum_{n=1}^{\infty} r^n \cos n\theta \cdot r'^n \cos n\theta'$$

$$+ 2\sum_{n=1}^{\infty} r^n \sin n\theta \cdot r'^n \sin n\theta'$$

$$= \frac{1 - |\overline{z'}|^2 |z|^2}{|1 - \overline{z'}z|^2}. \tag{50}$$

Hence, the images $u(r, \theta)$ of (49) are expressible in the form

$$u(r, \theta) = c_0 + 2\sum_{n=1}^{\infty} A_n r^n \cos n\theta + 2\sum_{n=1}^{\infty} B_n r^n \sin n\theta \tag{51}$$

for some real constants c_0, A_n and B_n satisfying

$$c_0^2 + 2\sum_{n=1}^{\infty} A_n^2 + 2\sum_{n=1}^{\infty} B_n^2 < \infty.$$

164

These constants are uniquely determined by

$$c_0 = \frac{1}{2\pi} \int_{-\pi}^{\pi} u(r, \theta) d\theta,$$

$$A_n = \frac{1}{2\pi r^n} \int_{-\pi}^{\pi} u(r, \theta) \cos n\theta \, d\theta,$$

and

$$B_n = \frac{1}{2\pi r^n} \int_{-\pi}^{\pi} u(r, \theta) \sin n\theta \, d\theta,$$

for any fixed $0 < r < 1$. Hence, we have the isometrical identity

$$\frac{1}{4\pi^2} \left(\int_{-\pi}^{\pi} u(r, \theta) d\theta \right)^2 + \frac{1}{2\pi^2} \sum_{n=1}^{\infty} \frac{1}{r^{2n}} \left(\int_{-\pi}^{\pi} u(r, \theta) \cos n\theta \, d\theta \right)^2$$

$$+ \frac{1}{2\pi^2} \sum_{n=1}^{\infty} \frac{1}{r^{2n}} \left(\int_{-\pi}^{\pi} u(r, \theta) \sin n\theta \, d\theta \right)^2 = \frac{1}{2\pi} \int_{-\pi}^{\pi} f(\theta)^2 d\theta, \qquad (52)$$

for any fixed r $(0 < r < 1)$. Then, we can determine all the values $u(r', \theta')$ on $|z| < 1$ by the values $u(r, \theta)$ for any fixed r.

(IV)   On the upper-half plane $\{\mathrm{Im}\, z > 0\}$, we consider the Poisson integral

$$u(x, y) = \frac{1}{\pi} \int_{-\infty}^{\infty} f(\xi) \frac{y}{y^2 + (x - \xi)^2} d\xi \quad (y > 0, -\infty < x < \infty), \qquad (53)$$

for functions $f \in L_2(\mathbb{R})$. We form the reproducing kernel (36)

$$K_H(x, y; x', y') = \frac{1}{\pi} \int_{-\infty}^{\infty} \frac{y}{y^2 + (x - \xi)^2} \frac{y'}{y'^2 + (x' - \xi)^2} d\xi$$

$$= \frac{y + y'}{(y + y')^2 + (x - x')^2}$$

$$= \int_0^{\infty} e^{-t(y+y')} \cos t(x - x') dt$$

$$= \int_0^{\infty} \left(e^{-ty} \cos tx \cdot e^{-ty'} \cos tx' + e^{-ty} \sin tx \cdot e^{-ty'} \sin tx' \right) dt.$$
$$(54)$$

Hence, the images $u(x, y)$ in (53) are expressible in the form

$$u(x, y) = \int_0^{\infty} F(t) e^{-ty} \cos tx \, dt + \int_0^{\infty} G(t) e^{-ty} \sin tx \, dt \qquad (55)$$

165

for some functions F and G satisfying $F, G \in L_2(0, \infty)$. For any fixed y, these functions are uniquely determined by using the Fourier cosine and sine transforms in the forms

$$F(t) = \frac{2}{\pi} e^{ty} \ s - \lim_{L \to \infty} \int_0^L u(x, y) \cos xt \, dx \tag{56}$$

and

$$G(t) = \frac{2}{\pi} e^{ty} \ s - \lim_{L \to \infty} \int_0^L u(x, y) \sin xt \, dx \tag{57}$$

in $L_2(0, \infty)$. Hence, we have the isometrical identity

$$\frac{1}{\pi} \int_{-\infty}^{\infty} f(\xi)^2 d\xi = \frac{4}{\pi^2} \int_0^{\infty} e^{2ty} \left(s - \lim_{L \to \infty} \int_0^L u(x, y) \cos xt \, dx \right)^2 dt$$

$$+ \frac{4}{\pi^2} \int_0^{\infty} e^{2ty} \left(s - \lim_{L \to \infty} \int_0^L u(x, y) \sin xt \, dx \right)^2 dt. \tag{58}$$

Meanwhile, from (55) we have, for any fixed x

$$F(t) \cos xt + G(t) \sin xt = \frac{1}{2\pi i} \int_{-i\infty}^{i\infty} u(x, y) e^{yt} \, dy \tag{59}$$

by the Laplace inversion formula. Here, we use the analytic extension of $u(x, y)$ to the right half plane with respect to the variable y. At this moment we can also use the real inversion formula of the Laplace transform in Theorem 4.2.2.

Note that $F(t) \cos xt + G(t) \sin xt \in L_2(0, \infty)$. Hence, we have

$$F(t) = \frac{1}{2\pi i} \int_{-i\infty}^{i\infty} u(0, y) e^{yt} \, dy \tag{60}$$

and

$$G(t) = \frac{1}{2\pi t i} \int_{-i\infty}^{i\infty} \frac{\partial u(x, y)}{\partial x} \bigg|_{x=0} e^{yt} \, dy. \tag{61}$$

Then we can obtain an alternative realization of the norm of $u(x, y)$ in the RKHS H_{K_H}.

166

In the two cases, we can determine all the values of $u(x, y)$ on $y > 0$ by the values $u(x, y)$ (for any fixed $y > 0$), and $u(0, y)$, $u_x(x, y)|_{x=0}$, in reasonable ways.

In the above examples, we know that there exist special sets E for which the RKHS induced from the original integral transform is simply and naturally determined. In miscellaneous concrete cases, it seems that to determine such sets E is an interesting and valuable problem.

5 A general Cauchy integral on the half plane

In this subsection, we shall give a fairly simple and beautiful isometrical identity between the Szegö space and the Bergman–Selberg space on the half plane by means of the Cauchy integral, following [By1]. For the Bergman–Selberg spaces H_{K_q} on the upper half plane U, see again (2.4.93)–(2.4.96).

Theorem 4 *For any fixed $q > 0$, let T_q be the integral transform of the Szegö space $H_{K_{1/2}}$ on the upper half plane U, for the principal value of $i^{q+1/2}$*

$$[T_q f](z) = \frac{\Gamma(q + \frac{1}{2})}{2\pi i^{q+\frac{1}{2}}} \int_{-\infty}^{\infty} f(\xi)(\xi - z)^{-(q+\frac{1}{2})} d\xi, \quad z \in U. \tag{62}$$

Then, $T_q f \in H_{K_q}$ and T_q is an isometrical mapping between $H_{K_{1/2}}$ and H_{K_q}. Furthermore, for $q > \frac{1}{2}$, the inversion T_q^{-1} of T_q is represented in the form

$$[T_q^{-1} g](w) = \frac{\Gamma(q + \frac{1}{2})}{\pi \Gamma(2q - 1) i^{q+\frac{1}{2}}} \iint_U g(z)(\bar{z} - w)^{-(q+\frac{1}{2})}$$
$$\times (2y)^{2q-2} dx\,dy, \quad w \in U. \tag{63}$$

Proof : For any $z \in U$, we have, by the Parseval–Plancherel theorem

$$[T_q f](z) = \frac{1}{\sqrt{2\pi}} \int_0^{\infty} \hat{f}(t) e^{izt} t^{q-\frac{1}{2}} dt.$$

As we saw in Chapter 2, 4.6, the Fourier transform $\hat{f}$ of f vanishes on $(-\infty, 0)$, $\hat{f} \in L_2(0, \infty)$ and any member of $L_2(0, \infty)$ is the Fourier transform of $f \in H_{K_{1/2}}$. Meanwhile, $K_q(z, \bar{u}) = (-1)^q \Gamma(2q)(z - \bar{u})^{-2q}$ and

$$K_q(z, \bar{u}) = \int_0^{\infty} e^{izt} e^{-i\bar{u}t} t^{2q-1} dt.$$

Hence, the image of T_q coincides with H_{K_q} and we have the isometrical identity

$$\|T_q f\|_{H_{K_q}}^2 = \frac{1}{2\pi} \int_0^{\infty} |\hat{f}(t)|^2 dt$$

167

because $\{e^{izt}t^{q-\frac{1}{2}}; z \in U\}$ is complete in $L_2(0,\infty)$. Hence, we have the desired isometrical identity

$$\|T_q f\|_{H_{K_q}} = \|f\|_{H_{K_{1/2}}}.$$

In order to invert T_q, notice since T_q is an isometry the inverse equals the adjoint $T_q^{-1} = T_q^*$. Hence, we have

$$
\begin{aligned}
[T_q^{-1}g](w) &= (T_q^{-1}g, K_{\frac{1}{2}}(\cdot, \overline{w}))_{H_{K_{1/2}}} \\
&= (T_q^* g, K_{\frac{1}{2}}(\cdot, \overline{w}))_{H_{K_{1/2}}} \\
&= (g, T_q K_{\frac{1}{2}}(\cdot, \overline{w}))_{H_{K_q}}.
\end{aligned}
$$

Here,

$$
\begin{aligned}
[T_q K_{\frac{1}{2}}(\cdot, \overline{w})](z) &= \int_0^\infty e^{-i(\overline{w}-z)t} t^{q-\frac{1}{2}} dt \\
&= \Gamma(q + \frac{1}{2}) i^{-(q+\frac{1}{2})} (\overline{w} - z)^{-(q+\frac{1}{2})}.
\end{aligned}
$$

Hence, we have the desired representation (63).

§5 The Meyer wavelets

The theory of wavelets is developing enormously in both the mathematical sciences and pure mathematics. See, for example, I. Daubechies [[D]] and C. Chui [[Chu]] as basic references.

See also the recent article [GL] for a good crash course in wavelet analysis with the aim of using it to solve the parabolic approximate wave equation. Furthermore, they develop the wavelet–Galerkin approximation for solving partial differential equations in the variational form and derive a range-depth adaptive wavelet approach for providing an algorithm for solving the propagation problem.

The theory of reproducing kernels will give a unified understanding of wavelet transforms, frames, multiresolution analysis and the sampling theorem. The general method for integral transforms needs to be treated case by case. In this section we shall examine typical wavelets of the Meyer wavelet type using the theory of reproducing kernels. For the general interrelationships between the theories of frames in wavelet transforms and of reproducing kernels, see ([[Sa]], Chapter 7). We will not refer to these general results in this book.

We shall use mainly the notation and theorems in Daubechies [[D]]; however, the integral parameters will be changed in order to be consistent with our notation.

168

1   Integral transform by the Meyer wavelets

The Mexican hat function $\psi(t)$ is defined by the second derivative of the Gaussian $e^{-\frac{t^2}{2}}$ with the normalization that its L_2 norm is set equal to one; that is,

$$\psi(t) = \frac{2}{\sqrt{3}} \pi^{-\frac{1}{4}} (1 - t^2) e^{-\frac{t^2}{2}}.$$

Its Fourier transform

$$(\mathcal{F}\psi)(\omega) = \frac{1}{\sqrt{2\pi}} \int_{-\infty}^{\infty} \psi(t) e^{-i\omega t} dt$$

is given by

$$(\mathcal{F}\psi)(\omega) = \hat{\psi}(\omega) = \frac{2}{\sqrt{3}} \pi^{-\frac{1}{4}} \omega^2 e^{-\frac{\omega^2}{2}}.$$

The Mexican hat function $\psi(t)$ has the properties

$$\int_{-\infty}^{\infty} \psi(t) dt = 0$$

and

$$C_{\psi} = 2\pi \int_{-\infty}^{\infty} |\hat{\psi}(\omega)|^2 |\omega|^{-1} d\omega = \frac{8}{3}\sqrt{\pi} < \infty$$

which are admitted by other functions appearing in wavelet theory.

The Meyer wavelets are

$$\psi^{a,x}(t) = |a|^{-\frac{1}{2}} \psi(\frac{t-x}{a}) \quad \text{for} \quad a, x \in \mathbb{R}, \quad a \neq 0,$$

and the wavelet transform with respect to this wavelet family is defined by

$$(T^{wav} f)(a, x) = \langle f, \psi^{a,x} \rangle = \int_{-\infty}^{\infty} f(t) |a|^{-\frac{1}{2}} \psi(\frac{t-x}{a}) dt \tag{1}$$

for $L^2(\mathbb{R})$ functions f. Of course,

$$|(T^{wav} f)(a, x)| \leq \|f\|,$$

where $\|f\| = \langle f, f \rangle$.

For the integral transform (1), we have the isometrical identity which is obtained formally using the Fourier transform: for any $f, g \in L^2(\mathbb{R})$,

$$\int_{-\infty}^{\infty} \int_{-\infty}^{\infty} (T^{wav} f)(a, x) \overline{(T^{wav} g)(a, x)} \frac{da\,dx}{a^2}$$

$$= \int_{-\infty}^{\infty} \int_{-\infty}^{\infty} \left[\int_{-\infty}^{\infty} \hat{f}(\omega) |a|^{\frac{1}{2}} e^{-ix\omega} \widehat{\psi}(a\omega) d\omega \right]$$

$$\cdot \left[\int_{-\infty}^{\infty} \overline{\hat{g}(\omega')} |a|^{\frac{1}{2}} e^{ix\omega'} \overline{\widehat{\psi}(a\omega')} d\omega' \right] \frac{da\,dx}{a^2}$$

$$= 2\pi \int_{-\infty}^{\infty} \frac{da}{|a|} \int_{-\infty}^{\infty} \hat{f}(\omega) \overline{\hat{g}(\omega)} |\widehat{\psi}(a\omega)|^2 d\omega$$

$$= \frac{8}{3} \sqrt{\pi} \langle f, g \rangle$$

and so,

$$\int_{-\infty}^{\infty} \int_{0}^{\infty} (T^{wav} f)(a, x) \overline{(T^{wav} g)(a, x)} \frac{da\,dx}{a^2} = \frac{4}{3} \sqrt{\pi} \langle f, g \rangle. \tag{2}$$

The finiteness of the norm in the left hand side in (2), however, does not characterize a Hilbert space formed by the images by (1). This means that the image space is a subspace in $L^2(\mathbb{R}^+ \times \mathbb{R}; \frac{3}{4\sqrt{\pi}} a^{-2} da\,dx)$ for $\mathbb{R}^+ = (0, \infty)$. So, we first wish to identify the image space.

By the symmetry in a, we shall consider the integral transform (1) for $a > 0$. However, we should consider the integral transform (1) in the complex form

$$(T^{wav} f)(A, z) = \int_{-\infty}^{\infty} f(t) A^{-\frac{1}{2}} \psi(\frac{t - z}{A}) dt \tag{3}$$

or

$$(T^{wav} f)(A, z) = \frac{2}{\sqrt{3}} \pi^{-\frac{1}{4}} A^{\frac{5}{2}} \int_{-\infty}^{\infty} \hat{f}(\omega) e^{i\omega z} \omega^2 e^{-\frac{A^2 \omega^2}{2}} d\omega \tag{4}$$

for the complex numbers $A = a + i\alpha$ and $z = x + iy$, in order to identify a natural image space of the integral transform (3) or (4).

In order to identify the natural image space of the integral transform (3) for

170

$L^2(\mathbb{R})$ functions, we calculate the kernel form

$$K(A, z; \overline{A'}, \overline{z'}) = A^{-\frac{1}{2}}\overline{A'}^{-\frac{1}{2}} \int_{-\infty}^{\infty} \psi(\frac{t-z}{A})\overline{\psi(\frac{t-z'}{A'})}dt$$

$$= \int_{-\infty}^{\infty} \left[\frac{2}{\sqrt{3}}\pi^{-\frac{1}{4}}A^{\frac{5}{2}}e^{i\omega z}\omega^2 e^{-\frac{A^2\omega^2}{2}}\right]$$

$$\cdot \left[\frac{2}{\sqrt{3}}\pi^{-\frac{1}{4}}A'^{\frac{5}{2}}e^{i\omega z'}\omega^2 e^{-\frac{A'^2\omega^2}{2}}\right]d\omega$$

$$= \frac{4\sqrt{2}}{3}\frac{A^{\frac{5}{2}}\overline{A'}^{\frac{5}{2}}}{(A^2+\overline{A'}^2)^{\frac{5}{2}}}\exp\left\{-\frac{(z-\overline{z'})^2}{2(A^2+\overline{A'}^2)}\right\}$$

$$\cdot \left\{3 - \frac{6(z-\overline{z'})^2}{A^2+\overline{A'}^2} + \frac{(z-\overline{z'})^4}{(A^2+\overline{A'}^2)^2}\right\} \tag{5}$$

for $A, A' \in \Delta(\frac{\pi}{4}) = \{|\arg A| < \frac{\pi}{4}\}$ and, $z, z' \in \mathbb{C}$. The identity (5) implies that $K(A, z; \overline{A'}, \overline{z'})$ are analytic in (A, z) on $\Delta(\frac{\pi}{4}) \times \mathbb{C}$, and anti-analytic in (A', z') on $\Delta(\frac{\pi}{4}) \times \mathbb{C}$. In particular, we see that the images $(T^{wav}f)(a, x)$ of (1) by $L^2(\mathbb{R})$ functions f can be extended analytically onto the product space $\Delta(\frac{\pi}{4}) \times \mathbb{C}$ in the form $(T^{wav}f)(A, z)$, and the images $(T^{wav}f)(A, z)$ are characterized as the members of the Hilbert space H_K admitting the reproducing kernel (5). Furthermore, since

$$\left\{A^{-\frac{1}{2}}\psi(\frac{t-z}{A}); A \in \Delta(\frac{\pi}{4}), \quad z \in \mathbb{C}\right\}$$

is complete in $L^2(\mathbb{R})$, we have the isometrical identity

$$\|(T^{wav}f)(A, z)\|_{H_K}^2 = \int_{-\infty}^{\infty}|\hat{f}(\omega)|^2 d\omega = \|f\|^2. \tag{6}$$

In particular, we obtain, from (2) and (6), the very curious identity

$$\|(T^{wav}f)(A, z)\|_{H_K}^2 = \frac{3}{4\sqrt{\pi}}\int_{-\infty}^{\infty}\int_{0}^{\infty}|(T^{wav}f)(a, x)|^2\frac{da\,dx}{a^2} \tag{7}$$

from the viewpoint of the complicated reproducing kernel (5).

2   Characterization as functions in space x

We first examine the wavelet transform (3) for any fixed $A \in \Delta(\frac{\pi}{4})$. Then, we have the corresponding reproducing kernel by setting $A = A'$ in (5)

$$K(A, z; \overline{A}, \overline{z'}) = K_A(z, \overline{z'}) = \frac{1}{3}\exp\left\{-\frac{(z-\overline{z'})^2}{4(a^2-\alpha^2)}\right\}$$

$$\cdot \left\{3 - \frac{3(z-\overline{z'})^2}{a^2-\alpha^2} + \frac{(z-\overline{z'})^4}{4(a^2-\alpha^2)^2}\right\}. \tag{8}$$

171

This implies that for any fixed $A \in \Delta(\frac{\pi}{4})$, the images $(T^{wav}f)(A,z) = (T^{wav}f)_A(z)$ of (3) for $L^2(\mathbb{R})$ functions f are characterized as the members of the Hilbert space H_{K_A} composed of entire functions admitting the reproducing kernel (8).

We shall determine the space H_{K_A}. We note the identity

$$K_A(z, \overline{z'}) = \frac{4(a^2 - \alpha^2)^2}{3} \frac{\partial^4}{\partial z^2 \partial \overline{z'}^2} \exp\left\{-\frac{(z - \overline{z'})^2}{4(a^2 - \alpha^2)}\right\} \tag{9}$$

and that

$$\frac{4(a^2 - \alpha^2)^2}{3} \exp\left\{-\frac{(z - \overline{z'})^2}{4(a^2 - \alpha^2)}\right\}$$

is the reproducing kernel for the Fock space $\mathcal{F}(A)$ composed of all entire functions $h(z)$ with finite norms

$$\left\{\frac{3}{8\pi(a^2 - \alpha^2)^3} \iint_{\mathbb{C}} |h(z)|^2 \exp(\frac{-y^2}{a^2 - \alpha^2}) dx dy\right\}^{\frac{1}{2}} < \infty \tag{10}$$

(see (2.4.9)). Then, from (9) we have the isometrical identity

$$\|h\|_{\mathcal{F}(A)} = \|h''(z)\|_{H_{K_A}} \quad \text{for} \quad h \in \mathcal{F}(A). \tag{11}$$

Note that for any fixed $A \in \Delta(\frac{\pi}{4})$

$$\left\{A^{-\frac{1}{2}} \psi(\frac{t - z}{A}); z \in \mathbb{C}\right\}$$

is complete in $L^2(\mathbb{R})$. Hence, for the twice integrals $(T^{wav}f)_A^{(-2)}(z)$ of $(T^{wav}f)_A(z)$ under the normalization that the values at infinity are zero, we have the following characterization of $(T^{wav}f)_A(z)$.

Theorem 1 *For any fixed $A \in \Delta(\frac{\pi}{4})$, the images $(T^{wav}f)_A(z)$ of (3) for $L^2(\mathbb{R})$ functions f are characterized by the properties that $(T^{wav}f)_A(z)$ are entire functions and*

$$\iint_{\mathbb{C}} |(T^{wav}f)_A^{(-2)}(z)|^2 \exp(\frac{-y^2}{a^2 - \alpha^2}) dx dy < \infty.$$

Furthermore, then we have the isometrical identity

$$\frac{3}{8\pi(a^2 - \alpha^2)^3} \iint_{\mathbb{C}} |(T^{wav}f)_A^{(-2)}(z)|^2 \exp(\frac{-y^2}{a^2 - \alpha^2}) dx dy = \|f\|^2. \tag{12}$$

172

Meanwhile, we shall derive a characterization of the images $(T^{wav}f)_A(x)$ as real variable functions. In (4), by Plancherel's theorem, we have, for any fixed $A \in \Delta(\frac{\pi}{4})$

$$\int_{-\infty}^{\infty} |\hat{f}(\omega)|^2 d\omega = \frac{3\sqrt{\pi}}{4(a^2-\alpha^2)^{\frac{5}{2}}}$$

$$\cdot \int_{-\infty}^{\infty} \left| \frac{1}{2\pi} \int_{-\infty}^{\infty} (T^{wav}f)_A(x)e^{-i\omega x} dx \right|^2 \frac{1}{\omega^4} e^{\omega^2(a^2-\alpha^2)} d\omega \tag{13}$$

and so, we obtain

Theorem 2 *For any fixed $A \in \Delta(\frac{\pi}{4})$, the images $(T^{wav}f)_A(x)$ of (3) for $L^2(\mathbb{R})$ functions f are characterized by the properties that they are of class $C^\infty(\mathbb{R})$ and have finite integrals (13). Then, we furthermore obtain the isometrical identity*

$$\|f\|^2 = \frac{3}{8\sqrt{\pi}(a^2-\alpha^2)^{\frac{5}{2}}} \sum_{n=0}^{\infty} \frac{(a^2-\alpha^2)^n}{n!} \int_{-\infty}^{\infty} \left| \frac{\partial^{n-2}(T^{wav}f)_A(x)}{\partial x^{n-2}} \right|^2 dx \tag{14}$$

where

$$\frac{\partial^{-1}}{\partial x^{-1}}(T^{wav}f)_A(x) = \int_{-\infty}^{x} (T^{wav}f)_A(\xi)d\xi$$

and

$$\frac{\partial^{-2}}{\partial x^{-2}} = \frac{\partial^{-1}}{\partial x^{-1}}(\frac{\partial^{-1}}{\partial x^{-1}}).$$

From the identities (12) and (14), we obtain the inversion formulas for (3), in a general argument.

Theorem 3 *In the wavelet transform (3) for any fixed $A \in \Delta(\frac{\pi}{4})$, we have the inversion formulas*

$$f = s - \lim_{N \to \infty}$$

$$\cdot \left[\frac{3\overline{A}^{-\frac{1}{2}}}{8\pi(a^2-\alpha^2)^3} \iint_{E_N} (T^{wav}f)_A^{(-2)}(z)\psi(\frac{t-\overline{z}}{\overline{A}})^{(-2)} \exp\left\{ \frac{-y^2}{a^2-\alpha^2} \right\} dx dy \right]_{(15)}$$

for any compact exhaustion $\{E_N\}_{N=1}^{\infty}$ of $\mathbb{C}$, and

$$f = s - \lim_{N \to \infty} \left[\frac{3\overline{A}^{-\frac{1}{2}}}{8\sqrt{\pi}(a^2-\alpha^2)^{\frac{5}{2}}} \right]$$

$$\cdot \sum_{n=0}^{N} \frac{(a^2-\alpha^2)^n}{n!} \int_{-\infty}^{\infty} \frac{\partial^{n-2}(T^{wav}f)_A(x)}{\partial x^{n-2}} \left[\frac{\partial^{n-2}}{\partial x^{n-2}}\psi(\frac{t-x}{A}) \right] dx \tag{16}$$

in the sense of strong convergence in $L^2(\mathbb{R})$.

3   Characterization as functions in scaling a

We next examine the wavelet transform (3) for any fixed z. Then, we have the corresponding reproducing kernel by setting $z = z'$ in (5)

$$K(A, z; \overline{A'}, \overline{z}) = K_z(A, \overline{A'})$$

$$= \frac{4\sqrt{2}}{3} \frac{A^{\frac{5}{2}}\overline{A'}^{\frac{5}{2}}}{(A^2 + \overline{A'}^2)^{\frac{5}{2}}} \exp\left(\frac{2y^2}{A^2 + \overline{A'}^2}\right)$$

$$\cdot \left\{3 + \frac{24y^2}{A^2 + \overline{A'}^2} + \frac{16y^4}{(A^2 + \overline{A'}^2)^2}\right\}. \tag{17}$$

We wish to identify the Hilbert space admitting the reproducing kernel (17) which characterizes the images $(T^{wav}f)_z(A)$ as the functions in A. But, it is quite involved. However, when z is a real number x, the reproducing kernel (17) is very simple; that is,

$$K_x(A, \overline{A'}) = \frac{4\sqrt{2}A^{\frac{5}{2}}\overline{A'}^{\frac{5}{2}}}{(A^2 + \overline{A'}^2)^{\frac{5}{2}}}. \tag{18}$$

So, we shall consider the integral transform (9) for any fixed real x.

In (9), note that for the even part of the functions f with respect to x; that is,

$$f_{e,x}(t) = \frac{1}{2}\{f(t - x) + f(x - t)\},$$

$$(T^{wav}f)_x(A) = \int_{-\infty}^{\infty} f_{e,x}(t) A^{-\frac{1}{2}} \psi(\frac{t - x}{A}) dt \tag{19}$$

and, furthermore, for any fixed x

$$\{A^{-\frac{1}{2}}\psi(\frac{t - x}{A}); A \in \Delta(\frac{\pi}{4})\}$$

is complete in the subspace of $L^2(\mathbb{R})$ comprising even functions with respect to the point x. This implies that for any fixed x, the images $(T^{wav}f)_x(A)$ for $L^2(\mathbb{R})$ functions f are characterized as the members of the Hilbert space H_{K_x} composed of all analytic functions $h(z)$ on $\Delta(\frac{\pi}{4})$ admitting the reproducing kernel (18) and we have the isometrical identity

$$\|(T^{wav}f)_x\|_{H_{K_x}}^2 = \int_{-\infty}^{\infty} |f_{e,x}(t)|^2 dt \tag{20}$$

174

for the even part $f_{e,x}$ of f with respect to the point x.

We shall determine the Hilbert space H_{K_x} admitting the reproducing kernel (18).

First note that for any $q > \frac{1}{2}$

$$K_{(q)}(z, \overline{u}) = \frac{\Gamma(2q)}{(z + \overline{u})^{2q}}$$

is the reproducing kernel for the Bergman–Selberg space $H_{K_{(q)}}$ comprising all analytic functions $f(z)$ on the right half plane $R^+ = \{\text{Re} z > 0\}$ with finite norms

$$\|f\|_{H_{K_{(q)}}}^2 = \frac{2^{2q-2}}{\pi \Gamma(2q-1)} \iint_{R^+} |f(z)|^2 x^{2q-2} dx dy$$

$$= \frac{1}{\Gamma(2q-1)\pi} \iint_{R^+} |f(z)|^2 K_B(z, \overline{z})^{1-q} dx dy$$

for the Bergman kernel $K_B(z, \overline{u}) = K_{(1)}(z, \overline{u})$ for the space $H_{K_{(1)}}$ on R^+. The members f in $H_{K_{(q)}}$ are, in fact, analytic differentials of order q, and for any conformal mapping $z^* = \varphi(z)$ from R^+ onto a domain R^* we have the isometrical identity

$$\frac{1}{\Gamma(2q-1)\pi} \iint_{R^+} |f(z)|^2 K_B(z, \overline{z})^{1-q} dx dy$$

$$= \frac{1}{\Gamma(2q-1)\pi} \iint_{R^*} |f(\varphi^{-1}(z^*))|^2 K(\varphi^{-1}(z^*), \overline{\varphi^{-1}(z^*)})^{1-q} dx^* dy^*,$$

$$z^* = x^* + iy^*$$

with the relations

$$f(z)(dz)^q = f(\varphi^{-1}(z^*))(dz^*)^q$$

and

$$K_{(q)}(z, \overline{u})(dz)^q (d\overline{u})^q = K_{(q)}(z^*, \overline{u^*})(dz^*)^q \overline{(du^*)^q}$$

for the Bergman–Selberg kernels on R^+ and R^*, respectively.

By the conformal mapping $z = A^2$, we have

$$K_B(A, \overline{A'}) = \frac{4A\overline{A'}}{(A^2 + \overline{A'}^2)^2}$$

which is the Bergman kernel on the sector $\Delta(\frac{\pi}{4})$. Hence, we obtain

$$\|f\|_{H_{K_{(q)}}}^2 = \frac{1}{\Gamma(2q-1)\pi} \iint_{\Delta(\frac{\pi}{4})} |f(A)|^2 \left\{ \frac{a^2 + \alpha^2}{(a^2 - \alpha^2)^2} \right\}^{1-q} da d\alpha$$

175

and

$$K_q(A, \overline{A'}) = \frac{\Gamma(2q)2^{2q}A^q\overline{A'}^q}{(A^2 + \overline{A'^2})^{2q}}$$

on $\Delta(\frac{\pi}{4})$. From these identities we have, as in Section 2

Theorem 4 *For any fixed real x, the images $(T^{wav}f)_x(A)$ of (3) for $L^2(\mathbb{R})$ functions f are characterized by the properties that $(T^{wav}f)_x(A)$ are analytic on $\Delta(\frac{\pi}{4})$, and*

$$\iint_{\Delta(\frac{\pi}{4})} |(T^{wav}f)_x(A)|^2 \frac{(a^2 - \alpha^2)^{\frac{1}{2}}}{(a^2 + \alpha^2)^{\frac{3}{2}}} da\, d\alpha < \infty.$$

Then, we have, furthermore, the isometrical identity for the even part $f_{e,x}$ of f with respect to the point x

$$\int_{-\infty}^{\infty} |f_{e,x}(t)|^2 dt = \frac{3}{2\pi} \iint_{\Delta(\frac{\pi}{4})} |(T^{wav}f)_x(A)|^2 \frac{(a^2 - \alpha^2)^{\frac{1}{2}}}{(a^2 + \alpha^2)^{\frac{3}{2}}} da\, d\alpha.$$

From Theorem 4, we have the inversion formula, as in Theorem 3

Theorem 5 *For the integral transform (19) for the even part $f_{e,x}$ of $L^2(\mathbb{R})$ functions f with respect to the point x, we have the inversion formula*

$$f_{e,x} = s - \lim_{N \to \infty} \frac{3}{\pi^{\frac{5}{4}}} \iint_{F_N} (T^{wav}f)_x(A)$$

$$\cdot \overline{A}^{-\frac{1}{2}} \left\{ 1 - (\frac{t-x}{\overline{A}})^2 \right\} \exp\left\{ \frac{-(t-x)^2}{2\overline{A}^2} \right\} \frac{(a^2 - \alpha^2)^{\frac{1}{2}}}{(a^2 + \alpha^2)^{\frac{3}{2}}} da\, d\alpha$$

for any compact exhaustion $\{F_N\}_{N=1}^{\infty}$ of $\Delta(\frac{\pi}{4})$.

4   Multiresolution analysis and the sampling theorem

One typical aspect of the theory of wavelets is that it provides a multiresolution analysis; that is, in our case, the family of functions

$$\Psi_{j,k}(t) = \Psi^{2^j, 2^j k}(t) = 2^{-\frac{j}{2}}\Psi(2^{-j}t - k); \qquad j, k \in \mathbb{Z}, \quad \text{integers} \qquad (21)$$

constitutes a complete orthonormal system of $L^2(\mathbb{R})$ (cf. [[D]], Chapters 4 and 5). This fact implies surprisingly that the image space H_K of the wavelet transform

176

(3) for $L^2(\mathbb{R})$ functions has a very simple realization, and for the space H_K, the sampling theorem is valid at the points

$$\{(2^j, 2^j k); j, k \in \mathbb{Z}\}$$

on the real half plane (a, x) $(a > 0)$.

Indeed, since (21) is a complete orthonormal system in $L^2(\mathbb{R})$ we obtain the expansion of $f \in L^2(\mathbb{R})$ in (3)

$$f(t) = \sum_{j,k \in \mathbb{Z}} c_{j,k} \Psi_{j,k}(t) \tag{22}$$

with

$$\sum_{j,k \in \mathbb{Z}} |c_{j,k}|^2 < \infty. \tag{23}$$

Here, in particular,

$$
\begin{aligned}
c_{j,k} &= \langle f, \Psi_{j,k} \rangle \\
&= \int_{-\infty}^{\infty} f(t) a^{-\frac{1}{2}} \Psi(2^{-j}t - k) dt \\
&= (T^{wav} f)(2^j, 2^j k).
\end{aligned} \tag{24}
$$

Then, by inserting f in (3) we obtain

$$(T^{wav} f)(A, z) = \sum_{j,k \in \mathbb{Z}} (T^{wav} f)(2^j, 2^j k) K(A, z; 2^j, 2^j k), \tag{25}$$

just as in the sampling theorem. We thus obtain

Theorem 6   *The image space H_K of the wavelet transform (3) for $L^2(\mathbb{R})$ functions is characterized by the properties that for any l^2 numbers $c_{j,k}$ with (23), there exists a unique member $g(A, z)$ in H_K satisfying*

$$c_{j,k} = g(2^j, 2^j k); j, k \in \mathbb{Z} \tag{26}$$

and it is expressible in the form

$$g(A, z) = \sum_{j,k \in \mathbb{Z}} c_{j,k} K(A, z; 2^j, 2^j k). \tag{27}$$

Here, the sum in (27) is taken in the Hilbert space H_K, but on the real half space (a, x) $(a > 0)$ the convergence in (27) is uniform.

In Theorem 6, the uniform convergence of (27) on the real half space is derived immediately from the boundedness of the reproducing kernel $K(a, x; a', x')$.

When we take a finite point set of (j, k) in the expression (27), we can give an estimate of the error, in a general argument. See, Theorem 4.2.3. In any case, multiresolution analysis in the theory of wavelets gives the connection between the very simple realization of the image spaces in the wavelet transforms and the sampling theorem.

4. Applications to the approximation of functions

In this chapter, we shall give applications to the best approximation of functions using the functions in a reproducing kernel Hilbert space. We provide approximations in reproducing kernel Hilbert spaces with several typical concrete examples.

§1 Best approximations by the functions in a RKHS

In this section, we shall establish a fundamental theorem for the best approximation by the functions in a RKHS. As a concrete application we discuss the analytic extension problem for functions on the real line to entire functions and approximations of functions by solutions of the heat equation. These materials are taken from [AS], [BS2], [BS3], [BS4], [Y]. For the hypercircle inequality and related topics, see, for example [[Da]].

1 A fundamental best approximation theorem

Let E be an arbitrary set, and let H_K be a RKHS admitting the reproducing kernel $K(p, q)$. Meanwhile, for any subset X of E we consider a Hilbert space $H(X)$ comprising functions F on X. In the relationship of two Hilbert spaces H_K and $H(X)$, we assume that

(a)   for the restriction $f|_X$ of the members f of H_K to the set X, $f|_X$ belongs to the Hilbert space $H(X)$,

and

(b)   the restriction operator $Lf = f|_X$ is continuous from H_K into $H(X)$.

Then, we shall consider the fundamental problem

$$\inf_{f \in H_K} \|Lf - F\|_{H(X)} \tag{1}$$

for a member F of $H(X)$.

For the sake of the nice properties of L and its adjoint L^* in our situation, we can obtain *'algorithms'* to decide whether best approximations f^* of F in the sense of

$$\inf_{f \in H_K} \|Lf - F\|_{H(X)} = \|Lf^* - F\|_{H(X)}$$

exist. Furthermore, when there exist best approximations f^*, we will be able to get the best approximation f^* constructively in a reasonable way. Indeed, we can obtain intrinsic representations of the best approximation in terms of F and the reproducing kernel $K(p,q)$.

Because of its importance, we give a general, abstract theorem for the existence of a solution to our problem (1).

For any Hilbert (possibly finite dimensional) space $\mathcal{H}$, let T denote a bounded linear operator of H_K into $\mathcal{H}$. For the adjoint operator T^* of T, we form the positive matrix

$$k(p,q) = (T^*TK(\cdot,q), T^*TK(\cdot,p))_{H_K} \quad \text{on} \quad E \times E. \qquad (2)$$

Then, we have

Theorem 1 *For a member* $\mathbf{d}$ *of* $\mathcal{H}$, *there exists a function* $\tilde{f}$ *in* H_K *such that*

$$\inf_{f \in H_K} \|Tf - \mathbf{d}\|_{\mathcal{H}} = \|T\tilde{f} - \mathbf{d}\|_{\mathcal{H}} \qquad (3)$$

if and only if, for the RKHS H_k

$$T^*\mathbf{d} \in H_k. \qquad (4)$$

Furthermore, if the existence of the best approximation $\tilde{f}$ *satisfying (3) is ensured, then there exists a unique extremal function* $\check{f}$ *with the minimum norm in* H_K, *and the function* $\check{f}$ *is expressible in the form*

$$\check{f}(p) = (T^*\mathbf{d}, T^*TK(\cdot,p))_{H_k} \quad \text{on} \quad E. \qquad (5)$$

Proof: For any $f \in H_K$, by using the reproducing property of $K(\cdot,p)$ in H_K, T^*Tf is expressible in the form

$$[T^*Tf](p) = (T^*Tf, K(\cdot,p))_{H_K} = (f, T^*TK(\cdot,p))_{H_K}.$$

Hence, by Theorem 2.1.2, the range of T^*T coincides with the RKHS H_k. Let P be the orthogonal projection of H_K onto $H_K \ominus N(T^*T)$. Then, we have, by the same theorem

$$\|T^*Tf\|_{H_k} = \|Pf\|_{H_K}.$$

We assume that best approximations $\tilde{f}$ satisfying (3) exist. Then, we have

$$\|T\tilde{f} - \mathbf{d}\|_{\mathcal{H}} \leq \|\mathbf{h} - \mathbf{d}\|_{\mathcal{H}}$$

179

for all $\mathbf{h}$ in $\overline{R(T)}$ ($\mathcal{H}$ - closure of the range of T). Hence, $\mathbf{d} = T\tilde{f} + \mathbf{g}$ for some $\mathbf{g} \in \mathcal{H} \ominus \overline{R(T)}$. Since $N(T^*) = \mathcal{H} \ominus \overline{R(T)}$, $T^*T\tilde{f} = T^*\mathbf{d}$, and so, $T^*\mathbf{d}$ is a member of H_k.

In order to show the converse, let f_1 be any member of H_K with $T^*Tf_1 = T^*\mathbf{d}$. In addition, we choose a member $\mathbf{h}_1$ in $\overline{R(T)}$ such that

$$\|\mathbf{h}_1 - \mathbf{d}\|_{\mathcal{H}} \leqq \|\mathbf{h} - \mathbf{d}\|_{\mathcal{H}} \tag{6}$$

for all $\mathbf{h} \in \overline{R(T)}$. Then,

$$T^*Tf_1 = T^*\mathbf{h}_1$$

and

$$Tf_1 = \mathbf{h}_1$$

because T^* is one-to-one on $\overline{R(T)}$. Hence, we have, from (6)

$$\|Tf_1 - \mathbf{d}\|_{\mathcal{H}} = \inf_{f \in H_K} \|Tf - \mathbf{d}\|_{\mathcal{H}}.$$

By setting $\check{f} = Pf_1$, we see that $\check{f}$ is a unique member in H_K such that

$$\|T\check{f} - \mathbf{d}\|_{\mathcal{H}} = \inf_{f \in H_K} \|Tf - \mathbf{d}\|_{\mathcal{H}},$$

and $\check{f}$ has the minimum norm in H_K because the family of functions f_1 satisfying (3) is exactly $\check{f} + N(T^*T)$.

Finally we shall derive the expression (5). Since T^*T is an isometry of $H_K \ominus N(T^*T)$ onto H_k, its adjoint S is the inversion of T^*T. Hence, we have

$$\check{f}(p) = [ST^*\mathbf{d}](p)$$
$$= (ST^*\mathbf{d}, K(\cdot, p))_{H_K}$$
$$= (T^*\mathbf{d}, T^*TK(\cdot, p))_{H_k}.$$

In Theorem 1, note that

$$(T^*\mathbf{d})(p) = (T^*\mathbf{d}, K(\cdot, p))_{H_K} = (\mathbf{d}, TK(\cdot, p))_{\mathcal{H}};$$

that is, $T^*\mathbf{d}$ is expressible in terms of $\mathbf{d}, T, K(p, q)$ and $\mathcal{H}$.

2   Approximations of functions on the real line by entire functions

Following Theorem 1, we examine best approximations of functions on the real line by entire functions. Since we need a concrete form of the reproducing kernel in

Theorem 1, as a typical reproducing kernel Hilbert space for entire functions, we consider the Fischer space $\mathcal{F}_a$ normed by

$$\|f\|_{\mathcal{F}_a}^2 = \frac{a^2}{\pi} \iint_{\mathbf{C}} |f(z)|^2 e^{-a^2|z|^2} \, dx \, dy$$

for fixed $a > 0$, whose reproducing kernel is

$$K_a(z, \overline{u}) = e^{a^2 \overline{u} z},$$

as stated in Chapter 2, 4.6 ((2.4.38) and (2.4.39)).

Meanwhile, as a function space approximated by the Fischer space $\mathcal{F}_a$ we shall first determine an $L_2(\mathbb{R}, W(x)dx)$ space with a natural weight $W(x)(\geq 0)$

$$\|F\|_{L_2(W)}^2 = \int_{\mathbb{R}} |F(x)|^2 W(x) \, dx$$

in connection with the Fischer space $\mathcal{F}_a$ and Theorem 1. Under these situations we examine the best approximation problem in the sense that for $F \in L_2(W)$

$$\inf_{f \in \mathcal{F}_a} \|Tf - F\|_{L_2(W)}.$$

In this case, $\{Tf; f \in \mathcal{F}_a\}$ will be complete in $L_2(W)$ and so,

$$\inf_{f \in \mathcal{F}_a} \|Tf - F\|_{L_2(W)} = 0.$$

Therefore, the condition for the existence of the best approximation f^* in the sense

$$\|Tf^* - F\|_{L_2(W)} = 0$$

will become the condition that F can be extended analytically to the member $f^* \in \mathcal{F}_a$ except for a null Lebesgue measure set on the real line $\mathbb{R}$. Furthermore, we can get a constructive sequence $\{f_n\}_{n=0}^{\infty} (f_n \in \mathcal{F}_a)$ such that

$$\lim_{n \to \infty} \|Tf_n - F\|_{L_2(W)} = 0$$

for any function F in $L_2(W)$.

We first look for a natural weight $W(x)$ such that the restriction operator T is bounded from $\mathcal{F}_a$ into $L_2(W)$.

Note that for any member $f \in \mathcal{F}_a$, the integrals exist by Fubini's theorem

$$\int_{-\infty}^{\infty} |f(x + iy)|^2 e^{-a^2(x^2 + y^2)} \, dx$$

181

for almost all $y \in \mathbb{R}$. In this connection, we have

Theorem 2 *For any $f \in \mathcal{F}_a$, we have the inequality*

$$\int_{\mathbb{R}} |f(x)|^2 e^{-a^2 x^2} dx \le \frac{\sqrt{2\pi}}{a} \|f\|_{\mathcal{F}_a}^2;$$

that is, for $W(x) = W_a(x) = e^{-a^2 x^2}$ the restriction operator T from $\mathcal{F}_a$ into $L_2(W_a)$ is bounded.

Proof : Recall the identity

$$K_a(z, \bar{u}) = \frac{1}{\sqrt{2\pi} a} e^{-\frac{a^2 z^2}{2}} e^{-\frac{a^2 \bar{u}^2}{2}} \int_{\mathbb{R}} e^{z\xi} e^{\bar{u}\xi} e^{-\frac{\xi^2}{2a^2}} d\xi.$$

This representation of $K_a(z, \bar{u})$ implies that any $f \in \mathcal{F}_a$ is expressible in the form

$$f(z) = \frac{1}{\sqrt{2\pi} a} e^{-\frac{a^2 z^2}{2}} \int_{\mathbb{R}} F(\xi) e^{z\xi} e^{-\frac{\xi^2}{2a^2}} d\xi \tag{7}$$

for some (of course, uniquely determined) function F satisfying

$$\int_{\mathbb{R}} |F(\xi)|^2 e^{-\frac{\xi^2}{2a^2}} d\xi < \infty,$$

and we have the isometrical identity

$$\|f\|_{\mathcal{F}_a}^2 = \frac{1}{\sqrt{2\pi} a} \int_{\mathbb{R}} |F(\xi)|^2 e^{-\frac{\xi^2}{2a^2}} d\xi. \tag{8}$$

Meanwhile, by the Parseval–Plancherel identity, we have, from (7)

$$\int_{\mathbb{R}} |f(iy)|^2 e^{-a^2 y^2} dy = \frac{1}{a^2} \int_{\mathbb{R}} |F(\xi)|^2 e^{-\frac{\xi^2}{a^2}} d\xi$$

$$\le \frac{1}{a^2} \int_{\mathbb{R}} |F(\xi)|^2 e^{-\frac{\xi^2}{2a^2}} d\xi = \frac{\sqrt{2\pi}}{a} \|f\|_{\mathcal{F}_a}^2.$$

For $f \in \mathcal{F}_a$, we set $f_1(z) = f(-iz)$. Then, $f_1 \in \mathcal{F}_a$ and $\|f_1\|_{\mathcal{F}_a} = \|f\|_{\mathcal{F}_a}$. Hence, we have the desired result

$$\int_{\mathbb{R}} |f(x)|^2 e^{-a^2 x^2} dx = \int_{\mathbb{R}} |f_1(ix)|^2 e^{-a^2 x^2} dx$$

$$= \int_{\mathbb{R}} |f_1(iy)|^2 e^{-a^2 y^2} dy$$

$$\le \frac{\sqrt{2\pi}}{a} \|f_1\|_{\mathcal{F}_a}^2 = \frac{\sqrt{2\pi}}{a} \|f\|_{\mathcal{F}_a}^2.$$

182

We shall determine the condition for the existence of the best approximations $f^* \in \mathcal{F}_a$ of a function $F \in L_2(W_a)$ in the sense

$$\inf_{f \in \mathcal{F}_a} \|Tf - F\|_{L_2(W_a)} = \|Tf^* - F\|_{L_2(W_a)}. \tag{9}$$

By Theorem 1, there exist the best approximations f^* in (9) if and only if

$$(T^*F)(z) = \int_{\mathbb{R}} F(\xi) e^{a^2 \xi z} e^{-a^2 \xi^2} d\xi \in \mathcal{R}(T^*T) \tag{10}$$

and $\mathcal{R}(T^*T)$ is characterized as the RKHS H_k admitting the reproducing kernel

$$
\begin{aligned}
k(z, \overline{u}) &= (T^*TK(\cdot, \overline{u}), T^*TK(\cdot, \overline{z}))_{\mathcal{F}_a} \\
&= (TK(\cdot, \overline{u}), TT^*TK(\cdot, \overline{z}))_{L_2(W_a)} \\
&= \frac{\sqrt{\pi}}{a} \int_{\mathbb{R}} e^{a^2 \overline{u} \xi} e^{\frac{a^2 (\xi + z)^2}{4}} e^{-a^2 \xi^2} d\xi \\
&= \frac{2\pi}{\sqrt{3} a^2} e^{\frac{1}{3} a^2 z^2} e^{\frac{1}{3} a^2 \overline{u}^2} e^{\frac{1}{3} a^2 \overline{u} z}.
\end{aligned} \tag{11}
$$

Note that the RKHS H_k is composed of all entire functions $f(z)$ with finite norms

$$\|f\|_{H_k}^2 = \frac{a^4}{2\sqrt{3}\pi^2} \iint_{\mathbb{C}} |f(z)|^2 e^{-a^2 x^2 + \frac{a^2 y^2}{3}} dx dy, \tag{12}$$

as we see from the technique in Corollary 2.3.3 (see also (2.4.9)). Of course,

$$\{Tf; f \in \mathcal{F}_a\} \quad \text{and, in particular,} \quad \{e^{a^2 \overline{u} \xi}; u \in \mathbb{C}\}$$

are complete in $L_2(W_a)$ and so $Tf^* = F$ in $L_2(W_a)$ in (9). Hence, from Theorem 1 we have

Theorem 3 *For $F \in L_2(W_a)$, there exists an analytic function $f^* \in \mathcal{F}_a$ satisfying $Tf^* = F$ in $L_2(W_a)$ if and only if*

$$\iint_{\mathbb{C}} \left| \int_{\mathbb{R}} F(\xi) e^{a^2 \xi z} e^{-a^2 \xi^2} d\xi \right|^2 e^{-a^2 x^2 + \frac{a^2 y^2}{3}} dx dy < \infty. \tag{13}$$

By Theorem 1, we can get an explicit representation of f^* in terms of F. Of course, f^* is uniquely determined. In order to use Theorem 1, note that

$$
\begin{aligned}
[T^*TK_a(\cdot, \overline{u})](z) &= (TK_a(\cdot, \overline{u}), TK_a(\cdot, \overline{z}))_{L_2(W_a)} \\
&= \frac{\sqrt{\pi}}{a} e^{\frac{a^2 z^2}{4}} e^{\frac{a^2 \overline{u}^2}{4}} e^{\frac{a^2 \overline{u} z}{2}}.
\end{aligned} \tag{14}
$$

Therefore, in particular, we have

Theorem 4 *For any $f \in \mathcal{F}_a$, $f(z)$ is expressible in terms of the trace $f(x)$ to the real line in the form*

$$f(z) = \frac{a^3}{2\sqrt{3}\pi^{\frac{3}{2}}} e^{\frac{a^2 z^2}{4}} \iint_{\mathbf{C}} \left(\int_{\mathbf{R}} f(\xi) e^{a^2 Z\xi} e^{-a^2\xi^2} d\xi \right)$$
$$\cdot \left(e^{\frac{a^2\overline{Z}}{4}} e^{\frac{a^2 z \overline{Z}}{2}} \right) e^{-a^2 X^2 + \frac{a^2}{3}Y^2} \, dX dY,$$

$$Z = X + iY.$$

For any $F \in L_2(W_a)$, we shall construct a sequence $\{f_n\}_{n=0}^{\infty}$ satisfying $f_n \in \mathcal{F}_a$ and

$$\lim_{n \to \infty} \|Tf_n - F\|_{L_2(W_a)} = 0.$$

First note that the images $T^* F = f(z)$; that is,

$$f(z) = (F(\cdot), TK(\cdot, \overline{z}))_{L_2(W_a)}$$

for $L_2(W_a)$ functions F are characterized as the RKHS $H_{\mathbb{K}}$ admitting the reproducing kernel

$$\mathbb{K}(z, \overline{u}) = (TK(\cdot, \overline{u}), TK(\cdot, \overline{z}))_{L_2(W_a)}$$
$$= \frac{\sqrt{\pi}}{a} e^{\frac{a^2 z^2}{4}} e^{\frac{a^2 \overline{u}^2}{4}} e^{\frac{a^2 \overline{u} z}{2}}. \tag{15}$$

As in the space H_k, we see that the space $H_{\mathbb{K}}$ is composed of all entire functions f with finite norms

$$\|f\|_{H_{\mathbb{K}}}^2 = \frac{a^3}{2\pi\sqrt{\pi}} \iint_{\mathbf{C}} |f(z)|^2 e^{-a^2 x^2} \, dx dy. \tag{16}$$

From the representation (15) of $\mathbb{K}(z, \overline{u})$ we see directly that the family of functions

$$\varphi_n(z) = \left(\frac{\sqrt{\pi} a^{2n-1}}{2^n n!} \right)^{\frac{1}{2}} e^{\frac{a^2 z^2}{4}} z^n; \tag{17}$$

$$n = 0, 1, 2, \cdots$$

is a complete orthonormal system in $H_{\mathbb{K}}$. Moreover, note that

$$\varphi_n \in H_k \quad \text{for any} \quad n. \tag{18}$$

For any $F \in L_2(W_a)$, we set

$$a_n = (T^*F, \varphi_n)_{H_\alpha} \qquad (19)$$

and

$$\widetilde{f_N}(z) = \sum_{n=0}^{N} a_n \varphi_n(z). \qquad (20)$$

Then, $\widetilde{f_N} \in H_k \subset H_\mathbf{K}$ and $\{\widetilde{f_N}\}$ converges to T^*F in $H_\mathbf{K}$ as N tends to infinity. By Theorem 1, we construct the functions f_N^* satisfying

$$T^*Tf_N^* = \widetilde{f_N}. \qquad (21)$$

Since the adjoint operator T^* is isometrical from $L_2(W_a)$ onto $H_\mathbf{K}$, we have

Theorem 5   *For the sequence $\{f_N^*\}_{N=0}^{\infty}$ constructed in (21) through (17), (19) and (20), and for any $F \in L_2(W_a)$, we have*

$$\lim_{N\to\infty} \|Tf_N^* - F\|_{L_2(W_a)} = 0.$$

To compare Theorems 2–5 with the following beautiful and surprising theorem by Ohsawa ([Oh2]) will be very interesting and valuable. See, also [Oh1], [Oh3] and [OT] and furthermore see [A] in connection with our arguments.

Theorem 6   *Let φ be a plurisubharmonic function on a bounded pseudoconvex domain $D \subset \mathbb{C}^n$, let $H \subset \mathbb{C}^n$ be a complex hyperplane, and let f be a holomorphic function on $D \cap H$ satisfying*

$$\int_{D\cap H} |f(z)|^2 e^{-\varphi(z)} dV_{n-1} < \infty.$$

Here, dV_k denotes the 2k-dimensional Lebesgue measure. Then there exists a holomorphic function F on D satisfying $F|_{D\cap H} = f$ and

$$\int_D |F(z)|^2 e^{-\varphi(z)} dV_n \leqq C_D \int_{D\cap H} |f(z)|^2 e^{-\varphi(z)} dV_{n-1}.$$

Here, C_D is a constant depending only on the diameter of D.

3   Approximation by solutions of the heat equation

We consider the heat equation

$$u_{xx}(x, t) = u_t(x, t) \quad \text{on} \quad \mathbb{R} \times T_+ \qquad (22)$$

185

satisfying the initial condition

$$u(x,0) = F(x) \quad \text{on} \quad \mathbb{R} \tag{23}$$

for $L_2(\mathbb{R})$ functions F. Let $u_F(x,t)$ be the solution of the Cauchy problem (22) and (23). For $h \in L_2(\mathbb{R})$ and for any fixed $t(> 0)$, we shall consider the approximation problem

$$\inf \int_{\mathbb{R}} |u_F(x,t) - h(x)|^2 dx, \tag{24}$$

where F runs over $L_2(\mathbb{R})$. In fact, the family of the solutions u_F is dense in $L_2(\mathbb{R})$. Hence, we shall determine the condition for h to exist, the solution u_F satisfying

$$\int_{\mathbb{R}} |u_F(x,t) - h(x)|^2 dx = 0;$$

that is,

$$u_F(x,t) = h(x) \quad a.e. \quad \text{on} \quad \mathbb{R}. \tag{25}$$

Further, for any $h \in L_2(\mathbb{R})$ we shall construct the minimizing sequence $\{u_{F_n}\}$ satisfying

$$\lim_{n \to \infty} \int_{\mathbb{R}} |u_{F_n}(x,t) - h(x)|^2 dx = 0. \tag{26}$$

The identity (25) will mean that the problem (24) can be connected to the analytic extension problem of h, as in Section 1.2.

In Theorem 3.2.1, we saw that the family of the solutions $u_F(x,t)$ forms the RKHS $H_{K(t)}$ admitting the reproducing kernel (3.2.9) equipped with the norm (3.2.10).

As we see from the identity (3.2.17), the restriction operator Tf of $H_{K(t)}$ into $L_2(\mathbb{R})$ is bounded, as in Theorem 2. Then, as in Theorem 3, we have

Theorem 7 *For $h \in L_2(\mathbb{R})$, there exists a function $F \in L_2(\mathbb{R})$ satisfying (25) if and only if*

$$\iint_{\mathbb{C}} \left| \int_{\mathbb{R}} h(\xi) \exp\left\{ -\frac{\xi^2}{8t} + \frac{z\xi}{4t} \right\} d\xi \right|^2 \exp\left\{ \frac{-3x^2 + y^2}{12t} \right\} dx dy < \infty. \tag{27}$$

Proof : Since $\{K(\cdot, \overline{z}; t) | z \in \mathbb{C}\}$ is complete in $L_2(\mathbb{R})$, the adjoint operator T^* is one-to-one. Hence, there exists a function $F \in L_2(\mathbb{R})$ satisfying (25) if and only if T^*h belongs to the range of $H_{K(t)}$ by the operator T^*T, by Theorem 1.

186

We compute the kernel form

$$k(z, \overline{u}; t) = (T^* T K(\cdot, \overline{u}; t), T^* T K(\cdot, \overline{z}; t))_{H_{K(t)}}$$
$$= (T K(\cdot, \overline{u}; t), T T^* T K(\cdot, \overline{z}; t))_{L_2(\mathbb{R})}.$$

Here,

$$[T^* T K(\cdot, \overline{u}; t)](z) = (T K(\cdot, \overline{u}; t), T K(\cdot, \overline{z}; t))_{L_2(\mathbb{R})}$$
$$= \frac{1}{4\sqrt{\pi t}} \exp\left\{\frac{-z^2}{16t}\right\} \exp\left\{\frac{-\overline{u}^2}{16t}\right\} \exp\left\{\frac{z\overline{u}}{8t}\right\}. \qquad (28)$$

Hence, we obtain

$$k(z, \overline{u}; t) = \frac{1}{2\sqrt{6\pi t}} \exp\left\{\frac{-z^2}{24t}\right\} \exp\left\{\frac{-\overline{u}^2}{24t}\right\} \exp\left\{\frac{z\overline{u}}{12t}\right\}.$$

We see that $k(z, \overline{u}; t)$ is the reproducing kernel for the Hilbert space $H_{k(t)}$ comprising all entire functions g equipped with the norm

$$\|g\|^2_{H_{k(t)}} = \frac{1}{\sqrt{6\pi t}} \iint_{\mathbb{C}} |g(z)|^2 \exp\left\{\frac{-y^2}{6t}\right\} dx\, dy.$$

Hence, from the representation of T^*, we have the desired result.

As in Theorem 4, we have

Theorem 8 *For a function h satisfying (27), the analytic extension $h(z) \in H_{K(t)}$ is expressible in the form*

$$h(z) = \frac{1}{16\sqrt{3}(\pi t)^{\frac{3}{2}}} \exp\left\{\frac{-z^2}{16t}\right\}$$
$$\cdot \iint_{\mathbb{C}} \left[\int_{\mathbb{R}} h(\xi) \exp\left\{\frac{-\xi^2}{8t} + \frac{Z\xi}{4t}\right\} d\xi \right]$$
$$\cdot \exp\left\{\frac{19X^2}{16t} + \frac{-65Y^2}{48t} + \frac{\overline{Z}z + iXY}{8t}\right\} dX\, dY$$
$$= \sum_{n=0}^{\infty} \frac{(2t)^n}{n!} \int_{\mathbb{R}} \partial_\xi^n h(\xi) \overline{\partial_\xi^n K(\xi, \overline{z}; t)} d\xi.$$

Proof : For the latter representation of $h(z)$, use the identity (3.2.17).

187

For any $h \in L_2(\mathbb{R})$, we shall construct a sequence $\{u_n(z)\}$ satisfying $u_n \in H_{K(t)}$ and

$$\lim_{n \to \infty} \int_{\mathbb{R}} |u_n(x) - h(x)|^2 dx = 0.$$

The images

$$f(z) = [T^* g](z) = (g(\cdot), TK(\cdot, \overline{z}; t))_{L_2(\mathbb{R})}$$

for $L_2(\mathbb{R})$ functions g are characterized as the RKHS $K_{\mathbb{K}(t)}$ admitting the reproducing kernel

$$\mathbb{K}(z, \overline{u}; t) = (TK(\cdot, \overline{u}; t), TK(\cdot, \overline{z}; t))_{L_2(\mathbb{R})}.$$

From the expression (28) of $\mathbb{K}(z, \overline{u}; t)$ we see that the family of functions

$$\varphi_n(z) = \frac{1}{2} \left[n!(8t)^n \sqrt{\pi t} \right]^{-\frac{1}{2}} \exp\left\{ \frac{-z^2}{16t} \right\} z^n, \tag{29}$$

$$n = 0, 1, 2, \cdots$$

is a complete orthonormal system in $H_{\mathbb{K}(t)}$. Note that

$$\varphi_n \in H_{k(t)}, \quad n = 0, 1, 2, \cdots.$$

We set

$$a_n = (T^* h, \varphi_n)_{H_{\mathbb{K}(t)}} \tag{30}$$

and

$$\widehat{f}_N(z) = \sum_{n=0}^{N} a_n \varphi_n(z). \tag{31}$$

Then, $\widehat{f}_N \in H_{k(t)} \subset H_{\mathbb{K}(t)}$ and the sequence $\{\widehat{f}_N\}$ converges to $T^* h$ in $H_{\mathbb{K}(t)}$ as N tends to infinity. We construct the sequence u_N^* satisfying

$$T^* T u_N^* = \widehat{f}_N. \tag{32}$$

Since the adjoint operator T^* is an isometry of $L_2(\mathbb{R})$ onto $H_{\mathbb{K}(t)}$, we have, as in Theorem 5

Theorem 9 *For any $h \in L_2(\mathbb{R})$, define the sequence $\{u_N^*\}_{N=0}^{\infty}$ through (29), (30), (31) and (32). Then, we have*

$$\lim_{N \to \infty} \int_{\mathbb{R}} |[Tu_N^*](x) - h(x)|^2 dx = 0.$$

In Theorem 9, furthermore, for u_N^* we can look for the initial function F_N satisfying

$$u_{F_N}(z,t) = u_N^*(z)$$

with (22) and (23), by Theorem 3.2.2.

§2 Approximations in reproducing kernel Hilbert spaces

In this section, we shall examine typical approximations in reproducing kernel Hilbert spaces. The main theme of this section is sampling theorems and related topics.

1 Approximations in sampling theorems

By Theorem 2.1.2, we can get the essence of the WKS (Whittakers–Kotel'nikov–Shannon) sampling theorem from the following theorem, as given in 2.4.16. For general references for sampling theorems, see [J] and [Hig]. See also the recent book [[Hi]] for the sampling theorem and the related topics.

Theorem 1 *Let* **h** *be a Hilbert space* $\mathcal{H}$*-valued function from an abstract set* E *into a Hilbert space* $\mathcal{H}$*. If for some points* $\{p_j\}_{j \in I}$ *of* E,

$$\{\mathbf{h}(p_j); j \in I\} \tag{1}$$

is a complete orthonormal system in $\mathcal{H}$*, then for the RKHS* H_K *admitting the reproducing kernel*

$$K(p,q) = (\mathbf{h}(q), \mathbf{h}(p))_{\mathcal{H}} \quad on \quad E \times E, \tag{2}$$

the sampling property

$$f(p) = \sum_{j \in I} f(p_j) K(p, p_j) \quad on \quad E \quad for \ all \quad f \in H_K \tag{3}$$

is valid both in the senses of strong and absolute convergences in H_K *and on* E, *respectively.*

In Theorem 1, if $K(p, p)$ is bounded on E, then the convergence in (3) is uniform on E, as we see from Theorem 2.2.3. The converse of Theorem 1 will be referred to in Theorem 4.

In practical applications of Theorem 1, we use only a finite number of sample points of $\{p_j\}_{j \in I}$. In order to obtain an upper bound on the truncation error when a finite number of sample points is taken in Theorem 1 we need the following hypercircle inequality of Golomb and Weinberger ([[Da]], Theorem 9.4.7) :

189

Theorem 2   *For a Hilbert space H, let L be a linear functional on H. For a complete orthonormal system $\{v_j; j \in I\}$ of H and for some fixed constants $\{b_j; j \in I'\}$ for some subset I' of I, we take the element x_0 of H with the minimum norm satisfying*

$$(x, v_j)_H = b_j, \quad j \in I' \subset I.$$

Then, for $H'' = \{x \in H; \|x\|_H^2 \leq E\}$ we have the inequality

$$|Lx - Lx_0|^2 \leq \left[E - \|x_0\|_H^2\right]\left[\sum_{j \in I \setminus I'} |Lv_j|^2\right]. \tag{4}$$

Moreover, if $\|x_0\|_H^2 \leq E$, there exists an element in H satisfying the equality in (4).

Proof :   In the representation

$$x = \sum_{j \in I}(x, v_j)_H v_j = \sum_{j \in I'} b_j v_j + \sum_{j \in I \setminus I'}(x, v_j)_H v_j,$$

we have

$$x_0 = \sum_{j \in I'} b_j v_j.$$

Hence,

$$x - x_0 = \sum_{j \in I \setminus I'}(x, v_j)_H v_j$$

and so,

$$Lx - Lx_0 = \sum_{j \in I \setminus I'}(x, v_j)_H L(v_j).$$

By the Schwarz inequality we have

$$|Lx - Lx_0|^2 \leq \|x - x_0\|_H^2 \sum_{j \in I \setminus I'} |L(v_j)|^2.$$

Since, $x - x_0 \perp x_0$,

$$\|x - x_0\|_H^2 = \|x\|_H^2 - \|x_0\|_H^2 \leq E - \|x_0\|_H^2,$$

and so we have the desired result (4).

Next we assume that $\|x_0\|_H^2 \leq E$.

190

We set

$$z = \sum_{j \in I \setminus I'} \overline{L(v_j)} v_j.$$

If $z = o$, of course, $L(v_j) = 0$ for $j \in I \setminus I'$. In this case, the element $x' = x_0$ satisfies the equality trivially.

If $z \neq o$, set

$$x' = x_0 + \lambda z,$$

where

$$|\lambda| = \{E - \|x_0\|_H^2\}^{\frac{1}{2}} / \|z\|_H.$$

Then,

$$(x', v_j)_H = (x_0, v_j)_H = b_j, \quad j \in I'$$

and

$$\|x'\|_H^2 = \|x_0\|_H^2 + |\lambda|^2 \|z\|_H^2 = E.$$

Furthermore,

$$|L(x' - x_0)| = |\lambda| |L(z)| = |\lambda| \sum_{j \in I \setminus I'} |L(v_j)|^2$$

$$= |\lambda| \|z\|_H^2 = (E - \|x_0\|_H^2)^{\frac{1}{2}} \|z\|_H$$

$$= (E - \|x_0\|_H^2)^{\frac{1}{2}} \left(\sum_{j \in I \setminus I'} |L(v_j)|^2 \right)^{\frac{1}{2}},$$

which implies the desired equality.

In Theorem 1, we set, for some subset I' of I, the truncation error as

$$E_{I'}(p) = \sum_{j \in I \setminus I'} f(p_j) K(p, p_j). \tag{5}$$

Then, we have

Theorem 3   *In Theorem 1 and in the truncation error (5), we have*

$$|E_{I'}(p)| \leqq \left[E - \sum_{j \in I'} |f(p_j)|^2 \right]^{\frac{1}{2}} \left[\sum_{j \in I \setminus I'} |K(p, p_j)|^2 \right]^{\frac{1}{2}}, \tag{6}$$

191

for any $f \in H_K''$, where

$$H_K'' = \{f \in H_K; \|f\|_{H_K}^2 \leq E\}.$$

Furthermore, if $E - \sum_{j \in I'} |f(p_j)|^2 \neq 0$, then there exists a member of H_K'' attaining the equality in (6).

Proof : In Theorem 2, we set

$$H = H_K, \quad H'' = H_K''; \quad Lf = f(p),$$

and

$$b_j = f(p_j) = (f(\cdot), K(\cdot, p_j))_{H_K}.$$

Then,

$$L\mathbf{x}_0 = f_0(p) = \sum_{j \in I'} f(p_j) K(p, p_j),$$

and

$$L\mathbf{x} = f(p) = \sum_{j \in I} f(p_j) K(p, p_j).$$

We thus obtain Theorem 3 from Theorem 2.

In general, in (3)

$$f_0(p) = \sum_{j \in I'} f(p_j) K(p, p_j)$$

satisfies the following two properties

$$\text{Interpolation Property :} \quad f_0(p_j) = f(p_j), \quad j \in I' \tag{7}$$

and

$$\text{Minimum Energy Property :} \quad \|f_0\|_{H_K}^2 = \min_{\substack{h \in H_K \\ h(p_j) = f(p_j)(j \in I')}} \|h\|_{H_K}^2. \tag{8}$$

We shall determine the condition such that for other distinct points $\{q_j\}_{j \in I'}$ of E the two properties are satisfied.

Theorem 4   *For some constant c and points $\{q_j\}_{j \in I'}$ of E,*

$$g(p) = c \sum_{j \in I'} f(q_j) K(p, q_j)$$

satisfies properties (7) and (8) for $\{q_j\}_{j \in I'}$ if and only if $c = 1/\lambda_n$, $f(q_j) = \theta_{n,j}$, where λ_n is the largest eigenvalue and θ is the corresponding eigenvector of the matrix equation

$$K\theta = \lambda\theta,$$
$$K = [K(q_j, q_{j'})]_{j,j'=1,2,\cdots,n},$$
$$\theta^T = [\theta_1, \theta_2, \cdots, \theta_n].$$

Proof :   Let $f(q_j) = \theta_{n,j}$ and $c = 1/\lambda_n$. Then,

$$g(q_{j'}) = \frac{1}{\lambda_n} \sum_{j \in I'} \theta_{n,j} K(q_{j'}, q_j)$$
$$= \theta_{n,j'} = f(q_{j'}).$$

Meanwhile, for any $h \in H_K$

$$\left[\sum_{j \in I'} |h(q_j)|^2 \right]^2 = \left[\sum_{j \in I'} \overline{h(q_j)}(h(\cdot), K(\cdot, q_j))_{H_K} \right]^2$$

$$= \left| (h(\cdot), \sum_{j \in I'} h(q_j) K(\cdot, q_j))_{H_K} \right|^2$$

$$\leqq \|h\|_{H_K}^2 \sum_{j,j' \in I'} h(q_j) \overline{h(q_{j'})} K(q_{j'}, q_j). \tag{9}$$

Hence

$$\min_{\substack{h \in H_K \\ h(q_j) = \theta_{n,j} (j \in I')}} \|h\|_{H_K}^2 = \frac{1}{\lambda_n} \sum_{j \in I'} |\theta_{n,j}|^2. \tag{10}$$

By direct calculation, we have

$$\|g\|_{H_K}^2 = \frac{1}{\lambda_n} \sum_{j \in I'} |\theta_{n,j}|^2. \tag{11}$$

Conversely, we assume that g satisfies (7); that is, for $j' \in I'$

$$g(q_{j'}) = f(q_{j'}) = c \sum_{j \in I'} f(q_j) K(q_{j'}, q_j).$$

Then,

$$K\hat{\mathbf{f}} = \frac{1}{c}\hat{\mathbf{f}} \tag{12}$$

has an eigenvalue $\frac{1}{c}$ and the corresponding eigenvector $\hat{\mathbf{f}}$, where $\hat{\mathbf{f}}^T = [f(q_1), f(q_2), \cdots, f(q_n)]$. Further we assume that g satisfies (8). Then,

$$\frac{\|g\|_{H_K}^2}{\sum_{j \in I'} |f(q_j)|^2} = \min_{\substack{h \in H_K, \\ h(q_j)=f(q_j) \\ (j \in I')}} \frac{\|h\|_{H_K}^2}{\sum_{j \in I'} |h(q_j)|^2} = \frac{1}{\lambda_n}. \tag{13}$$

From (10), (11), (12) and (13) we have the desired results $c = \frac{1}{\lambda_n}$ and $\hat{\mathbf{f}} = \theta_n$.

Corollary 1 *For the images*

$$f(t) = \frac{1}{\sqrt{2\pi}} \int_{-\pi}^{\pi} F(w)e^{iwt}dw \tag{14}$$

for $L_2(-\pi, \pi)$ functions $F(w)$, we have the sampling property

$$f(t) = \sum_{j=-\infty}^{\infty} f(j)\frac{\sin \pi(t-j)}{\pi(t-j)} \quad on \quad \mathbb{R}, \tag{15}$$

which converges absolutely and uniformly.
   We take $I' = \{j_0(t) - M \leq j \leq j_0(t) + N\}$, where $j_0(t)$ is the nearest integer to time t, and M and N are positive integers. The truncation error is, for all integers I

$$E_{M,N}(t) = \sum_{I \setminus I'} f(j)\frac{\sin \pi(t-j)}{\pi(t-j)}. \tag{16}$$

For the RKHS S formed by the images in (14), we set $H'' = \{f \in S; \|f\|_S^2 \leqq E\}$. Then, we have

$$|E_{M,N}(t)| \leq \begin{cases} \dfrac{E_0^{\frac{1}{2}}}{\pi} \left[\dfrac{1}{M} + \dfrac{2}{2N-1}\right]^{\frac{1}{2}}, t \in [j_0(t), j_0(t) + \frac{1}{2}] \\[2ex] \dfrac{E_0^{\frac{1}{2}}}{\pi} \left[\dfrac{2}{2M-1} + \dfrac{1}{N}\right]^{\frac{1}{2}}, t \in [j_0(t) - \frac{1}{2}, j_0(t)], \end{cases}$$

194

where

$$E_0 = \sum_{j \in I \setminus I'} |f(j)|^2 < E < \infty.$$

Proof :   Since the family

$$\left\{ \frac{1}{\sqrt{2\pi}} e^{ijw} \right\}_{j \in I}$$

forms a complete orthonormal system in $L_2(-\pi, \pi)$, we have the sampling property directly by Theorem 1. The uniform convergence follows from the boundedness of the reproducing kernel

$$\int_{-\pi}^{\pi} \frac{1}{\sqrt{2\pi}} e^{itw} \frac{1}{\sqrt{2\pi}} e^{-itw} \, dw \leqq 1$$

and from Theorem 2.2.3.

For the upper bound of $E_{M,N}(t)$, note that $|\sin \pi(t - j)| \leqq 1$ and

$$\sum_{j \in I \setminus I'} \frac{1}{(t-j)^2} < \left[\frac{1}{M} + \frac{2}{2N-1} \right], t \in \left[j_0(t), j_0(t) + \frac{1}{2} \right],$$

$$\sum_{j \in I \setminus I'} \frac{1}{(t-j)^2} < \left[\frac{2}{2M-1} + \frac{1}{N} \right], t \in \left[j_0(t) - \frac{1}{2}, j_0(t) \right].$$

Hence we have the desired bound directly from Theorem 3.

2   Extremal functions in a RKHS with the constraint of a finite norm

First, note that if $K(q, q) \neq 0$, then

$$f^*(p) = \frac{K(p, q)}{K(q, q)^{\frac{1}{2}}}$$

is the extremal function in the problem

$$\max_{\|f\|_{H_K} \leq 1} f(q), \tag{17}$$

as we see directly from the reproducing property of $K(p, q)$ for H_K and from the Schwarz inequality. We shall consider a general extremal problem of (17) for a bounded linear operator T from a RKHS H_K on E into a Hilbert space $\mathcal{H}$ in the form

$$\max_{\|f\|_{H_K} \leq 1} \|Tf\|_{\mathcal{H}}. \tag{18}$$

For this problem, we have

Theorem 5 *The following identities are valid :*

$$\max_{f \in H_K} \frac{(T_q f(q), T_q(Tf(\cdot), TK(\cdot, q))_{\mathcal{H}})_{\mathcal{H}}}{\|Tf\|_{\mathcal{H}}^2}$$
$$= \text{the maximum eigenvalue } \mu \text{ of } TT^*$$
$$= \|T^*\|^2 = \|T\|^2. \tag{19}$$

Then, we have

$$\max_{\|f\|_{H_K} \leqq 1} \|Tf\|_{\mathcal{H}}^2 \leqq \mu \tag{20}$$

and the extremal function $h(p)$ attaining the equality in (20) is given by

$$h(p) = \frac{(T\hat{f}(\cdot), TK(\cdot, p))_{\mathcal{H}}}{(T_q \hat{f}(q), T_q(T\hat{f}(\cdot), TK(\cdot, q))_{\mathcal{H}})_{\mathcal{H}}^{\frac{1}{2}}} \tag{21}$$

for the eigenfunction (extremal function in (19)) $\hat{f}$.

Proof : For general properties of the linear operator T, see Chapter 2, 3.3. The identity (19) follows directly from the identity

$$(T_q f(q), T_q(Tf(\cdot), TK(\cdot, q))_{\mathcal{H}})_{\mathcal{H}} = (Tf, TT^* \cdot Tf)_{\mathcal{H}}.$$

Furthermore, we have

$$TT^* \cdot T\hat{f} = \mu T\hat{f}$$

and so,

$$T^*T \cdot (T^*T\hat{f}) = \mu T^*T\hat{f}.$$

We thus have (21).

In particular, we have

Corollary 2 *Let $\{q_j\}_{j=1}^n$ be any distinct points of E and set*

$$H_K' = \{f \in H_K; \|f\|_{H_K} \leqq 1\}.$$

Then

$$f_0(p) = \left[(1/\lambda_n) \sum_{j=1}^n |\theta_{n,j}|^2 \right]^{-\frac{1}{2}} \sum_{j=1}^n \theta_{n,j} K(p, q_j)$$

196

is the extremal function satisfying

$$\max_{f \in H_{K'}} \sum_{j=1}^{n} |f(q_j)|^2 = \sum_{j=1}^{n} |f_0(q_j)|^2 = \lambda_n$$

where λ_n is the largest eigenvalue and θ_n is the corresponding eigenvector of the matrix equation

$$K\theta = \lambda\theta;$$
$$K = [K(q_{j'}, q_j)]_{j,j'=1,2,\cdots,n},$$
$$\theta_j^T = [\theta_{j,1}, \theta_{j,2}, \cdots, \theta_{j,n}].$$

3   Best approximations of $g \in H_K$ by H_K functions taking given values

For a subset X of E and for a function F on X we consider the subfamily of H_K such that

$$H_K(X, F) = \{f \in H_K; f(p) = F(p) \quad \text{on} \quad X\}.$$

Then, we shall determine the best approximation of any $g \in H_K$ by $H_K(X, F)$ functions in the sense of

$$\min_{f \in H_K(X,F)} \|f - g\|_{H_K}. \tag{22}$$

For this purpose, recall the construction of the Hilbert space $H_{K|X}$ admitting the reproducing kernel $K(p, q)|_X$ restricted to the subset X (Theorem 2.3.1).

For any $f \in H_K$, we have the expression, by the reproducing property of $K(p, q)|_X$,

$$f(p) = (f|_X(\cdot), K(\cdot, p)|_X)_{H_{K|X}} \quad \text{on} \quad X. \tag{23}$$

Then, for $f|_X = F$, we define the function f^* on E by

$$f^*(p) = (F(\cdot), TK(\cdot, p))_{H_{K|X}} \tag{24}$$

for the restriction operator T of H_K to $H_{K|X}$. Of course, T is a bounded linear operator and we can define its adjoint operator T^*. Then, from (23) and (24) we see that

$$Tf^* = F \quad \text{on} \quad X. \tag{25}$$

Furthermore,

$$f^* \in H_K \quad \text{and} \quad \|f^*\|_{H_K} = \|F\|_{H_{H|X}}. \tag{26}$$

We thus see that f^* is the extremal function in the sense that

$$\|f^*\|_{H_K} = \min_{f \in H_K(X,F)} \|f\|_{H_K}. \tag{27}$$

By the translation $f - g$, we have

Theorem 6 *For any $g \in H_K$, the best approximation f^* in the sense*

$$\min_{f \in H_K(X,F)} \|f - g\|_{H_K} = \|f^* - g\|_{H_K}$$

is given by

$$f^*(p) = g(p) + (F(\cdot) - Tg(\cdot), TK(\cdot, p))_{H_{K|X}}. \tag{28}$$

If $K(q, q) \neq 0$, then for one point q of E and, in particular, for $f(q) = 1, g \equiv 0$, we have

$$f^*(p) = \frac{K(p, q)}{K(q, q)}$$

satisfying

$$\min_{\substack{f \in H_K \\ f(q)=1}} \|f\|_{H_K} = \|f^*\|_{H_K} = \frac{1}{K(q, q)^{\frac{1}{2}}}.$$

Note that this problem is a dual for the problem (17).

As a general form, we have

Corollary 3 *Let $\{q_j\}_{j=1}^n$ be distinct points of E, the values $\{M_j\}_{j=1}^n$ be arbitrary, and let $H_K(\{q_j\}, \{M_j\})$ be the subfamily of H_K satisfying*

$$\{f \in H_K; f(q_j) = M_j, \quad j = 1, 2, \cdots, n\}.$$

Then, for any $g \in H_K$

$$f^*(p) = g(p) + \sum_{j=1}^n d_j D_j(p)$$

is the unique best approximation in the sense that

$$\min_{f \in H_K(\{q_j\}, \{M_j\})} \|f - g\|_{H_K}^2 = \|f^* - g\|_{H_K}^2 = \sum_{j=1}^n |d_j|^2,$$

where, for $j = 1, 2, \cdots, n$,

$$d_j = \frac{1}{(G_{j-1}G_j)^{\frac{1}{2}}} \begin{vmatrix} K(q_1, q_1) & \cdots & K(q_n, q_1) \\ \vdots & & \vdots \\ K(q_1, q_{n-1}) & \cdots & K(q_n, q_{n-1}) \\ m_1 & \cdots & m_n \end{vmatrix},$$

$$m_j = M_j - g(q_j),$$
$$G_j = \det[K(q_{j''}, q_{j'})]_{j', j''=1,2,\cdots,j},$$
$$G_0 = 1,$$

and

$$D_j(p) = \frac{1}{(G_{j-1}G_j)^{\frac{1}{2}}} \begin{vmatrix} K(q_1,q_1) & \cdots & K(q_n,q_1) \\ \vdots & & \vdots \\ K(q_1,q_{n-1}) & \cdots & K(q_n,q_{n-1}) \\ K(q_1,q_n) & \cdots & K(q_n,q_n) \end{vmatrix}.$$

For Corollaries, see [Y], directly.

5. Applications to analytic extension formulas and real inversion formulas for the Laplace transform

By using the representations of the norms in Bergman–Selberg spaces on the right half plane in terms of the trace of analytic functions to the real positive axis, we shall establish very natural real inversion formulas for the Laplace transform.

§1 Representations of the norms in Bergman–Selberg spaces

We shall give very simple representations of the norms in Bergman–Selberg spaces on strips and sectors on the complex plane in terms of the values on the real line and on the real positive axis, respectively. At the same time, analytic extension formulas of functions on the real line and on the real positive axis to the members of the Bergman–Selberg spaces on the strips and the sectors are established, respectively. The special results for the right half complex plane will be needed to derive the natural real inversion formulas for the Laplace transform. These materials are taken from [Sa22] and [AHS2]. See also [AHOS] for analytic extension formulas on multidimensional spaces.

1 Results

For the Bergman–Selberg spaces $H_{K_q}(S_r)$ $(q > 0)$ for the strip $S_r = \{z; |\mathrm{Im} z| < r\}$, we shall establish

Theorem 1 *For $f \in H_{K_q}(S_r)$ we have the identity*

$$
\|f\|^2_{H_{K_q}(S_r)} = \left(\frac{1}{\Gamma(2q-1)\pi^q} \iint_{S_r} |f(z)|^2 K_{S_r}(z,\overline{z})^{1-q} dx dy, \quad q > \frac{1}{2} \right)
$$

$$
= \frac{1}{\Gamma(q)^2} \sum_{n=0}^{\infty} \left(\frac{2r}{\pi} \right)^{2n+2q-1}
$$

$$
\cdot \left(\sum_{j_1 > j_2 > \cdots > j_n \geq 0} \Pi^n_{k=1} \frac{1}{(q+j_k)^2} \right) \int_{\mathbb{R}} |f^{(n)}(x)|^2 dx, \tag{1}
$$

where $K_{S_r}(z,\overline{u})$ is the usual Bergman kernel on S_r and the summation $\sum_{j_1 > j_2 > \cdots > j_n \geq 0}$ in (1) is understood as one for $n = 0$.

Conversely, any C^∞ function $f(x)$ on the real line with convergent summation in (1) can be extended analytically onto the strip S_r, and the analytic extension $f(z)$ belongs to $H_{K_q}(S_r)$ and satisfies the identity (1).

Theorem 2   *For the right half plane $R^+ = \{z; Re\, z > 0\}$ we have the identity*

$$\|f\|^2_{H_{K_q}(R^+)} = \left(\frac{1}{\Gamma(2q-1)\pi} \iint_{R^+} |f(z)|^2 (2x)^{2q-2} dx\, dy, \quad q > \frac{1}{2} \right)$$

$$= \sum_{n=0}^{\infty} \frac{1}{n!\Gamma(n+2q+1)} \int_0^{\infty} |\partial_x^n (xf'(x))|^2 x^{2n+2q-1} dx. \qquad (2)$$

Conversely, any C^∞ function $f(x)$ on the real positive line with convergent summation in (2) can be extended analytically onto the right half plane R^+. The analytic extension $f(z)$ satisfying $\lim_{x\to\infty} f(x) = 0$ belongs to $H_{K_q}(R^+)$ and the identity (2) is valid.

In Theorems 1 and 2, for $q > 0$ the Bergman–Selberg space $H_{K_q}(\cdot)$ can be defined as the RKHS admitting the reproducing kernel

$$K_q(z, \overline{u}) = \Gamma(2q)\pi^q K_{(\cdot)}(z, \overline{u})^q \qquad (3)$$

for the usual Bergman kernel $K_{(\cdot)}(z, \overline{u})$. For $q > \frac{1}{2}$ the Bergman–Selberg space admits the norms as in Theorems 1 and 2. For $q = \frac{1}{2}$ we have the classical Szegö space stated in Chapter 2, 4.6. For $0 < q < \frac{1}{2}$ we do not have the representations of the norms in the Bergman–Selberg spaces. We can, however, use the isometrical mapping between $H_{K_{1/2}}$ and H_{K_q} in Theorem 3.4.4 for some special cases to express the norms in H_{K_q}.

The special cases of $q = 1$ and $1/2$ in Theorem 1 were given in [HS1] and the results were applied to the investigations of analyticity of solutions of nonlinear partial differential equations in [HS1], [HS2], [Hay1], [Hay2], [BHK] and [HK].

Meanwhile, a similar type of result to Theorem 2 for the sector $\{|\arg z| < \frac{\pi}{4}\}$ and for $q = 1$ was obtained incidentally in [AHS1] from both the viewpoints of our general method for integral transforms and the property of the solutions for the one-dimensional heat equation on $x > 0$ as stated in Chapter 2, (22). After that, for the general sector $\Delta(r) = \{|\arg z| < r\}$ $(0 < r < \frac{\pi}{2})$, similar results to Theorem 2 were derived for the special cases of $q = 1$ and $1/2$ in [AHS2] as follows:

Theorem 3   *For an analytic function $f(z)$ on the sector $\Delta(r)$ $(0 < r < \frac{\pi}{2})$ we have the identity*

$$\iint_{\Delta(r)} |f(z)|^2 dx\, dy = \sin(2r) \sum_{j=0}^{\infty} \frac{(2\sin r)^{2j}}{(2j+1)!} \int_0^{\infty} x^{2j+1} |f^{(j)}(x)|^2 dx. \qquad (4)$$

Conversely, if any C^∞ function $f(x)$ has a convergent sum in the right hand side in (4), then the function $f(x)$ can be extended analytically onto the sector $\Delta(r)$ in the form $f(z)$ and the identity (4) is valid.

Theorem 4 *For any member $f(z)$ of the Szegö space on the sector $\Delta(r)$ $(0 < r < \frac{\pi}{2})$ we have the identity*

$$\int_{\partial\Delta(r)} |f(z)|^2 |dz| = 2\cos r \sum_{j=0}^{\infty} \frac{(2\sin r)^{2j}}{(2j)!} \int_0^{\infty} x^{2j} |f^{(j)}(x)|^2 dx, \qquad (5)$$

where $f(z)$ means the nontangential Fatou limit on $\partial\Delta(r)$. Conversely, if any C^{∞} function $f(x)$ on $(0,\infty)$ has a convergent sum in the right hand side in (5), then the function $f(x)$ can be extended analytically onto the sector $\Delta(r)$ and the identity (5) is valid.

These results were applied to the investigations of analyticity of the solutions of nonlinear partial differential equations in [Hay1], [Hay2], [BHK] and [HK].

In particular, note that the special cases $q = 1$ and $1/2$ are essential in Theorems 3 and 4 for the sake of using the identities

$$\Gamma(1 + it)\Gamma(1 - it) = \frac{\pi t}{\sinh \pi t} \qquad (6)$$

and

$$\Gamma(\frac{1}{2} + it)\Gamma(\frac{1}{2} - it) = \frac{\pi}{\cosh \pi t}. \qquad (7)$$

Of course, Theorems 3 and 4 are not valid for the typical case R^+; that is, for $r = \frac{\pi}{2}$. See [Ai] for further interesting investigations in connection with this fact. For this we obtain Theorem 2 for any positive q which is different from Theorems 3 and 4 even for $q = 1$ and $1/2$, and which gives a quite simple representation of the norms $\|f\|_{H_{K_q}(R^+)}$. This representation will be needed essentially to establish the natural and analytical real inversion formula for the Laplace transform.

2   Proof of Theorem 1

Note first that the members $f(z)$ in $H_{K_q}(D)$, for a domain D, are, in fact, analytic differentials of order q and for any conformal mapping $\varphi(z)$ on D onto $\varphi(D)$ on the w-plane we have the transformation formula such that for $f \in H_{K_q}(D)$ and for the corresponding function $\tilde{f}(w) \in H_{K_q}(\varphi(D))$,

$$f(z)(dz)^q = \tilde{f}(w)(dw)^q;$$

that is,

$$f(z) = \tilde{f}(\varphi(z))\varphi'(z)^q. \qquad (8)$$

202

Furthermore, this corresponding mapping is isometrical from $H_{K_q}(D)$ onto H_{K_q} $(\varphi(D))$, as we see from Theorem 2.3.4 or Theorem 2.3.6. Further we have the transformation formula

$$K_q(z, \overline{u})(dz)^q \overline{(du)^q} = K_q(\varphi(z), \overline{\varphi(u)})(d\varphi(z))^q \overline{(d\varphi(u))^q}. \tag{9}$$

By using the conformal mapping $\varphi(z) = \tanh \frac{\pi z}{4r}$ from S_r onto the unit disc $|w| < 1$, the explicit form

$$K_\Delta(z, \overline{u}) = \frac{1}{\pi(1 - \overline{u}z)^2}, \quad \Delta = \{|z| < 1\},$$

and (9), we obtain the explicit form of $K_q(z, \overline{u})$ on S_r

$$K_q(z, \overline{u}) = \left(\frac{\pi}{4r}\right)^{2q} \frac{\Gamma(2q)}{\left[\cosh \frac{\pi(z-\overline{u})}{4r}\right]^{2q}}. \tag{10}$$

In order to obtain expressions of the norms in $H_{K_q}(S_r)$ on the real line, we consider the representation of $K_q(z, \overline{u})$ by means of Fourier integrals. By using the integral formula

$$\frac{2^{\nu-2}}{a\Gamma(\nu)}\Gamma\left(\frac{\nu}{2} + \frac{iy}{2a}\right)\Gamma\left(\frac{\nu}{2} - \frac{iy}{2a}\right) = \int_0^\infty [\operatorname{sech}(ax)]^\nu \cos xy\, dx \quad (y > 0),$$

$$\text{for} \quad Re\,\nu > 0, Re\,a > 0, \tag{11}$$

([[EMOT]], p.30, (5); [[GR]], p.506, 3.985; [[PBM]], p.353, 2.4.4.4), and by using the inverse cosine integral transform and transform of parameters, we obtain the representation

$$K_q(z, \overline{u}) = \frac{\pi^{2q-2}}{2^{2q}r^{2q-1}}\int_{\mathbb{R}} e^{it(z-\overline{u})} \left|\Gamma(q + \frac{2rit}{\pi})\right|^2 dt, \tag{12}$$

$$\text{on} \quad S_r \times \overline{S_r}.$$

This representation implies that any member $f(z)$ of $H_{K_q}(S_r)$ is expressible in the form

$$f(z) = \frac{\pi^{2q-2}}{2^{2q}r^{2q-1}}\int_{\mathbb{R}} F(t)e^{itz} \left|\Gamma(q + \frac{2rit}{\pi})\right|^2 dt \tag{13}$$

for a uniquely determined function F satisfying

$$\int_{\mathbb{R}} |F(t)|^2 \left|\Gamma(q + \frac{2rit}{\pi})\right|^2 dt < \infty, \tag{14}$$

203

and we have the isometrical identity

$$\|f\|^2_{H_{K_q}(S_r)} = \frac{\pi^{2q-2}}{2^{2q}r^{2q-1}} \int_{\mathbb{R}} |F(t)|^2 \left|\Gamma(q + \frac{2rit}{\pi})\right|^2 dt. \tag{15}$$

From (13) we have, by using the Fourier transform in the framework of the $L_2(\mathbb{R}, |\Gamma(q + \frac{2rit}{\pi})|^{-2}dt)$ space,

$$F(t) \left|\Gamma(q + \frac{2\pi it}{\pi})\right|^2 = \frac{2^{2q-1}r^{2q-1}}{\pi^{2q-1}} \int_{\mathbb{R}} f(x)e^{-itx}dx \tag{16}$$

and so, from (15)

$$\|f\|^2_{H_{K_q}(S_r)} = \frac{2^{2q-2}r^{2q-1}}{\pi^{2q}} \int_{\mathbb{R}} \left|\int_{\mathbb{R}} f(x)e^{-itx}dx\right|^2 \frac{dt}{\left|\Gamma(q + \frac{2rit}{\pi})\right|^2}. \tag{17}$$

From the identity

$$\frac{|\Gamma(x)|^2}{|\Gamma(x+iy)|^2} = \prod_{k=0}^{\infty} \left\{1 + \frac{y^2}{(x+k)^2}\right\} \tag{18}$$

([[AS]], p.256, 6.1.25;[[GR]], p.936, 8.325), we have the expansion

$$\frac{1}{\left|\Gamma(q + \frac{2rit}{\pi})\right|^2} = \frac{1}{\Gamma(q)^2} \sum_{n=0}^{\infty} \left(\frac{2r}{\pi}\right)^{2n} \left\{\sum_{j_1>j_2>\cdots>j_n\geq 0} \prod_{k=1}^{n} \frac{1}{(q+j_k)^2}\right\} t^{2n}, \tag{19}$$

which converges absolutely. Hence, from the Parseval–Plancherel identity, we obtain the desired identity.

Conversely, for any C^∞ function on $\mathbb{R}$ with convergent summation in (1), we define the function $F(t)$ by (16) which satisfies (14). Then, the function defined by (13) gives the desired analytic extension which satisfies the identity (1).

From (6), (7), (18) and (19) we have the simple identities

$$\sum_{j_1>j_2>\cdots>j_n\geq 0} \prod_{k=1}^{n} \frac{1}{(q+j_k)^2} = \begin{cases} \frac{\pi^{2n}}{(2n+1)!} & \text{for} \quad q=1 \\ \frac{\pi^{2n}}{(2n)!} & \text{for} \quad q=\frac{1}{2}. \end{cases}$$

We do not know whether there exist such simple expressions for general $q > 0$.

3   Proof of Theorem 2

We shall start with (17). By setting $u = e^x$, we have

$$\int_{\mathbb{R}} f(x) e^{-itx} dx = \int_0^\infty f(\log u) u^{-it-1} du. \tag{20}$$

On the other hand, by the conformal mapping $w = e^z$ of S_r onto $\Delta(r)' = \{|\arg w| < r\}$, for the corresponding function $\tilde{f}(w)$ to $f(z)$, we have, from the transformation formula (8)

$$f(z) = w^q \tilde{f}(w)$$

and so, for $\mathrm{Re}\, w = u$,

$$f(\log u) = u^q \tilde{f}(u). \tag{21}$$

Since the norm in $H_{K_q}(S_r)$ is conformally invariant by the transformation (8), we obtain, from (17), (20) and (21), the identity

$$\|f\|_{H_{K_q}(\Delta(r))}^2 = \frac{2^{2q-2} r^{2q-1}}{\pi^{2q}} \int_{\mathbb{R}} |Mf(q - it)|^2 \frac{dt}{\left|\Gamma\left(q + \frac{2rit}{\pi}\right)\right|^2} \tag{22}$$

for the Mellin transform

$$Mf(q - it) = \int_0^\infty f(x) x^{q-it-1} dx.$$

Now we shall look for an explicit representation of (22) in terms of $f(x)$. By the famous Gauss formula

$$F(a, b; c; 1) = \frac{\Gamma(c)\Gamma(c - a - b)}{\Gamma(c - a)\Gamma(c - b)} \quad c \neq 0, -1, -2, \cdots,$$

$$\mathrm{Re}(c - a - b) > 0$$

with

$$F(a, b; c; 1) = \frac{\Gamma(c)}{\Gamma(a)\Gamma(b)} \sum_{n=0}^\infty \frac{\Gamma(a + n)\Gamma(b + n)}{n!\Gamma(c + n)}$$

(see, for example, [[AS]], p.556, 15.1.10 and 15.1.1), we obtain the expansion, by setting $a = q + \frac{2rit}{\pi}$, $b = q - \frac{2rit}{\pi}$ and $c = 2q + 1$,

$$\frac{1}{\left|\Gamma\left(q + \frac{2rit}{\pi}\right)\right|^2} = \left\{q^2 + \left(\frac{2rt}{\pi}\right)^2\right\}$$

$$\cdot \sum_{n=0}^\infty \frac{\left\{q^2 + \left(\frac{2rt}{\pi}\right)^2\right\}\left\{(q+1)^2 + \left(\frac{2rt}{\pi}\right)^2\right\}\cdots\left\{(q+n-1)^2 + \left(\frac{2rt}{\pi}\right)^2\right\}}{n!\Gamma(n + 2q + 1)}. \tag{23}$$

Here, the term for $n = 0$ will be understood as one.

Now, in the case $r = \frac{\pi}{2}$; that is, in the case of $\Delta(\frac{\pi}{2}) = R^+$, we shall prepare the Parseval–Plancherel identity in the Mellin transform in the framework of the $L_2(\mathbb{R}, |\Gamma(q+it)|^{-2}dt)$ space which matches the expression (23) with (22).

We shall start with the Parseval–Plancherel identity in the form

$$\int_{\mathbb{R}} |Mf(q - it)|^2 dt = 2\pi \int_0^\infty |f(x)|^2 x^{2q-1} dx. \tag{24}$$

See, for example, [[Ti]], pp.94–95 for details.

This identity, however, can be derived immediately from the identity

$$\frac{1}{\tau + \overline{\tau'}} = \int_0^\infty x^{-q-\overline{\tau'}} x^{-q-\tau} x^{2q-1} dx,$$

$$\operatorname{Re} \tau, \operatorname{Re} \tau' > 0,$$

as in Theorem 3.1.1 for the Laplace transform.

We shall proceed formally by means of integration by parts and multiplication by x as follows.

By the transformation rule in the Mellin transform

$$-(q - it - 1)Mf(q - it - 1) = Mf'(q - it)$$

and

$$-(q - it)Mf(q - it) = M(xf'(x))(q - it)$$

(see, for example, [[Sn]], pp.262–263; 266–267), we have

$$\int_{\mathbb{R}} |Mf(q - it)|^2 (q^2 + t^2) dt = 2\pi \int_0^\infty |xf'(x)|^2 x^{2q-1} dx. \tag{25}$$

Similarly, from

$$(q - it)^2 Mf(q - it) = M(x(xf'(x))'(q - it),$$

we have

$$\int_{\mathbb{R}} |Mf(q - it)|^2 (q^2 + t^2) dt = 2\pi \int_0^\infty |(xf'(x))'|^2 x^{2q+1} dx. \tag{26}$$

Next, from the transformations

$$-(q - it - 1)^2 Mf(q - it - 1) = M(xf'(x))'(q - it),$$
$$(q - it - 1)(q - it - 2)^2 Mf(q - it - 2) = M(xf'(x))''(q - it),$$

and

206

$$- (q + 1 - it)(q - it)^2 M f(q - it) = M(x^2(xf'(x))'')(q - it),$$

we have

$$\int_{\mathbb{R}} |M f(q - it)|^2 (q^2 + t^2)^2 \{(q + 1)^2 + t^2\} dt = 2\pi \int_0^\infty |(xf'(x))''|^2 x^{2q+3} dx. \quad (27)$$

In general, we have

$$\int_{\mathbb{R}} |M f(q - it)|^2 (q^2 + t^2)^2 \{(q + 1)^2 + t^2\} \cdots \{(q + n - 1)^2 + t^2\} dt$$

$$= 2\pi \int_0^\infty |\partial_x^n (xf'(x))|^2 x^{2n+2q-1} dx. \quad (28)$$

Now we shall justify the Parseval–Plancherel identity (28). In the identity (22), note that the integral exists in the sense of the $L_2(\mathbb{R}, |\Gamma(q + it)|^{-2} dt)$ space as we see from the theory of Fourier integrals in the framework of the L_2 space. Hence, the formal transformation rules in the Mellin transform by means of integration by parts as in (25)–(27) are not valid. In order to justify (28), we need a completion procedure in the framework of the L_2 space by sufficiently smooth functions in which the transformation rules are valid in the Mellin transform. As a result, for the justification of (28) it is sufficient to show that for $f \in H_{K_q}(R^+)$ and for any nonnegative integer n

$$\int_0^\infty |\partial_x^n (xf'(x))|^2 x^{2n+2q-1} dx < \infty. \quad (29)$$

Then, the Mellin transform $M f(q - it)$ exists in the sense of the $L_2(\mathbb{R}, |\Gamma(q+it)|^{-2} dt)$ space and the identity (28) is valid. See, for example, [AHS2] and [[Sn]], pp.94–95. Under the condition (28), we can apply the arguments in [AHS2] for $-M f(q - it)(q - it), xf'(x)$ and for $k = j + q$ there.

We shall prove (29) directly.

For any $f \in H_{K_q}(R^+)$ $(q > \frac{1}{2})$ and $\xi > 0$, by the Cauchy integral formula

$$\xi f'(\xi) = \frac{1}{2\pi i} \int_{|z - \xi| = \hat{r}} \frac{\xi f(z)}{(z - \xi)^2} dz \quad \text{for} \quad 0 < \hat{r} < \frac{\xi}{\sqrt{2}}, \quad (30)$$

and we have

$$|\partial_\xi^n (\xi f'(\xi))| \leqq \frac{C_1(n)}{\hat{r}^{n+1}} \int_0^{2\pi} |f(\xi + \hat{r} e^{i\theta})| d\theta, \quad (31)$$

for a constant $C_1(n)$, independent of f and ξ. Hence,

$$\int_{\frac{\xi}{2\sqrt{2}}}^{\frac{\xi}{\sqrt{2}}} \left| \partial_\xi^n (\xi f'(\xi)) \right| d\hat{r}$$

$$\leq C_1(n) \int_{\frac{\xi}{2\sqrt{2}}}^{\frac{\xi}{\sqrt{2}}} \frac{d\hat{r}}{\hat{r}^{n+1}} \int_0^{2\pi} |f(\xi + \hat{r}e^{i\theta})| d\theta$$

$$= C_1(n) \iint_{\frac{\xi}{2\sqrt{2}} < |z-\xi| < \frac{\xi}{\sqrt{2}}} \hat{r}^{-(n+2)+1-q} |f(\xi + \hat{r}e^{i\theta})| \hat{r}^{q-1} \hat{r} \, d\hat{r} \, d\theta$$

$$= C_1(n) \left(\iint \hat{r}^{-2n-2q-1} \hat{r} \, d\hat{r} \, d\theta \right)^{\frac{1}{2}} \left(\iint |f(\xi + \hat{r}e^{i\theta})|^2 \hat{r}^{2q-2} \hat{r} \, d\hat{r} \, d\theta \right)^{\frac{1}{2}};$$

that is,

$$\left| \partial_\xi^n (\xi f'(\xi)) \right| \leq C_2(n) \xi^{-n-q-1} \left(\iint_{|z-\xi| < \frac{\xi}{\sqrt{2}}} |f(z)|^2 x^{2q-2} \, dx \, dy \right)^{\frac{1}{2}},$$

for a constant $C_2(n)$, independent of f and ξ. Hence, we have

$$\int_0^\infty \left| \partial_\xi^n (\xi f'(\xi)) \right|^2 \xi^{2n+2q-1} d\xi \leq C_2(n)^2 \int_0^\infty \frac{d\xi}{\xi} \iint_{|z-\xi| < \frac{\xi}{\sqrt{2}}} |f(z)|^2 x^{2q-2} \, dx \, dy.$$

From $|z - \xi| < \frac{\xi}{\sqrt{2}}, \frac{|z|}{1+\frac{1}{\sqrt{2}}} < \xi < \frac{|z|}{1-\frac{1}{\sqrt{2}}}$, and we have from Fubini's theorem

$$\int_0^\infty \left| \partial_\xi^n (\xi f'(\xi)) \right|^2 \xi^{2n+2q-1} d\xi \leq C_2(n)^2 \log \frac{1+\frac{1}{\sqrt{2}}}{1-\frac{1}{\sqrt{2}}} \iint_{R^+} |f(z)|^2 x^{2q-2} \, dx \, dy.$$

Hence, we have the desired result (29) for $q > \frac{1}{2}$. For $q = \frac{1}{2}$ we need only replace the norm in $H_{K_q}(R^+)$ by the Szegö norm in the above arguments.

Note the identity, however, for any $q > 0$

$$\|f\|_{H_{K_q}(R^+)} = \|f'\|_{H_{K_{q+1}}(R^+)}. \tag{32}$$

Then, we substitute (30) and (31) in the above arguments by

$$\xi f'(\xi) = \frac{1}{2\pi i} \int_{|z-\xi|=\hat{r}} \frac{\xi f'(z)}{z-\xi} dz \quad \text{for} \quad 0 < \hat{r} < \frac{\xi}{\sqrt{2}}$$

and

$$\left| \partial_\xi^n (\xi f'(\xi)) \right| \leq \frac{C_1(n)}{\hat{r}^n} \int_0^{2\pi} |f'(\xi + \hat{r}e^{i\theta})| d\theta,$$

respectively. Then, we obtain the desired result (29) for any $0 < q \leq \frac{1}{2}$.

For the identity (32), recall the representation

$$K_q(z, \overline{u}) = \frac{\Gamma(2q)}{(z + \overline{u})^{2q}} = \int_0^\infty e^{-tz} e^{-t\overline{u}} t^{2q-1} dt. \tag{33}$$

This identity implies that any member $f(z)$ of $H_{K_q}(R^+)$ is expressible in the form

$$f(z) = \int_0^\infty F(t) e^{-tz} t^{2q-1} dt \tag{34}$$

for a uniquely determined function F satisfying

$$\int_0^\infty |F(t)|^2 t^{2q-1} dt < \infty$$

and we have the identity

$$\|f\|^2_{H_{K_q}(R^+)} = \int_0^\infty |F(t)|^2 t^{2q-1} dt.$$

From (34), we have

$$f'(z) = \int_0^\infty F(t) e^{-tz} (-t) t^{2q-1} dt.$$

Then, the reproducing kernel

$$\int_0^\infty e^{-tz} (-t) e^{-t\overline{u}} (-t) t^{2q-1} dt$$

coincides with $K_{q+1}(z, \overline{u})$. Hence, we obtain the desired result (32).

As in Theorem 1, we can see that for any C^∞ function $f(x)$ on $(0, \infty)$ with convergent summation in (2), there exists an analytic function $f(z) \in H_{K_q}(R^+)$ satisfying $\lim_{x \to \infty} f(x) = 0$ and the identity (2).

4 Proof of Theorem 3

We shall start with (22) for $q = 1$, using the identity (6)

$$\|f\|^2_{H_{K_1}(\Delta(r))} = \iint_{\Delta(r)} |f(z)|^2 dx\, dy$$

$$= \frac{1}{2\pi} \int_R |Mf(1 - it)|^2 \frac{\sinh(2rt)}{t} dt. \tag{35}$$

209

In order to get an expansion of $\sinh(2rz)/z$ into successive polynomials $(j^2 + z^2) \cdots (1^2 + z^2)$, we use the identity

$$F(1 + z, 1 - z; \frac{3}{2}; \sin^2 r) = \frac{\sin(2rz)}{z \sin(2r)}$$

(cf. [[AS]], 15.1.16). Replacing z by iz and developing the left hand side into the series as above, we obtain

$$\frac{\sinh(2rz)}{z} = \sin(2r) \left\{ 1 + \sum_{j=1}^{\infty} \frac{(2 \sin r)^{2j}}{(2j + 1)!} (j^2 + z^2) \cdots (1^2 + z^2) \right\}. \qquad (36)$$

In order to derive the corresponding Parseval–Plancherel identity

$$\int_0^{\infty} x^{2j+1} |f^{(j)}(x)|^2 dx = \frac{1}{2\pi} \int_{\mathbb{R}} (j^2 + t^2) \cdots (1^2 + t^2) |Mf(1 + it)|^2 dt, \qquad (37)$$

as in (28), we shall show that

$$\int_0^{\infty} x^{2j+1} |f^{(j)}(x)|^2 dx < \infty. \qquad (38)$$

From Cauchy's integral formula,

$$|f^{(j)}(x)| \leq \frac{j!}{2\pi \hat{r}^j} \int_0^{2\pi} |f(x + \hat{r}e^{i\theta})| d\theta$$

for $0 < \hat{r} < x \sin r$. Hence,

$$|f^{(j)}(x)| \leq \frac{2}{x \sin r} \frac{j!}{2\pi} \int_{(1/2)x \sin r}^{x \sin r} |f(x + \hat{r}e^{i\theta})| d\theta$$

$$\leq \frac{j!}{\sqrt{\pi}} \left(\frac{2}{x \sin r} \right)^{j+1} \left(\iint_{|u+iv-x|<x \sin r} |f(u + iv)|^2 du dv \right)^{\frac{1}{2}}.$$

Since $|u + iv - x| < x \sin r$ implies $|(u + iv)|/(1 + \sin r) < x < |u + iv|/(1 - \sin r)$, it follows from Fubini's theorem that

$$\int_0^{\infty} x^{2j+1} |f^{(j)}(x)|^2 dx \leq \frac{(j!)^2}{\pi} \left(\frac{2}{\sin r} \right)^{2j+2} \int_0^{\infty} \frac{dx}{x}$$

$$\cdot \iint_{|u+iv-x|<x \sin r} |f(u + iv)|^2 du dv$$

$$\leq \frac{(j!)^2}{\pi} \left(\frac{2}{\sin r} \right)^{2j+2} \log \frac{1 + \sin r}{1 - \sin r} \iint_{\Delta(r)} |f(u + iv)|^2 du dv,$$

210

which implies the desired result.

From (35), (36) and (37) we have Theorem 3 as in Theorem 2.

5   Proof of Theorem 4

We start with (22) for $q = \frac{1}{2}$, using the identity (7)

$$\int_{\partial \Delta(r)} |f(z)|^2 |dz| = \frac{1}{\pi} \int_{\mathbb{R}} |Mf(\frac{1}{2} - it)|^2 \cosh(2rt) dt. \tag{39}$$

Recall the identity

$$\cosh(2rz) = \cos r \left\{ 1 + \sum_{j=1}^{\infty} \frac{(2\sin r)^{2j}}{(2j)!} \left((j - \frac{1}{2})^2 + z^2 \right) \cdots \left((\frac{1}{2})^2 + z^2 \right) \right\} \tag{40}$$

(cf. [[AS]], 15.1.18).

For $f \in H_{K_{1/2}}(\Delta(r))$ we have

$$\iint_{\Delta(r)} |f(u + iv)|^2 \frac{du\,dv}{|u + iv|} \le 2r \sup_{|\theta| < r} \int_0^{\infty} |f(\hat{r}e^{i\theta})|^2 d\hat{r} < \infty.$$

Hence, as in the previous case we have

$$\int_0^{\infty} x^{2j} |f^{(j)}(x)|^2 dx < \infty \quad \text{for} \quad j = 0, 1, \cdots.$$

Hence we have the Parseval–Plancherel identity

$$\int_0^{\infty} x^{2j} |f^{(j)}(x)|^2 dx = \frac{1}{2\pi} \int_{\mathbb{R}} |Mf(1 - it)|^2$$

$$\cdot \left((j - \frac{1}{2})^2 + t^2 \right) \cdots \left((\frac{1}{2})^2 + t^2 \right) dt. \tag{41}$$

Here, $((j - \frac{1}{2})^2 + t^2) \cdots ((\frac{1}{2})^2 + t^2)$ is understood to be 1 for $j = 0$. Hence, we have
Theorem 4 as in Theorem 2.

§2   Real inversion formulas for the Laplace transform

As an application of the representations of norms in the Bergman–Selberg spaces on
the right half plane in terms of the values on the real positive axis, we shall derive
natural inversion formulas for the Laplace transforms, following [BS2].

1   Results

For any $q > 0$, let L_q^2 be the class of all square integrable functions with respect to the measure $t^{1-2q}dt$ on the positive real line $(0, \infty)$. Then, we consider the Laplace transform

$$[\mathcal{L}F](z) = \int_0^\infty F(t)e^{-zt}dt,$$

$$(z \in R^+ = \{Re z > 0\})$$

for $F \in L_q^2$. In Theorem 3.1.1, we saw that the image of L_q^2 under the Laplace transform $\mathcal{L}$ coincides with the RKHS $H_q = H_{K_q}(R^+)$ and $\mathcal{L}$ is an isometry of L_q^2 onto H_q.

The inversion formula of the Laplace transform $\mathcal{L}$ is, in general, given by complex forms. The observation in many cases however gives us real data $[\mathcal{L}F](x)$ only, and so it is important to establish its inversion formula in terms of real data $[\mathcal{L}F](x)$. Such a formula was given for $L^1[(0, \infty), dt]$ functions F by R. P. Boas and D. V. Widder about fifty years ago (see [[Wi]], p.386). See also Chapter 1, (11) for Ramm's inversion formula. See also [[Ra2]], pp.230–233 for some real inversion formulas for the Laplace transform. By use of the representation (1.2) of H_q norms in Theorem 1.2, we shall establish the natural real and analytical inversion formula

Theorem 1   *For any fixed $q > 0$ and for any function of L_2^q, the inversion formula for the Laplace transform $f = \mathcal{L}F$*

$$F(t) = s - \lim_{N \to \infty} \int_0^\infty f(x)e^{-xt}P_{N,q}(xt)dx \quad (t > 0)$$

is valid, where the limit is taken in the space L_q^2 and the polynomials $P_{N,q}$ are given by the formula

$$P_{N,q}(\xi) = \sum_{0 \le \nu \le n \le N} \frac{(-1)^{\nu+1}\Gamma(2n + 2q)}{\nu!(n - \nu)!\Gamma(n + 2q + 1)\Gamma(n + \nu + 2q)}\xi^{n+\nu+2q-1}$$

$$\cdot \left\{ \frac{2(n + q)}{n + \nu + 2q}\xi^2 - \left(\frac{2(n + q)}{n + \nu + 2q} + 3n + 2q \right)\xi + (n + \nu + 2q) \right\}.$$

Moreover, the series

$$\sum_{n=0}^\infty \frac{1}{n!\Gamma(n + 2q + 1)}\int_0^\infty |\partial_x^n[xf'(x)]|^2 x^{2n+2q-1}dx$$

212

converges and the truncation error is estimated by the inequality

$$\left\| F(t) - \int_0^\infty f(x) e^{-xt} P_{N,q}(xt)\,dx \right\|_{L_2^q}^2$$

$$\leqq \sum_{n=N+1}^\infty \frac{1}{n!\,\Gamma(n+2q+1)} \int_0^\infty |\partial_x^n [x f'(x)]|^2 x^{2n+2q-1}\,dx.$$

Note that, even if $q = \frac{1}{2}$, our polynomial $P_{N,\frac{1}{2}}$ is different from the one of R. P. Boas and D. W. Widder.

2 Preliminaries

In order to prove Theorem 1, we prepare three lemmas.

Lemma 1 *For any fixed $q > 0$, and for any function $f \in H_q$,*

$$f_N(z) = \sum_{n=0}^N \frac{1}{n!\,\Gamma(n+2q+1)} \int_0^\infty \partial_\xi^n [\xi f'(\xi)] \overline{\partial_\xi^n [\xi \partial_\xi K_q(\xi,\overline{z})]} \xi^{2n+2q-1}\,d\xi$$

belongs to H_q and $\{f_N\}_{N=0}^\infty$ converges to f in H_q.

Proof : By using the representation (1.2) of the norm in H_q and the reproducing property of $K_q(\cdot,\overline{z})$, we have

$$f(z) = \sum_{n=0}^\infty \frac{1}{n!\,\Gamma(n+2q+1)} \int_0^\infty \partial_\xi^n [\xi f'(\xi)] \overline{\partial_\xi^n [\xi \partial_\xi K_q(\xi,\overline{z})]} \xi^{2n+2q-1}\,d\xi.$$

Hence, we see that f_N is a member in H_q and

$$\|f - f_N\|_q^2 = \left\| \sum_{n=N+1}^\infty \frac{1}{n!\,\Gamma(n+2q+1)} \int_0^\infty \partial_\xi^n [\xi f'(\xi)] \overline{\partial_\xi^n [\xi \partial_\xi K_q(\xi,\overline{z})]} \xi^{2n+2q-1}\,d\xi \right\|_q^2$$

$$\leqq \sum_{n=N+1}^\infty \frac{1}{n!\,\Gamma(n+2q+1)} \int_0^\infty |\partial_\xi^n [\xi f'(\xi)]|^2 \xi^{2n+2q-1}\,d\xi.$$

Therefore, our claim is true.

Lemma 2 *For any fixed $q > 0$, let the function f be a member of the space H_q and set, for any nonnegative integer N,*

$$F_N(t) = \sum_{n=0}^N \frac{t^{2q-1}}{n!\,\Gamma(n+2q+1)} \int_0^\infty \partial_x^n [x f'(x)] \partial_x^n [x \partial_x (e^{-tx})] x^{2n+2q-1}\,dx$$

213

for $t \in (0, \infty)$. *Then, the function* F_N *belongs to the space* L^2_q, *and furthermore, for the functions* f_N *defined in Lemma 1,* $\mathcal{L}F_N = f_N$.

Proof : We first prove that, for any n, the function g_n defined by

$$g_n(t) = t^{q-\frac{1}{2}} \int_0^\infty \partial_x^n [x f'(x)] \partial_x^n [x \partial_x (e^{-tx})] x^{2n+2q-1} dx$$

belongs to the space $L^2[(0, \infty), dt]$. By the Leibniz rule,

$$\partial_x^n [x \partial_x (e^{-tx})] t^{q-\frac{1}{2}} = (-1)^n t^{n+q-\frac{1}{2}} (n - tx) e^{-tx},$$

and we have

$$g_n(t) = (-1)^n n t^{n+q-\frac{1}{2}} \int_0^\infty \partial_x^n [x f'(x)] e^{-tx} x^{2n+2q-1} dx$$
$$- (-1)^n t^{n+q+\frac{1}{2}} \int_0^\infty \partial_x^n [x f'(x)] e^{-tx} x^{2n+2q} dx.$$

Moreover, the expression (1.2) implies that the functions defined by

$$\partial_x^n [x f'(x)] x^{2n+2q-1} \quad \text{and} \quad \partial_x^n [x f'(x)] x^{2n+2q}$$

are contained in the spaces L^2_{n+q} and L^2_{n+q+1}, respectively. Hence the function g_n is the restriction of a member in the family

$$\left\{ \tau^{n+q-\frac{1}{2}} h_1(\tau) + \tau^{n+q+\frac{1}{2}} h_2(\tau) : h_1 \in H_{n+q} \quad \text{and} \quad h_2 \in H_{n+q+1}(\tau = t + i\hat{t}) \right\}$$

to the half-axis $(0, \infty)$, and it is represented by

$$g_n(t) = t^{n+q-\frac{1}{2}} \hat{h}_1(t) + t^{n+q+\frac{1}{2}} \hat{h}_2(t)$$

for some functions $\hat{h}_1 \in H_{n+q}$ and $\hat{h}_2 \in H_{n+q+1}$. If $n = 0$, we have $g_n(t) = t^{q+\frac{1}{2}} h_3(t)$ for some function $h_3 \in H_{q+1}$. Furthermore, for $n \neq 0$ we have the representation

$$g_n(t) = t^{n+q-\frac{1}{2}} k_1'(t) + t^{n+q+\frac{1}{2}} k_2'(t)$$

for some functions $k_1 \in H_{n+q-1}$ and $k_2 \in H_{n+q}$, following (1.32). Hence, from (1.2) we get the relations

$$\int_0^\infty |t^{n+q-\frac{1}{2}} k_1'(t)|^2 dt = \int_0^\infty |t k_1'(t)|^2 t^{2n+2q-3} dt < \infty$$

214

and

$$\int_0^\infty |t^{n+q+\frac{1}{2}}k_2'(t)|^2 dt = \int_0^\infty |tk_2'(t)|^2 t^{2n+2q-1} dt < \infty,$$

and so the function g_n ($n \neq 0$) belongs to the space $L^2[(0,\infty), dt]$. Likewise, the function g_0 is also a member of the space $L^2[(0,\infty), dt]$. By virtue of the isometry $s(t) \longmapsto s(t)t^{q-\frac{1}{2}}$ of the space $L^2[(0,\infty), dt]$ onto the space L_q^2 we conclude that the function F_N belongs to the space L_q^2.

Next, in order to prove that $\mathcal{L}F_N = f_N$, we examine, for a fixed number $\xi > 0$, the integrability of the functions

$$\varphi(x,t;n,\xi) = \partial_x^n\{xf'(x)\}\partial_x^n\{x\partial_x(e^{-tx})\}e^{-\xi t}t^{2q-1}x^{2n+2q-1} \quad (n=0,1,2,\cdots)$$

with respect to the Lebesgue measure on the set $(0,\infty) \times (0,\infty)$. We first have the estimate

$$\begin{aligned}
|\varphi(x,t;n,\xi)| &= |\partial_x^n\{xf'(x)\}||\partial_x^n\{x\partial_x(e^{-tx})\}|e^{-\xi t}t^{2q-1}x^{2n+2q-1} \\
&= |\partial_x^n\{xf'(x)\}||(-t)^n xe^{-tx} + n(-t)^{n-1}e^{-tx}|t^{2q}e^{-\xi t}x^{2n+2q-1} \\
&\leqq |\partial_x^n\{xf'(x)\}| \left\{ t^{n+2q}xe^{-(x+\xi)t} + nt^{n+2q-1}e^{-(x+\xi)t} \right\} x^{2n+2q-1}.
\end{aligned}$$

Therefore, since the functions defined by

$$x\int_0^\infty t^{n+2q}e^{-(x+\xi)t} dt = \Gamma(n+2q+1)x(x+\xi)^{-(n+2q+1)}$$

and

$$n\int_0^\infty t^{n+2q-1}e^{-(x+\xi)t} dt = n\Gamma(n+2q)(x+\xi)^{-(n+2q)}$$

belong to the space $L^2[(0,\infty), x^{2n+2q-1}dx]$, we see by the Schwarz inequality that the function $\varphi(x,t;n,\xi)$ is integrable for all n. By Fubini's theorem, the following

215

sequence of equalities is therefore valid:

$$\int_0^\infty F_N(t)e^{-\xi t}\,dt = \sum_{n=0}^N \frac{1}{n!\Gamma(n+2q+1)}$$

$$\times \int_0^\infty \left[\int_0^\infty \partial_x^n\{xf'(x)\}\partial_x^n\{x\partial_x(e^{-tx})\}x^{2n+2q-1}\,dx \right] t^{2q-1}e^{-\xi t}\,dt$$

$$= \sum_{n=0}^N \frac{1}{n!\Gamma(n+2q+1)}$$

$$\times \int_0^\infty \partial_x^n[xf'(x)] \left[\int_0^\infty \partial_x^n\{x\partial_x(e^{-tx})\}e^{-\xi t}t^{2q-1}\,dt \right] x^{2n+2q-1}\,dx$$

$$= \sum_{n=0}^N \frac{1}{n!\Gamma(n+2q+1)}$$

$$\times \int_0^\infty \partial_x^n[xf'(x)]\partial_x^n \left[x\partial_x \int_0^\infty e^{-tx}e^{-\xi t}t^{2q-1}\,dt \right] x^{2n+2q-1}\,dx$$

$$= \sum_{n=0}^N \frac{1}{n!\Gamma(n+2q+1)}$$

$$\times \int_0^\infty \partial_x^n[xf'(x)]\partial_x^n[x\partial_x K_q(x,\xi)]x^{2n+2q-1}\,dx = f_N(\xi).$$

Thus the assertions of the lemma are proved.

Lemma 3   For any fixed $q > 0$, let the function f be a member of the space H_q. Then the following statements are true.

(i) If $n \geq 1$ and $0 \leq m \leq n-1$, then $\partial_x^m[xf'(x)]x^{n+m+2q} = o(1)$ as $x \longrightarrow 0+$.
(ii) $f(x)x^q = O(1)$ as $x \longrightarrow 0+$.

Proof :   By the Leibniz rule, we have the equality

$$\partial_x^m[xf'(x)] = x\partial_x^{m+1}f(x) + m\partial_x^m f(x).$$

We also see that the function $\partial_x^{m+1}f$ belongs to the space H_{q+m+1} following (1.32), and from the Schwarz inequality the following estimate is valid:

$$|\partial_x^{m+1}f(x)| = \left| \left(\partial_\xi^{m+1}f(\xi), K_{q+m+1}(\xi,x) \right)_{q+m+1} \right|$$

$$\leq \|\partial_x^{m+1}f\|_{q+m+1} K_{q+m+1}(x,x)^{\frac{1}{2}}$$

$$= \|\partial_x^{m+1}f\|_{q+m+1} \Gamma(2q+2m+2)^{\frac{1}{2}}2^{-(q+m+1)}x^{-(q+m+1)}.$$

Likewise, the estimates

$$|\partial_x^m f(x)| \le \|\partial_x^m f\|_{q+m} \Gamma(2q+2m)^{\frac{1}{2}} 2^{-(q+m)} x^{-(q+m)}$$

and

$$|f(x)| \le \|f\|_q \Gamma(2q)^{\frac{1}{2}} 2^{-q} x^{-q}$$

are valid. Therefore, our lemma is obtained.

3   Proof of Theorem 1

From Lemma 3, and by integration by parts we have, for any nonnegative integer n,

$$\int_0^\infty \partial_x^n[xf'(x)]\partial_x^n[x\partial_x(e^{-tx})]x^{2n+2q-1}dx$$

$$= t^n \int_0^\infty xf'(x)\partial_x^n[(n-tx)e^{-tx}x^{2n+2q-1}]dx$$

$$= -t^n \int_0^\infty f(x)\partial_x[x\partial_x^n\{(n-tx)e^{-tx}x^{2n+2q-1}\}]dx.$$

Meanwhile, for $n \ge 1$ we also have

$$-t^n\partial_x[x\partial_x^n\{(n-tx)e^{-tx}x^{2n+2q-1}\}]$$

$$= -e^{-tx}t^n\left[\sum_{\nu=0}^n \binom{n}{\nu}(-t)^\nu\{n\partial_x^{n-\nu}x^{2n+2q-1} - t\partial_x^{n-\nu}x^{2n+2q}\}\right.$$

$$- tx\sum_{\nu=0}^n \binom{n}{\nu}(-t)^\nu\{n\partial_x^{n-\nu}x^{2n+2q-1} - t\partial_x^{n-\nu}x^{2n+2q}\}$$

$$\left.+ x\sum_{\nu=0}^n \binom{n}{\nu}(-t)^\nu\{n\partial_x^{n-\nu+1}x^{2n+2q-1} - t\partial_x^{n-\nu+1}x^{2n+2q}\}\right]$$

$$= e^{-tx}\sum_{\nu=0}^n (-1)^{\nu+1}\binom{n}{\nu}(xt)^{n+\nu}\frac{\Gamma(2n+2q)}{\Gamma(n+\nu+2q)}$$

$$\times \left\{\frac{2n+2q}{n+\nu+2q}t^2x^{2q+1} - \left(\frac{2n+2q}{n+\nu+2q}+3n+2q\right)tx^{2q} + n(n+\nu+2q)x^{2q-1}\right\}.$$

Applying Lemma 1 and Lemma 2 to the isometry $\mathcal{L}$, we therefore obtain the inversion formula of $\mathcal{L}$. Also, the inequality in Theorem 1 gives the estimate of the truncation error.

Some characteristics of the strong singularity of the polynomial $P_{N,1}(\xi)$ in Theorem 2.1 as $N \longrightarrow \infty$ and some effective algorithms for the real inversion formula in Theorem 2.1 are examined by [KT1,2] and [T] using computers.

217

6.   Applications to source inverse problems

As a typical source inverse problem, we shall establish analytical solutions for Poisson's equation and Helmholtz's equation in some typical settings.

§1   Inverse source formulas in Poisson's equation

We shall consider Poisson's equation

$$\Delta u = -\rho(\mathbf{r}) \quad \text{in} \quad \mathbb{R}^1, \mathbb{R}^2 \quad \text{and} \quad \mathbb{R}^3 \tag{1}$$

for a real-valued $L_2(d\mathbf{r})$ source function ρ whose support is contained in either the interior of a sphere $r < a$ or in the exterior; r denotes the distance $r = |\mathbf{r}|$ from the origin. We shall first give the characterization and natural representation of the potential u exterior to the support of ρ. As an application, we shall give a surprisingly simple expression of ρ^* in terms of u exterior to the support of ρ, which has the minimum $L_2(d\mathbf{r})$ norm among the source functions ρ satisfying (1) on the outside of the support of ρ. These representations have a practical application for determining the source function ρ^* from the potential u. These inverse problems were proposed by Laplace some 200 years ago and many mathematicians have worked on these problems. We shall present solutions for these problems in the framework of $L_2(d\mathbf{r})$ spaces based on the work [Sa28]. See Foreword and References of [[I]] and, for general references on inverse source problems, see in particular [[I]] and [[Che]].

1   Case with compact support on $\mathbb{R}^3$

We assume that the support of ρ is contained in the sphere $r < a$. So, for $\mathbf{r} \in \mathbb{R}^3$ and $|\mathbf{r}| = r$, we shall examine the integral representation of the solution u of Poisson's equation (1)

$$u(\mathbf{r}') = \frac{1}{4\pi} \int_{r<a} \frac{1}{|\mathbf{r}' - \mathbf{r}|} \rho(\mathbf{r}) d\mathbf{r} \tag{2}$$

in $\mathbb{R}^3$, where the source function ρ satisfies

$$\int_{r<a} \rho(\mathbf{r})^2 d\mathbf{r} < \infty. \tag{3}$$

In order to determine some properties of the potential u for $r > a$, we consider the kernel

$$K_{i,a}(\mathbf{r}', \mathbf{r}'') = \frac{1}{4\pi} \int_{r<a} \frac{1}{|\mathbf{r}' - \mathbf{r}||\mathbf{r}'' - \mathbf{r}|} d\mathbf{r},$$

where $r', r'' > a$. In order to calculate $K_{i,a}(\mathbf{r}', \mathbf{r}'')$ we shall use the identity

$$\frac{1}{|\mathbf{r}' - \mathbf{r}|} = \sum_{n=0}^{\infty} \sum_{m=0}^{n} \frac{\varepsilon_m (n-m)!}{(n+m)!} P_n^m(\cos\theta') P_n^m(\cos\theta) \cos m(\varphi' - \varphi) \frac{r^n}{r'^{n+1}},$$

$$\text{for} \quad r' > r \tag{4}$$

where (r, θ, φ) and (r', θ', φ') are spherical coordinates (cf. [[MF]], p.1274). Here, $\varepsilon_m = 2 - \delta_{mo}$ is the Neumann factor. By using the two orthogonality relations

$$\int_0^{2\pi} \cos m(\varphi' - \varphi) \cos m'(\varphi'' - \varphi) d\varphi = 2\pi \delta_{mm'}(\varepsilon_m)^{-1} \cos m(\varphi' - \varphi'') \tag{5}$$

and

$$\int_0^{\pi} P_n^m(\cos\theta) P_{n'}^m(\cos\theta) \sin\theta d\theta$$

$$= \int_{-1}^{1} P_n^m(x) P_{n'}^m(x) dx$$

$$= \delta_{nn'} \frac{2}{2n+1} \frac{(n+m)!}{(n-m)!}, \tag{6}$$

we obtain the formal expansion

$$K_{i,a}(\mathbf{r}', \mathbf{r}'') = \sum_{n=0}^{\infty} \frac{a^{2n+3}}{(2n+1)(2n+3)} \frac{1}{r'^{n+1}} \frac{1}{r''^{n+1}}$$

$$\times \sum_{m=0}^{n} \frac{\varepsilon_m (n-m)!}{(n+m)!} P_n^m(\cos\theta') P_n^m(\cos\theta'')$$

$$\times (\cos m\varphi' \cos m\varphi'' + \sin m\varphi' \sin m\varphi'')$$

$$\text{for} \quad r', r'' > a. \tag{7}$$

This series converges absolutely on $r', r'' > a$. The kernel $K_{i,a}(\mathbf{r}', \mathbf{r}'')$ is a positive matrix on $r > a$ and so there exists a uniquely determined Hilbert space $H_{K_{i,a}}$ admitting the reproducing kernel $K_{i,a}(\mathbf{r}', \mathbf{r}'')$. Furthermore, the images $u(\mathbf{r}')$ of (2) belong just to the Hilbert space $H_{K_{i,a}}$. The expansion (7) implies that the images

$u(\mathbf{r}')$ are expressible in the form

$$u(\mathbf{r}') = \sum_{n=0}^{\infty} \frac{a^{2n+3}}{(2n+1)(2n+3)} \frac{1}{r'^{n+1}}$$

$$\times \sum_{m=0}^{n} \frac{\varepsilon_m (n-m)!}{(n+m)!} P_n^m(\cos\theta')(A_n^m \cos m\varphi' + B_n^m \sin m\varphi')$$

(8)

for some constants $\{A_n^m, B_n^m\}_{n,m=0}^{\infty}$ satisfying

$$\sum_{n=0}^{\infty} \frac{a^{2n+3}}{(2n+1)(2n+3)} \sum_{m=0}^{n} \frac{\varepsilon_m (n-m)!}{(n+m)!} \{(A_n^m)^2 + (B_n^m)^2\} < \infty.$$

(9)

Conversely, any $u(\mathbf{r}')$ defined by (8) with (9) belongs to $H_{K_{i,a}}$. Since the family

$$\{P_n^m(\cos\theta)\cos m\varphi, P_n^m(\cos\theta)\sin m\varphi\}_{n,m=0}^{\infty}$$

is complete in the Hilbert space consisting of the functions $f(\theta,\varphi)$ with the finite norm

$$\left\{ \int_0^{\pi} \int_0^{2\pi} f(\theta,\varphi)^2 \sin\theta d\theta d\varphi \right\}^{\frac{1}{2}} < \infty,$$

we have the representation of the norm $\|u\|_{H_{K_{i,a}}}$ in the form

$$\|u\|_{H_{K_{i,a}}}^2 = \sum_{n=0}^{\infty} \frac{a^{2n+3}}{(2n+1)(2n+3)} \sum_{m=0}^{n} \frac{\varepsilon_m (n-m)!}{(n+m)!} \{(A_n^m)^2 + (B_n^m)^2\}.$$

(10)

Furthermore, we have the isometrical identity

$$\|u\|_{H_{K_{i,a}}}^2 = \min \int_{r<a} \rho(\mathbf{r})^2 d\mathbf{r}$$

$$= \int_{r<a} \rho^*(\mathbf{r})^2 d\mathbf{r}.$$

(11)

Here, the minimum is taken over all $\rho(\mathbf{r})$ satisfying (2) and (3) for $r' > a$ and ρ^* is the uniquely determined function with the minimum norm.

In (8), by using the orthogonality relation

$$\int_0^{\pi} \int_0^{2\pi} P_n^m(\cos\theta)\cos m\varphi P_{n'}^{m'}(\cos\theta) \left\{ \begin{array}{c} \cos m'\varphi \\ \text{or} \\ \sin m'\varphi \end{array} \right\} \sin\theta d\theta d\varphi$$

$$= \delta_{mm'}\delta_{nn'} \frac{4\pi(\varepsilon_m)^{-1}}{2n+1} \frac{(n+m)!}{(n-m)!},$$

220

we have the expressions, for any fixed $b > a$,

$$A_n^m = \frac{(2n+1)^2(2n+3)b^{n+1}}{4\pi a^{2n+3}} \int_0^\pi \int_0^{2\pi} u(b,\theta,\varphi)$$
$$\times P_n^m(\cos\theta)\cos m\varphi \sin\theta \, d\theta \, d\varphi \qquad (12)$$

and

$$B_n^m = \frac{(2n+1)^2(2n+3)b^{n+1}}{4\pi a^{2n+3}} \int_0^\pi \int_0^{2\pi} u(b,\theta,\varphi)$$
$$\times P_n^m(\cos\theta)\sin m\varphi \sin\theta \, d\theta \, d\varphi. \qquad (13)$$

These expressions with (8) imply that for any point $\mathbf{r}'$ ($r' > a$), the potentials $u(\mathbf{r}')$ are expressible in terms of

$$u(b,\theta,\varphi) \quad \text{for any fixed} \quad b(b > a). \qquad (14)$$

We shall derive the inversion formula representing ρ^* in terms of (14). Using the reproducing property of $K_{i,a}(\mathbf{r},\mathbf{r}')$ in $H_{K_{i,a}}$, we have

$$u(\mathbf{r}') = (u(\mathbf{r}), K_{i,a}(\mathbf{r},\mathbf{r}'))_{H_{K_{i,a}}}$$
$$= \left(u(\mathbf{r}), \frac{1}{4\pi}\int_{r_1<a} \frac{1}{|\mathbf{r}-\mathbf{r}_1||\mathbf{r}'-\mathbf{r}_1|}d\mathbf{r}_1\right)_{H_{K_{i,a}}}$$
$$= \frac{1}{4\pi}\int_{r_1<a} \frac{d\mathbf{r}_1}{|\mathbf{r}'-\mathbf{r}_1|}(u(\mathbf{r}), \frac{1}{|\mathbf{r}-\mathbf{r}_1|})_{H_{K_{i,a}}}$$
$$= \frac{1}{4\pi}\int_{r_1<a} \frac{d\mathbf{r}_1}{|\mathbf{r}'-\mathbf{r}_1|}\rho^*(\mathbf{r}_1) \qquad (15)$$

and so we have

$$\rho^*(\mathbf{r}_1) = (u(\mathbf{r}), \frac{1}{|\mathbf{r}-\mathbf{r}_1|})_{H_{K_{i,a}}}. \qquad (16)$$

Here, note that $\frac{1}{|\mathbf{r}-\mathbf{r}_1|}$ belongs to the Hilbert space $H_{K_{i,a}}$.

Indeed, from (8) and (4) the corresponding coefficients $\widetilde{A_n^m}$ and $\widetilde{B_n^m}$ of $\frac{1}{|\mathbf{r}-\mathbf{r}_1|}$ in the representation (8) are

$$\widetilde{A_n^m} = \frac{(2n+1)(2n+3)}{a^{2n+3}}P_n^m(\cos\theta_1)r_1^n \cos m\varphi_1$$

and

$$\widetilde{B_n^m} = \frac{(2n+1)(2n+3)}{a^{2n+3}}P_n^m(\cos\theta_1)r_1^n \sin m\varphi_1.$$

Then,

$$\sum_{n=0}^{\infty} \frac{a^{2n+3}}{(2n+1)(2n+3)} \sum_{m=0}^{n} \frac{\varepsilon_m (n-m)!}{(n+m)!} \{(\widetilde{A_n^m})^2 + (\widetilde{B_n^m})^2\}$$

$$= \sum_{n=0}^{\infty} \frac{(2n+1)(2n+3)}{a^{2n+3}} \sum_{m=0}^{n} \frac{\varepsilon_m (n-m)!}{(n+m)!} P_n^m (\cos\theta_1)^2 r_1^{2n}$$

$$= \sum_{n=0}^{\infty} \frac{(2n+1)(2n+3) r_1^{2n}}{a^{2n+3}} P_n(1)$$

$$= \sum_{n=0}^{\infty} \frac{(2n+1)(2n+3) r_1^{2n}}{a^{2n+3}} < \infty \quad \text{for} \quad r_1 < a.$$

See [[MF]], p.1274.

Hence, the formally obtained expressions in (15) and (16) are justified. We thus have

Theorem 1   *The source function ρ^* in the sense of (11) is expressible in terms of (14) in the form*

$$\rho^*(\mathbf{r}_1) = \frac{1}{4\pi} \sum_{n=0}^{\infty} \frac{(2n+1)^2(2n+3)}{a^{2n+3}} r_1^n b^{n+1}$$

$$\times \sum_{m=0}^{n} \frac{\varepsilon_m (n-m)!}{(n+m)!} P_n^m(\cos\theta_1)$$

$$\times \int_0^\pi \int_0^{2\pi} u(b,\theta,\varphi) P_n^m(\cos\theta) \cos m(\varphi - \varphi_1) \sin\theta\, d\theta\, d\varphi.$$

If we decompose ρ in (3) as

$$\rho = \rho^* + \rho_o,$$

where ρ^* is defined by (11) and given in Theorem 1, then ρ_o is expressible in the form

$$\rho_o = \Delta f, \quad f \in C_o^2(r < a)$$

(see, for example, [[I]] or [[Che]]). This null (invisible) part ρ_o of ρ, can of course, not be determined by the potential u in $\{r > a\}$. The Poisson equation (1) has physical interpretations, namely the correspondence

$$\rho \longleftrightarrow u$$

222

can be taken as

$$\text{mass distribution} \longleftrightarrow \text{gravitational potential}$$
$$\text{current} \longleftrightarrow \text{magnetic field}$$
$$\text{charge} \longleftrightarrow \text{electrostatic field}$$
$$\text{heat source} \longleftrightarrow \text{temperature,}$$

where we consider only steady state situations. Hence, we can apply our inverse formula for miscellaneous physical situations.

As we see from Theorem 1, the source functions ρ^* obtained are harmonic in $r < a$.

Indeed,

$$P_n^m(\cos\theta_1)r_1^n \cos m\varphi_1, \qquad P_n^m(\cos\theta_1)r_1^n \sin m\varphi_1$$

are harmonic functions; furthermore, we have

$$|\rho^*(\mathbf{r}_1)| \leq \|u\|_{H_{K_{i,a}}} \left\| \frac{1}{|\mathbf{r} - \mathbf{r}_1|} \right\|_{H_{K_{i,a}}}$$

$$= \|u\|_{H_{K_{i,a}}} \left\{ \sum_{n=0}^{\infty} \frac{(2n+1)(2n+3)r_1^{2n}}{a^{2n+3}} \right\}^{\frac{1}{2}}.$$

Hence, the expansion of $\rho^*(\mathbf{r}_1)$ in the right hand side in Theorem 1 converges absolutely and uniformly on $r \leq a' < a$ for any fixed a'.

The fact that ρ^* are harmonic means that the source functions obtained are smooth on the ball $r < a$.

In order to treat more general sources, we shall consider the integral transform

$$u(\mathbf{r}') = \frac{1}{4\pi} \int_{r<a} \frac{1}{|\mathbf{r}' - \mathbf{r}|} \rho_1(\mathbf{r}) D(r) d\mathbf{r},$$

where $D(r)$ is a nonnegative measurable function, depending only on r such that the integral

$$K_{i,a}(\mathbf{r}', \mathbf{r}''; D(r)) = \frac{1}{4\pi} \int_{r<a} \frac{1}{|\mathbf{r}' - \mathbf{r}||\mathbf{r}'' - \mathbf{r}|} D(r) d\mathbf{r}$$

$$(= K_{i,a}(D(r)))$$

exists for $r', r'' > a$. Then, we can obtain an inversion formula representing ρ_1^* in terms of $u(b, \theta, \varphi)$ satisfying

$$\|u\|_{H_{K_{i,a}(D(r))}}^2 = \frac{1}{4\pi} \int_{r<a} \rho_1^*(\mathbf{r})^2 D(r) d\mathbf{r},$$

223

as in Theorem 1. Hence,

$$\rho(\mathbf{r}) = \rho_1^*(\mathbf{r})D(r).$$

For this argument, see the following subsection 2 for $D(r) = r^{-2}$. By taking a suitable $D(r)$, we will be able to obtain a more general and useful source $\rho(\mathbf{r})$.

2 Case with unbounded support on $\mathbb{R}^3$

Next, we shall consider the case where the support of ρ is contained in the sphere $r > a$. In this case, in order to obtain the existence of the corresponding reproducing kernel we assume that

$$\int_{r>a} \rho(\mathbf{r})^2 r^2 d\mathbf{r} < \infty. \tag{17}$$

In order to examine the integral representation

$$u(\mathbf{r}') = \frac{1}{4\pi} \int_{r>a} \frac{1}{|\mathbf{r}' - \mathbf{r}|} \rho(\mathbf{r}) d\mathbf{r} \tag{18}$$

of u, we shall rewrite the integral transform in the form

$$u(\mathbf{r}') = \frac{1}{4\pi} \int_{r>a} \frac{1}{|\mathbf{r}' - \mathbf{r}|} \rho_1(\mathbf{r}) \frac{1}{r^2} d\mathbf{r}, \tag{19}$$

where the function $\rho_1 = r^2 \rho$ satisfies

$$\int_{r>a} \rho_1(\mathbf{r})^2 \frac{1}{r^2} d\mathbf{r} = \int_{r>a} \rho(\mathbf{r})^2 r^2 d\mathbf{r} < \infty. \tag{20}$$

We form the reproducing kernel

$$K_{o,a}(\mathbf{r}', \mathbf{r}'') = \frac{1}{4\pi} \int_{r>a} \frac{1}{|\mathbf{r}' - \mathbf{r}||\mathbf{r}'' - \mathbf{r}|} \frac{1}{r^2} d\mathbf{r}. \tag{21}$$

By using (4) and the orthogonality conditions (6) and (7), we have

$$K_{o,a}(\mathbf{r}', \mathbf{r}'') = \sum_{n=0}^{\infty} \frac{r'^n r''^n}{(2n+1)^2 a^{2n+1}} \sum_{m=0}^{n} \frac{\varepsilon_m (n-m)!}{(n+m)!}$$
$$\times P_n^m(\cos\theta') P_n^m(\cos\theta'')(\cos m\varphi' \cos m\varphi'' + \sin m\varphi' \sin m\varphi''),$$
$$\text{for} \quad r', r'' < a. \tag{22}$$

Hence, by arguments parallel to those of Theorem 1, we have

224

Theorem 2 *For the source function ρ^* satisfying (18) and (17) with the minimum norm in (17), we have the inverse formula, for any fixed $\hat{b}$ $(0 < \hat{b} < a)$,*

$$\rho^*(\mathbf{r}_1) = \frac{1}{4\pi r_1^2} \sum_{n=0}^{\infty} (2n+1)^3 a^{2n+1} r_1^{-(n+1)} \hat{b}^{-n}$$

$$\times \sum_{m=0}^{n} \frac{\varepsilon_m (n-m)!}{(n+m)!} P_n^m(\cos\theta_1) \int_0^{\pi} \int_0^{2\pi} u(\hat{b}, \theta, \varphi)$$

$$\times P_n^m(\cos\theta) \cos m(\varphi - \varphi_1) \sin\theta \, d\theta \, d\varphi.$$

3 $\mathbb{R}^2$ case

We shall consider the two-dimensional potential

$$u(\mathbf{r}') = \frac{1}{2\pi} \int_{r<a} \log\frac{1}{|\mathbf{r}' - \mathbf{r}|} \rho(\mathbf{r}) d\mathbf{r} \qquad (23)$$

for a source function ρ satisfying

$$\int_{r<a} \rho(\mathbf{r})^2 d\mathbf{r} < \infty \qquad (24)$$

whose support is contained in the disc $r < a$.

In order to examine the integral transform (23) with (24), we form the reproducing kernel

$$K_{i,a}^{(2)}(\mathbf{r}', \mathbf{r}'') = \frac{1}{2\pi} \int_{r<a} \log\frac{1}{|\mathbf{r}' - \mathbf{r}|} \log\frac{1}{|\mathbf{r}'' - \mathbf{r}|} d\mathbf{r}$$

$$\text{for} \quad r', r'' > a. \qquad (25)$$

By using the expansion

$$\log\frac{1}{|\mathbf{r}' - \mathbf{r}|} = \log\frac{1}{r'} + \sum_{n=1}^{\infty} \frac{1}{n}\left(\frac{r}{r'}\right)^n \cos n(\varphi' - \varphi),$$

$$\text{for} \quad r' > r \qquad (26)$$

in the polar coordinates (r, φ) and (r', φ') (cf. [[MF]], p.1188), we have

$$K_{i,a}^{(2)}(\mathbf{r}', \mathbf{r}'') = \frac{a^2}{2} \log\frac{1}{r'} \log\frac{1}{r''} + \frac{1}{4}$$

$$\times \sum_{n=1}^{\infty} \frac{a^{2n+2}}{n^2(n+1)} \frac{1}{r'^n r''^n} (\cos n\varphi' \cos n\varphi'' + \sin n\varphi' \sin n\varphi'')$$

$$\text{for} \quad r', r'' > a, \qquad (27)$$

225

which converges absolutely. Hence, as in Theorem 2 we have

Theorem 3 *The source function ρ^* with the minimum norm (24) satisfying (23) for $r' > a$ is expressible in the form, for any fixed b $(b > a)$*

$$\rho^*(\mathbf{r}_1) = \left(u(\mathbf{r}), \log \frac{1}{|\mathbf{r} - \mathbf{r}_1|} \right)_{H_{K_{i,a}^{(2)}}}$$

$$= \frac{1}{\pi a^2 \log \frac{1}{b}} \int_0^{2\pi} u(b, \varphi) d\varphi$$

$$+ \frac{4}{\pi} \sum_{n=1}^{\infty} \frac{n(n+1)b^n r_1^n}{a^{2n+2}} \int_0^{2\pi} u(b, \varphi) \cos n(\varphi - \varphi_1) d\varphi.$$

Next, we shall examine the case where ρ has unbounded support. In order to consider the potential

$$u(\mathbf{r}') = \frac{1}{2\pi} \int_{r>a} \log \frac{1}{|\mathbf{r}' - \mathbf{r}|} \rho(\mathbf{r}) d\mathbf{r}, \tag{28}$$

we shall examine the integral transform

$$u(\mathbf{r}') = \frac{1}{2\pi} \int_{r>a} \log \frac{1}{|\mathbf{r}' - \mathbf{r}|} \rho_1(\mathbf{r}) \frac{1}{r^3} d\mathbf{r} \tag{29}$$

for a function $\rho_1 = r^3 \rho$ satisfying

$$\int_{r>a} \rho_1(\mathbf{r})^2 \frac{1}{r^3} d\mathbf{r} = \int_{r>a} \rho(\mathbf{r})^2 r^3 d\mathbf{r} < \infty. \tag{30}$$

Then, the corresponding reproducing kernel $K_{o,a}^{(2)}(\mathbf{r}', \mathbf{r}'')$ can be calculated, using (26), as follows:

$$K_{o,a}^{(2)}(\mathbf{r}', \mathbf{r}'') = \frac{1}{2\pi} \int_{r>a} \log \frac{1}{|\mathbf{r}' - \mathbf{r}|} \log \frac{1}{|\mathbf{r}'' - \mathbf{r}|} \frac{1}{r^3} d\mathbf{r}$$

$$= \frac{1}{a} \left\{ 2 + (\log \frac{1}{a})^2 - 2 \log \frac{1}{a} \right\}$$

$$+ \frac{1}{2} \sum_{n=1}^{\infty} \frac{r'^n r''^n}{(2n+1)n^2 a^{2n+1}} (\cos n\varphi' \cos n\varphi'' + \sin n\varphi' \sin n\varphi'').$$
$$\tag{31}$$

Hence, we have

Theorem 4 *The source function ρ^* satisfying (28) and (29) with the minimum norm is expressible in the form, for any fixed $\hat{b}$ $(0 < \hat{b} < a)$,*

$$\rho^*(\mathbf{r}_1) = \frac{1}{r_1^3}\left(u(\mathbf{r}), \log\frac{1}{|\mathbf{r} - \mathbf{r}_1|}\right)_{H_{K_{i,a}^{(2)}}}$$

$$= \frac{1}{r_1^3}\left[\left\{\frac{1}{a}(2 + (\log\frac{1}{a})^2 - 2\log\frac{1}{a})\right\}^{-1}\right.$$

$$\times \left(\frac{1}{2\pi}\int_0^{2\pi} u(\hat{b}, \varphi)d\varphi\right)\log\frac{1}{r_1}$$

$$+ \frac{2}{\pi}\sum_{n=0}^{\infty}\frac{n(2n+1)a^{2n+1}}{\hat{b}^n r_1^n}\int_0^{2\pi} u(\hat{b}, \varphi)\cos n(\varphi - \varphi_1)d\varphi\Bigg].$$

4 $\mathbb{R}^1$ case

We consider the one-dimensional potential

$$u(x) = \int_{-a}^{a} -\frac{1}{2}|x - t|\rho(t)dt, \quad x > a > 0 \tag{32}$$

where the source function ρ satisfies

$$\int_{-a}^{a} \rho(t)^2\,dt < \infty. \tag{33}$$

In order to examine the integral transform (32) with (33), we form the reproducing kernel

$$K_r(x, y) = \frac{1}{4}\int_{-a}^{a}|x - t||y - t|dt = \frac{a}{2}(xy + \frac{1}{3}a^2) \quad \text{for} \quad x, y > a. \tag{34}$$

This expression means that the Hilbert space H_{K_r} admitting the reproducing kernel $K_r(x, y)$ is two dimensional, and any member $f(x)$ of H_{K_r} is expressible in the form

$$f(x) = \frac{a}{2}(c_1 x + c_2\frac{1}{3}a^2) \quad \text{for} \quad x > a, \tag{35}$$

for some constants c_1 and c_2; furthermore,

$$\|f\|_{H_{K_r}}^2 = \frac{a}{2}(c_1^2 + c_2^2\frac{1}{3}a^2). \tag{36}$$

227

The constants c_1 and c_2 are determined by any two point values, say $f(b_1)$ and $f(b_2)$ for $b_1, b_2 > a$, as follows:

$$c_1 = \frac{2}{a}\frac{f(b_1) - f(b_2)}{b_1 - b_2} \qquad (37)$$

and

$$c_2 = \frac{6}{a^3}\frac{b_1 f(b_2) - b_2 f(b_1)}{b_1 - b_2}. \qquad (38)$$

We thus have

Theorem 5   *In the potential (32) satisfying (33), the source function ρ^* with the minimum norm in (33) is expressible in terms of $u(b_1)$ and $u(b_2)$ $(b_1, b_2 > a, b_1 \neq b_2)$ in the form*

$$\rho^*(t) = (u(x), -\frac{1}{2}|x - t|)_{H_{K_r}}$$

$$= \frac{3(b_1 u(b_2) - b_2 u(b_1))}{a^3(b_1 - b_2)}t - \frac{u(b_1) - u(b_2)}{a(b_1 - b_2)}.$$

In order to determine a source function with unbounded support we shall examine the integral transform, for $a > 0, x < a$

$$u(x) = \int_a^\infty -\frac{1}{2}|x - t|\rho(t)dt$$

$$= \int_a^\infty -\frac{1}{2}|x - t|\rho_1(t)\frac{1}{t^4}dt \qquad (39)$$

for a source function $\rho = \rho_1 t^{-4}$ satisfying

$$\int_a^\infty \rho_1(t)^2 \frac{dt}{t^4} = \int_a^\infty \rho(t)^2 t^4 dt < \infty. \qquad (40)$$

We form the corresponding reproducing kernel

$$K_l(x, y) = \int_a^\infty -\frac{1}{2}|x - t| \cdot -\frac{1}{2}|y - t|\frac{dt}{t^4}$$

$$= \frac{1}{12a^3}(x - \frac{3}{2}a)(y - \frac{3}{2}a) + \frac{1}{16a}. \qquad (41)$$

Hence, by arguments parallel to those of Theorem 5, we have

228

Theorem 6 *In the potential u in (39) satisfying (40), the source function ρ^* with the minimum norm in (40) is expressible in terms of $u(b_1)$ and $u(b_2)$ for any two fixed points $b_1, b_2 < a$ as follows:*

$$\rho^*(t) = \frac{1}{t^4}(u(x), -\frac{1}{2}|x - t|)_{H_{K_l}}$$

$$= \frac{1}{t^4}\left[\frac{8a}{b_1 - b_2}\left\{(b_2 u(b_1) - b_1 u(b_2)) - \frac{3}{2}a(u(b_1) - u(b_2))\right\}t\right.$$

$$\left. + \frac{12a^2}{b_1 - b_2}\left\{(b_1 u(b_2) - b_2 u(b_1)) + 2a(u(b_1) - u(b_2))\right\}\right].$$

§2   Inverse source formulas for Helmholtz's equation

We shall consider Helmholtz's equation

$$\Delta u + k^2 u = -\rho(\mathbf{r}) \quad \text{on} \quad \mathbb{R}^3 \tag{1}$$

for a complex-valued $L_2(d\mathbf{r})$ source function ρ whose support is contained in a sphere $r < a$ or outside of the sphere. This equation is very fundamental in mathematical physics, in particular, mathematical acoustics and electromagnetics. See, for example, [[Bert]], [[CK]], [[MF]] and [[Ra2]]. We shall first give the characterization and natural representation of the wave $u(\mathbf{r})$ outside the support of ρ. As an application, we shall give a simple expression of ρ^* in terms of u outside the support of ρ, which has a minimum $L_2(d\mathbf{r})$ norm among the set of source functions ρ satisfying (1) outside the support of ρ. These representations provide a useful means for determining the source ρ^* from the field u. Secondly, in terms of the radiation pattern, that is,

$$g(\frac{\mathbf{r}'}{r'}) = \frac{1}{4\pi}\int_{r<a} e^{-ik(\frac{\mathbf{r}'}{r'},\mathbf{r})}\rho(\mathbf{r})d\mathbf{r}, \tag{2}$$

we give the solution of the inverse problem representing the visible component of ρ. See, for example, [[Bert]], pp.22–26 for this problem.

1   Case with compact support

We assume that the support of ρ is contained in the sphere $r < a$. Since the fundamental solution of (1) is

$$\frac{e^{ik|\mathbf{r}-\mathbf{r}'|}}{4\pi|\mathbf{r}-\mathbf{r}'|},$$

we shall examine the integral representation of the solution u of Helmholtz's equation (1) satisfying the Sommerfeld radiation condition

$$\lim_{r\to\infty} r\left(\frac{\partial u}{\partial r} - iku\right) = 0,$$

$$u(\mathbf{r}') = \frac{1}{4\pi} \int_{r<a} \frac{e^{ik|\mathbf{r}'-\mathbf{r}|}}{|\mathbf{r}'-\mathbf{r}|} \rho(\mathbf{r}) d\mathbf{r} \tag{3}$$

for a source function ρ satisfying

$$\int_{r<a} |\rho(\mathbf{r})|^2 d\mathbf{r} < \infty. \tag{4}$$

In order to determine the characteristic property of the field u in $r > a$, we calculate the kernel form

$$K_{a,i}(\mathbf{r}',\mathbf{r}'') = \frac{1}{4\pi} \int_{r<a} \frac{e^{ik|\mathbf{r}'-\mathbf{r}|} e^{-ik|\mathbf{r}''-\mathbf{r}|}}{|\mathbf{r}'-\mathbf{r}||\mathbf{r}''-\mathbf{r}|} d\mathbf{r}$$

for $r', r'' > a$. In order to calculate $K_{a,i}(\mathbf{r}',\mathbf{r}'')$, we shall use the expansion, in terms of the spherical coordinates (r, θ, φ) and (r', θ', φ')

$$\frac{e^{ik|\mathbf{r}'-\mathbf{r}|}}{|\mathbf{r}'-\mathbf{r}|} = ik \sum_{n=0}^{\infty} (2n+1) \sum_{m=0}^{n} \frac{\varepsilon_m (n-m)!}{(n+m)!}$$
$$\times P_n^m(\cos\theta) P_n^m(\cos\theta') j_n(kr) h_n^{(1)}(kr') \cos m(\varphi - \varphi'),$$
$$\text{for} \quad r \le r'. \tag{5}$$

This series converges absolutely and uniformly on compact sets in $\{r < r'\}$. Here, $h_n^{(1)}$ is the spherical Hankel function of the first kind of order n. See, for example, [[MF]], p.887 and [[CK]], p.29. By using the two orthogonality relations (1.5) and (1.6) we have the expansion

$$K_{a,i}(\mathbf{r}',\mathbf{r}'') = k^2 \sum_{n=0}^{\infty} (2n+1) C_n^{(i)}(a) \sum_{m=0}^{n} \frac{\varepsilon_m (n-m)!}{(n+m)!}$$
$$\times P_n^m(\cos\theta') P_n^m(\cos\theta'') h_n^{(1)}(kr') \overline{h_n^{(1)}(kr'')}$$
$$\times \{\cos m\varphi' \cos m\varphi'' + \sin m\varphi' \sin m\varphi''\} \tag{6}$$

for

$$C_n^{(i)}(a) := \int_0^a j_n(kr)^2 r^2 dr < \infty.$$

This series converges absolutely and uniformly on compact sets of $\{r > a\}$. The kernel $K_{a,i}(\mathbf{r}',\mathbf{r}'')$ is a positive matrix on $r > a$ and so there exists a uniquely determined Hilbert space $H_{K_{a,i}}$ admitting the reproducing kernel $K_{a,i}(\mathbf{r}',\mathbf{r}'')$. Furthermore, the images $u(\mathbf{r}')$ of (3) belong just to the Hilbert space $H_{K_{a,i}}$. The expansion (6) implies that the images $u(\mathbf{r}')$ are expressible in the form

$$u(\mathbf{r}') = k^2 \sum_{n=0}^{\infty} (2n+1) C_n^{(i)}(a) \sum_{m=0}^{n} \frac{\varepsilon_m (n-m)!}{(n+m)!}$$
$$\times P_n^m(\cos\theta') h_n^{(1)}(kr') (A_n^m \cos m\varphi' + B_n^m \sin m\varphi') \tag{7}$$

230

for some constants $\{A_n^m, B_n^m\}$ satisfying

$$\sum_{n=0}^{\infty}(2n+1)C_n^{(i)}(a) \sum_{m=0}^{n} \frac{\varepsilon_m(n-m)!}{(n+m)!}\{|A_n^m|^2 + |B_n^m|^2\} < \infty. \tag{8}$$

Conversely, any $u(\mathbf{r'})$ defined by (7) with (8) converges absolutely and uniformly on compact sets in $\{r' > a\}$ and belongs to the Hilbert space $H_{K_{a,i}}$. Since the family

$$\{P_n^m(\cos\theta)\cos m\varphi, P_n^m(\cos\theta)\sin m\varphi\}_{m,n=0}^{\infty}$$

is complete in the Hilbert space consisting of the functions $f(\theta, \varphi)$ with the finite norm

$$\left\{\int_0^{\pi} \int_0^{2\pi} |f(\theta, \varphi)|^2 \sin\theta \, d\theta \, d\varphi\right\}^{\frac{1}{2}} < \infty,$$

we have the representation of the norm $\|u\|_{H_{K_{a,i}}}$ in the form

$$\|u\|_{H_{K_{a,i}}}^2 = k^2 \sum_{n=0}^{\infty}(2n+1)C_n^{(i)}(a) \sum_{m=0}^{n} \frac{\varepsilon_m(n-m)!}{(n+m)!}\{|A_n^m|^2 + |B_n^m|^2\}. \tag{9}$$

Furthermore, we have the isometrical identity

$$\|u\|_{H_{K_{a,i}}}^2 = \min \frac{1}{4\pi} \int_{r<a} |\rho(\mathbf{r})|^2 d\mathbf{r}$$

$$= \frac{1}{4\pi} \int_{r<a} |\rho^*(\mathbf{r})|^2 d\mathbf{r}. \tag{10}$$

Here, the minimum is taken over all $\rho(\mathbf{r})$ satisfying (3) and (4) for $r' > a$ and ρ^* is the uniquely determined function with the minimum norm.

In (7), by using the two orthogonality conditions (1.5) and (1.6), we have for any fixed $b > a$

$$A_n^m = \frac{1}{4\pi k^2 h_n^{(1)}(kb)C_n^{(i)}(a)} \int_0^{\pi} \int_0^{2\pi} u(b,\theta,\varphi)P_n^m(\cos\theta)\cos m\theta \sin\theta \, d\theta \, d\varphi \tag{11}$$

and

$$B_n^m = \frac{1}{4\pi k^2 h_n^{(1)}(kb)C_n^{(i)}(a)} \int_0^{\pi} \int_0^{2\pi} u(b,\theta,\varphi)P_n^m(\cos\theta)\sin m\theta \sin\theta \, d\theta \, d\varphi. \tag{12}$$

Here, note that $h_n^{(1)}(kb)$ does not vanish. The expressions with (7) imply that for any point $\mathbf{r'}$ $(r' > a)$, the waves $u(\mathbf{r'})$ are expressible in terms of

$$u(b,\theta,\varphi) \quad \text{for any fixed} \quad b(b > a). \tag{13}$$

231

We shall derive the inverse formula representing ρ^* in terms of (13). Using the reproducing property of $K_{a,i}(\mathbf{r}, \mathbf{r}')$ in $H_{K_{a,i}}$, we have

$$u(\mathbf{r}') = (u(\mathbf{r}), K_{a,i}(\mathbf{r}, \mathbf{r}'))_{H_{K_{a,i}}}$$

$$= \left(u(\mathbf{r}), \frac{1}{4\pi} \int_{r_1 < a} \frac{e^{ik|\mathbf{r}-\mathbf{r}_1|} e^{-ik|\mathbf{r}'-\mathbf{r}_1|}}{|\mathbf{r} - \mathbf{r}_1||\mathbf{r}' - \mathbf{r}_1|} d\mathbf{r}_1 \right)_{H_{K_{a,i}}}$$

$$= \frac{1}{4\pi} \int_{r_1 < a} \frac{e^{ik|\mathbf{r}'-\mathbf{r}_1|}}{|\mathbf{r}' - \mathbf{r}_1|} d\mathbf{r}_1 \left(u(\mathbf{r}), \frac{e^{ik|\mathbf{r}-\mathbf{r}_1|}}{|\mathbf{r} - \mathbf{r}_1|} \right)_{H_{K_{a,i}}}$$

$$= \frac{1}{4\pi} \int_{r_1 < a} \frac{e^{ik|\mathbf{r}'-\mathbf{r}_1|} \rho^*(\mathbf{r}_1)}{|\mathbf{r}' - \mathbf{r}_1|} d\mathbf{r}_1$$

and so, we have

$$\rho^*(\mathbf{r}_1) = \left(u(\mathbf{r}), \frac{e^{ik|\mathbf{r}-\mathbf{r}_1|}}{|\mathbf{r} - \mathbf{r}_1|} \right)_{H_{K_{a,i}}}. \tag{14}$$

Here, note that for any fixed $\mathbf{r}_1$ ($r_1 < a$)

$$\frac{e^{ik|\mathbf{r}-\mathbf{r}_1|}}{|\mathbf{r} - \mathbf{r}_1|} \tag{15}$$

belongs to the Hilbert space $H_{K_{a,i}}$.

Indeed, from (7) and (5) the corresponding coefficients $\widetilde{A_n^m}$ and $\widetilde{B_n^m}$ of (15) in the representation (7) are

$$\widetilde{A_n^m} = \frac{P_n^m(\cos\theta_1) j_n(kr_1) \cos m\varphi_1}{-ikC_n^{(i)}(a)}$$

and

$$\widetilde{B_n^m} = \frac{P_n^m(\cos\theta_1) j_n(kr_1) \sin m\varphi_1}{-ikC_n^{(i)}(a)}.$$

Then,

$$k^2 \sum_{n=0}^{\infty}(2n+1)C_n^{(i)}(a) \sum_{m=0}^{n} \frac{\varepsilon_m (n-m)!}{(n+m)!}\{|\widetilde{A_n^m}|^2 + |\widetilde{B_n^m}|^2\}$$

$$= \sum_{n=0}^{\infty}(2n+1)C_n^{(i)}(a) \sum_{m=0}^{n} \frac{\varepsilon_m (n-m)!}{(n+m)!} \frac{P_n^m(\cos\theta_1)^2 j_n(kr_1)^2}{C_n^{(i)}(a)^2}$$

$$= \sum_{n=0}^{\infty}(2n+1)\frac{j_n(kr_1)^2}{C_n^{(i)}(a)} P_n(1)$$

$$= \sum_{n=0}^{\infty}(2n+1)\frac{j_n(kr_1)^2}{C_n^{(i)}(a)}. \tag{16}$$

See, for example, [[MF]], p.1274. As we see from the series representation of the spherical Bessel functions j_n, we have

$$j_n(t) = \frac{t^n}{1 \cdot 3 \cdots (2n+1)}(1 + O(\frac{1}{n})), n \to \infty$$

uniformly on compact subsets of $\mathbb{R}$. Hence, we have

$$\lim_{n \to \infty} \frac{\frac{(2n+3)j_{n+1}(kr_1)^2}{C_{n+1}^{(i)}(a)}}{\frac{(2n+1)j_n(kr_1)^2}{C_n^{(i)}(a)}} = \left(\frac{r_1}{a}\right)^2.$$

Hence, for $r_1 < a$ the series (16) converges and so (15) belongs to the Hilbert space $H_{K_{a,i}}$.

Therefore, the arguments in (14) will be justified. We thus obtain

Theorem 1 *The source function ρ^* in the sense of (10) in (1) is expressible in terms of (13) in the form*

$$\rho^*(\mathbf{r}_1) = \frac{-i}{4\pi k} \sum_{n=0}^{\infty} \frac{(2n+1)j_n(kr_1)}{C_n^{(i)}(a)h_n^{(1)}(kb)} \sum_{m=0}^{n} \frac{\varepsilon_m(n-m)!}{(n+m)!}$$

$$\times P_n^m(\cos\theta_1) \int_0^\pi \int_0^{2\pi} u(b,\theta,\varphi) P_n^m(\cos\theta) \cos m(\varphi - \varphi_1) \sin\theta \, d\theta \, d\varphi.$$

2   Case with noncompact support

Next, we shall consider the case where the support of ρ is contained in the outside of the sphere $r < a$. In this case, in order to obtain the existence of the corresponding reproducing kernel, we assume that

$$\int_{r>a} |\rho(\mathbf{r})|^2 r^2 d\mathbf{r} < \infty. \tag{17}$$

In order to examine the integral representation

$$u(\mathbf{r}') = \frac{1}{4\pi} \int_{r>a} \frac{e^{ik|\mathbf{r}'-\mathbf{r}|}}{|\mathbf{r}'-\mathbf{r}|} \rho(\mathbf{r}) d\mathbf{r} \tag{18}$$

of u, we shall consider the integral transform in the form

$$u(\mathbf{r}') = \frac{1}{4\pi} \int_{r>a} \frac{e^{ik|\mathbf{r}'-\mathbf{r}|}}{|\mathbf{r}'-\mathbf{r}|} \rho_1(\mathbf{r}) \frac{1}{r^2} d\mathbf{r} \tag{19}$$

233

for a function $\rho_1 = r^2 \rho$ satisfying

$$\int_{r>a} |\rho_1(\mathbf{r})|^2 \frac{d\mathbf{r}}{r^2} = \int_{r>a} |\rho(\mathbf{r})|^2 r^2 d\mathbf{r} < \infty. \tag{20}$$

We form the reproducing kernel

$$K_{a,o}(\mathbf{r}', \mathbf{r}'') = \frac{1}{4\pi} \int_{r>a} \frac{e^{ik|\mathbf{r}'-\mathbf{r}|} e^{-ik|\mathbf{r}''-\mathbf{r}|}}{|\mathbf{r}'-\mathbf{r}||\mathbf{r}''-\mathbf{r}|} \frac{d\mathbf{r}}{r^2}. \tag{21}$$

By using (5) and the orthogonality relations of (1.5) and (1.6), we have

$$K_{a,o}(\mathbf{r}', \mathbf{r}'') = k^2 \sum_{n=0}^{\infty} (2n+1) C_n^{(0)}(a) \sum_{m=0}^{n} \frac{\varepsilon_m (n-m)!}{(n+m)!}$$

$$\times P_n^m(\cos\theta') P_n^m(\cos\theta'') \overline{h_n^{(1)}(kr') h_n^{(1)}(kr'')}$$
$$\times \{\cos m\varphi' \cos m\varphi'' + \sin m\varphi' \sin m\varphi''\}, \tag{22}$$

where we note that

$$C_n^{(0)}(a) := \int_a^{\infty} j_n(kr)^2 dr \leq \frac{\pi}{2k} \int_0^{\infty} J_{n+\frac{1}{2}}(kr)^2 \frac{1}{r} dr < \infty,$$

(see, for example, [[WG]], pp.377 and 405 and [[W]], p.399). This series (22) converges absolutely and uniformly on compact subsets on $r > a$. Hence, by arguments parallel to those of Theorem 1, we have

Theorem 2 *For the source function ρ^* satisfying (18) and (17), we have the inverse formula, for any fixed $\hat{b}$ $(0 < \hat{b} < a)$,*

$$\rho^*(\mathbf{r}_1) = \frac{1}{r_1^2} \left(u(\mathbf{r}), \frac{e^{ik|\mathbf{r}-\mathbf{r}_1|}}{|\mathbf{r}-\mathbf{r}_1|} \right)_{H_{K_{a,i}}}$$

$$= \frac{-i}{4\pi kr^2} \sum_{n=0}^{\infty} \frac{(2n+1) j_n(kr_1)}{C_n^{(0)}(a) h_n^{(1)}(k\hat{b})} \sum_{m=0}^{n} \frac{\varepsilon_m (n-m)!}{(n+m)!}$$

$$\times P_n^m(\cos\theta_1) \int_0^{\pi} \int_0^{2\pi} u(\hat{b}, \theta, \varphi) P_n^m(\cos\theta) \cos m(\varphi - \varphi_1) \sin\theta \, d\theta \, d\varphi.$$

3   Use of radiation patterns

The behaviour at infinity of the solution $u(\mathbf{r})$ of (1) is

$$u(\mathbf{r}) = g(\frac{\mathbf{r}}{r}) \frac{e^{ikr}}{4\pi r} [1 + O(r^{-1})],$$

where the radiation pattern g is given by (2) (see, for example, [[CK]], p.20). We shall give the solution of the inverse problem representing the source function ρ in terms of the radiation pattern g.

In order to examine the integral transform (2), we form the reproducing kernel

$$K(\frac{\mathbf{r}'}{r'}, \frac{\mathbf{r}''}{r''}) = \frac{1}{4\pi} \int_{r<a} e^{-ik(\frac{\mathbf{r}'}{r'},\mathbf{r})} e^{ik(\frac{\mathbf{r}''}{r''},\mathbf{r})} d\mathbf{r}. \tag{23}$$

Recall here the Jacobi–Anger expansion

$$e^{-ik(\frac{\mathbf{r}'}{r'},\mathbf{r})} = 4\pi \sum_{n=0}^{\infty} \sum_{m=-n}^{n} \overline{i^m j_n(kr) Y_n^m(\frac{\mathbf{r}}{r})} Y_n^m(\frac{\mathbf{r}'}{r'}), \tag{24}$$

which converges uniformly on compact subsets of $\mathbb{R}^3$ (see, for example, [[CK]], p.31). By using the expansion (24) and the orthonormal property of the spherical harmonic functions

$$Y_n^m(\frac{\mathbf{r}}{r}) = P_n^m(\cos\theta) e^{im\varphi} = Y_n^m(\theta, \varphi)$$

in the L_2 space on the unit sphere (see, for example, [[CK]], p.24), we have

$$K(\frac{\mathbf{r}'}{r'}, \frac{\mathbf{r}''}{r''}) = 4\pi \sum_{n=0}^{\infty} C_n^{(i)}(a) \sum_{m=-n}^{n} Y_n^m(\frac{\mathbf{r}'}{r'}) \overline{Y_n^m(\frac{\mathbf{r}''}{r''})}. \tag{25}$$

Hence, by arguments parallel to those of Theorem 1, we see that for the Hilbert space H_K admitting the reproducing kernel (23), the norm $\|g\|_{H_K}$ in H_K of the images g of (2) is given by

$$\|g\|_{H_K}^2 = \frac{1}{4\pi} \sum_{n=0}^{\infty} \frac{1}{C_n^{(i)}(a)} \sum_{m=-n}^{n} \left| \int_0^{\pi} \int_0^{2\pi} g(\theta, \varphi) \overline{Y_n^m(\theta, \varphi)} \sin\theta d\theta d\varphi \right|^2. \tag{26}$$

Furthermore, we have the isometrical identity

$$\|g\|_{H_K}^2 = \min \frac{1}{4\pi} \int_{r<a} |\rho(\mathbf{r})|^2 d\mathbf{r} = \frac{1}{4\pi} \int_{r<a} |\rho^*(\mathbf{r})|^2 d\mathbf{r}. \tag{27}$$

Here, ρ^* is the function with the minimum norm satisfying (2). Of course, ρ can be decomposed in the form

$$\rho = \rho_o + \rho^* \tag{28}$$

where ρ_o belongs to the null space N (the invisible component) for the integral transform (2) and ρ^* belongs to $N^{\perp}$, the orthocomplement (the visible component). For the construction of the null space N, see, for example, [[Bert]], pp.23–26. By arguments parallel to those of Theorem 1, we obtain

Theorem 3 *The source function ρ^* in the sense of (27) or (28) is expressible in the form*

$$\rho^*(\mathbf{r}) = \left(g(\frac{\mathbf{r}'}{r'}), e^{-ik(\frac{\mathbf{r}'}{r'},\mathbf{r})} \right)_{H_K}$$

$$= \sum_{n=0}^{\infty} \frac{i^n j_n(kr)}{C_n^{(i)}(a)} \left(\int_0^\pi \int_0^{2\pi} g(\theta, \varphi) \overline{Y_n^m(\theta, \varphi)} \sin\theta \, d\theta \, d\varphi \right) Y_n^m(\frac{\mathbf{r}}{r}).$$

Appendix 1
Applications to representations of inverse functions

We shall consider an arbitrary mapping $p = \phi(\hat{p})$

$$\phi : \hat{E} \longrightarrow E \qquad (1)$$

from an abstract set $\hat{E}$ into an abstract set E. Then, we shall consider the formal problem of representing the inverse function ϕ^{-1} in terms of ϕ. Of course, the inverse is, in general, multivalued. By using Theorem 2.1.2 and Theorem 2.3.7, we shall first treat this general problem, and then we shall establish a general principle to solve this problem in some general and reasonable settings. We shall also give several typical concrete examples.

§1   A general approach

Let $K(p, q)$ be a positive matrix on E and H_K a uniquely determined functional Hilbert space consisting of functions on E and admitting the reproducing kernel $K(p, q)$. We shall assume that $K(p, q)$ is expressible in the form

$$K(p, q) = (\mathbf{h}(q), \mathbf{h}(p))_{\mathcal{H}} \quad \text{on} \quad E \times E \qquad (2)$$

in terms of a Hilbert space $\mathcal{H}$-valued function $\mathbf{h}(p)$ on E. We further assume that

$$\{\mathbf{h}(p); p \in E\} \quad \text{is complete in} \quad \mathcal{H}. \qquad (3)$$

Then, for the linear transform of $\mathcal{H}$

$$f(p) = (\mathbf{f}, \mathbf{h}(p))_{\mathcal{H}}, \quad \mathbf{f} \in \mathcal{H}, \qquad (4)$$

the images $f(p)$ form precisely the Hilbert space H_K; furthermore, we have the isometrical mapping

$$\|f\|_{H_K} = \|\mathbf{f}\|_{\mathcal{H}}. \qquad (5)$$

If the assumption (3) is not valid, then we have in (5), in general, the inequality

$$\|f\|_{H_K} \leqq \|\mathbf{f}\|_{\mathcal{H}}, \qquad (6)$$

by Theorem 2.1.2. Now, by using the mapping ϕ in (1) we shall define the function

$$K_\phi(\hat{p}, \hat{q}) = K(\phi(\hat{p}), \phi(\hat{q})) \qquad (7)$$
$$= (\mathbf{h}(\phi(\hat{q})), \mathbf{h}(\phi(\hat{p})))_{\mathcal{H}} \quad \text{on} \quad \hat{E} \times \hat{E}$$

237

and the linear mapping of $\mathcal{H}$

$$f(\phi(\hat{p})) = (\mathbf{f}, \mathbf{h}(\phi(\hat{p})))_{\mathcal{H}}, \quad \mathbf{f} \in \mathcal{H}. \tag{8}$$

Of course, $K_\phi(\hat{p}, \hat{q})$ is a positive matrix on $\hat{E}$, and so there exists a uniquely determined functional Hilbert space H_{K_ϕ} admitting the reproducing kernel $K_\phi(\hat{p}, \hat{q})$. Then, we obtain, from (5) and (6)

Theorem 2.3.6 *For an arbitrary mapping ϕ in (1), the functions $f(\phi(\hat{p}))$ in (8), form the reproducing kernel Hilbert space H_{K_ϕ} and we have the inequality*

$$\|f(\phi)\|_{H_{K_\phi}} \leq \|f\|_{H_K}.$$

Isometry holds here if and only if

$$\{\mathbf{h}(p); p \in Range \ \phi\} \quad is \ complete \ in \quad \mathcal{H}.$$

Furthermore, we have

Theorem 2.3.7 *In an arbitrary mapping ϕ in (1), for any $\hat{f} \in H_{K_\phi}$ we take the function $f^* \in H_K$ satisfying*

$$\hat{f} = f^*(\phi) \quad and \quad \|\hat{f}\|_{H_{K_\phi}} = \|f^*\|_{H_K}. \tag{9}$$

Then, we have

$$f^*(p) = (\hat{f}(\cdot), K(\phi(\cdot), p))_{H_{K_\phi}}. \tag{10}$$

Now, in Theorem 2.3.7, we shall assume that the mapping ϕ is onto E, and so an isometry between H_K and H_{K_ϕ}.

If we know the mapping $p = \phi(\hat{p})$ from $\hat{E}$ to E, then the isometrical mapping from H_K onto H_{K_ϕ} is given by

$$f(p) \in H_K \longrightarrow f(\phi(\hat{p})) \in H_{K_\phi}.$$

If we know the (in general, multivalued) inverse $\hat{p} = \phi^{-1}(p)$ of $p = \phi(\hat{p})$, then for $\hat{f} \in H_{K_\phi}$, the function $\hat{f}(\phi^{-1}(p))$ is a single-valued function on E and the isometrical mapping from H_{K_ϕ} onto H_K is given by

$$\hat{f} \in H_{K_\phi} \longrightarrow \hat{f}(\phi^{-1}(p)) \in H_K.$$

In general, Theorem 2.3.7 establishes the isometrical mapping

$$\hat{f} \in H_{K_\phi} \longrightarrow f^* \in H_K, \tag{11}$$

explicitly in terms of the reproducing kernel $K(p, q)$ on E, the mapping ϕ and the reproducing kernel Hilbert space H_{K_ϕ}. This fact will mean that Theorem 2.3.7 gives, in a sense, a method to construct the inverse ϕ^{-1}.

Indeed, in our general setting (1), for any point $p \in E$ and for any fixed function $\hat{f} \in H_{K_\phi}$ we first construct the function f^* by (10). Then, we have the inclusion relationship

$$\phi^{-1}(p) \subset \left\{ \hat{p} \in \hat{E}; \hat{f}(\hat{p}) = f^*(p) \right\}. \tag{12}$$

Hence, we will be able to look for all the inverses $\phi^{-1}(p)$ in the point set in the right hand side in (12), by using a suitable Hilbert space H_K and a suitable function $\hat{f} \in H_{K_\phi}$.

§2   Reasonable settings

We shall analyse the principle in Section 1 to represent the inverse ϕ^{-1} in (1) in terms of ϕ.

In order to use the formula (10) and (12) we need

(I)   a concrete structure for the Hilbert space H_{K_ϕ}, admitting its reproducing kernel $K_\phi(\hat{p}, \hat{q})$ defined by (7)

and

(II)   for a function $\hat{f}$ belonging to H_{K_ϕ}, its inverse $\hat{f}^{-1}$.

Then we have the inverse function of ϕ

$$\hat{p} = \hat{f}^{-1}(f^*(p)), \tag{13}$$

in (12).

For (I), if the Hilbert space H_K admitting the reproducing kernel $K(p, q)$ on E is concretely given, then the transformed Hilbert space H_{K_ϕ} will be constructed by a unified method stated in Theorem 2.3.6 and Theorem 2.3.7.

If the inner product in H_K is given by an integral form, then the inner product in H_{K_ϕ} will be given by some integral form induced by the mapping ϕ, as we shall see in examples.

For (II), if the identity mapping belongs to H_{K_ϕ}, then our situation will become, of course, extremely simple.

§3   Examples

Following our general principle, we shall give typical examples. These examples will show that concrete reproducing kernels have great value also from the viewpoint of representations of inverse functions.

1   The Riemann mapping function

Let Δ be the unit disc $\{|z| < 1\}$ and D an arbitrary bounded (for simplicity) domain on the Z-plane ($Z = X + iY$). Let $z = \varphi(Z)$ be a Riemann mapping function from D onto Δ which is analytic and univalent on D. Let $K_\Delta(z, \overline{u})$ be the Bergman kernel

$$K_\Delta(z, \overline{u}) = \frac{1}{\pi(1 - \overline{u}z)^2} \tag{14}$$

on Δ for the Bergman space H_{K_Δ} consisting of all analytic functions $f(z)$ on Δ with finite norms

$$\left\{ \iint_\Delta |f(z)|^2 dx dy \right\}^{\frac{1}{2}} < \infty \quad (z = x + iy). \tag{15}$$

Then, we can see directly that

$$K_{\Delta,\varphi}(\varphi(Z), \overline{\varphi(U)}) = \frac{1}{\pi(1 - \overline{\varphi(U)}\varphi(Z))^2} \tag{16}$$

is the reproducing kernel for the Hilbert space $H_{K_{\Delta,\varphi}}$ consisting of all analytic functions $\hat{f}(Z) = f(\varphi(Z))$ ($f \in H_{K_\Delta}$) with finite norms

$$\left\{ \iint_D |\hat{f}(Z)|^2 |\varphi'(Z)|^2 dX dY \right\}^{\frac{1}{2}} < \infty. \tag{17}$$

Hence, we have the inverse formula by (10) and by using the identity for $\hat{f}$ in (12)

$$\varphi^{-1}(z) = \frac{1}{\pi} \iint_D \frac{Z|\varphi'(Z)|^2}{(1 - z\overline{\varphi(Z)})^2} dX dY. \tag{18}$$

Note that when $Z = \varphi^{-1}(z)$ is not one to one, if we consider D as a Riemann surface $\varphi^{-1}(\Delta)$ spread over $\mathbb{C}$ counting its multiplicity, then the formula (18) is still valid. By this method, we can, in general, overcome the multivaluedness of the inverse functions.

When D is a bounded domain whose boundary ∂D is an analytic Jordan curve, we recall the Szegö reproducing kernel

$$\frac{1}{2\pi(1 - \overline{u}z)} \tag{19}$$

240

for the Hilbert space consisting of all analytic functions $f(z)$ on Δ with harmonic majorants $u(z)$ such that

$$|f(z)|^2 < u(z) \quad \text{on} \quad \Delta$$

and with finite norms

$$\left\{ \int_{\partial\Delta} |f(z)|^2 |dz| \right\}^{\frac{1}{2}} < \infty. \tag{20}$$

Here, $f(z)$ means the Fatou nontangential boundary values on $\partial\Delta$. Then, we obtain similarly the simple inverse formula

$$\varphi^{-1}(z) = \frac{1}{2\pi} \int_{\partial D} \frac{Z|\varphi'(Z)|}{1 - z\overline{\varphi(Z)}} |dZ|. \tag{21}$$

Note that the Riemann mapping function $\varphi(Z)$ satisfying $\varphi(Z_0) = 0 \quad (Z_0 \in D)$ is expressible in the form

$$\varphi(Z) = \sqrt{\frac{\pi}{K_D(Z_0, \overline{Z_0})}} \int_{Z_0}^{Z} K_D(\zeta, \overline{Z_0}) d\zeta$$

in terms of the Bergman kernel $K_D(\zeta, \overline{Z})$ on the domain D (see, for example, [[Ber]]).

2 Harmonic mappings

Note that on Δ

$$\frac{1 - |\overline{u}|^2 |z|^2 |2 - \overline{u}z|^2}{|1 - \overline{u}z|^4} \tag{22}$$

and

$$\frac{1 - |\overline{u}|^2 |z|^2}{|1 - \overline{u}z|^2} \tag{23}$$

are the reproducing kernels for the Hilbert spaces consisting of all harmonic functions $u(z)$ on Δ with finite norms

$$\left\{ \frac{1}{\pi} \iint_{\Delta} |u(z)|^2 \, dx dy \right\}^{\frac{1}{2}} < \infty \tag{24}$$

and

$$\left\{ \frac{1}{2\pi} \int_{\partial\Delta} |u(z)|^2 |dz| \right\}^{\frac{1}{2}} < \infty, \tag{25}$$

respectively. For the latter case we need the assumption that $|u(z)|^2$ has a harmonic majorant on Δ and $u(z)$ $(z \in \partial\Delta)$ means the Fatou nontangential boundary values as in the Szegö case. As in the Riemann mapping function we can obtain the representations of the inverse functions for harmonic mappings of D onto Δ.

3   Increasing functions

Note first that

$$\min(x, y) \quad (x, y > 0) \tag{26}$$

is the reproducing kernel for the Hilbert space $H(0, a)$ $(0 < a \leqq \infty)$ consisting of all real-valued functions $f(x)$ on $[0, a)$ such that $f(x)$ are absolutely continuous on $[0, a)$, $f(0) = 0$ and with finite norms

$$\left\{ \int_0^a f'(x)^2 dx \right\}^{\frac{1}{2}} < \infty. \tag{27}$$

Similarly, for an increasing function $x = \varphi(\hat{x})$ from $[0, b)$ $(0 < b \leqq \infty)$ onto $[0, a)$ of C^1-class satisfying

$$\varphi'(\hat{x}) > 0 \quad \text{on} \quad (0, b),$$

the function

$$\min(\varphi(\hat{x}), \varphi(\hat{y})) \quad \text{on} \quad [0, b) \times [0, b) \tag{28}$$

is the reproducing kernel for the Hilbert space H_φ consisting of all functions $\hat{f}(\hat{x}) = f(\varphi(\hat{x}))$ $(f \in H(0, a))$, such that $\hat{f}(\hat{x})$ are absolutely continuous on $[0, b)$, $\hat{f}(0) = 0$ and with finite norms

$$\left\{ \int_0^b \hat{f}'(\xi)^2 \frac{d\xi}{\varphi'(\xi)} \right\}^{\frac{1}{2}} < \infty, \tag{29}$$

as stated in Chapter 2, Section 4.3. Hence, by (10) and by the identity as $\hat{f}$ in (12) we have the inverse formula

$$
\begin{aligned}
\varphi^{-1}(x) &= \int_0^b \{\min(\varphi(\xi), x)\}' \frac{d\xi}{\varphi'(\xi)} \\
&= \int_0^b \left(\frac{2}{\pi} \int_0^\infty \frac{\sin(\varphi(\xi)t)\sin xt}{t^2} dt \right)' \frac{d\xi}{\varphi'(\xi)} \\
&= \frac{2}{\pi} \int_0^b \int_0^\infty \frac{\cos(\varphi(\xi)t)\sin xt}{t} dt\, d\xi.
\end{aligned} \tag{30}
$$

In particular, we have

$$\sqrt[n]{x} = \frac{2}{\pi} \int_0^\infty \int_0^\infty \frac{\cos(\xi^n t)\sin xt}{t} dt\, d\xi. \tag{31}$$

242

Appendix 2
Natural norm inequalities in nonlinear transforms

We shall see the existence of natural norm inequalities in some general nonlinear transforms of reproducing kernel Hilbert spaces and as its applications we shall derive typical concrete norm inequalities in the nonlinear transforms. Concrete norm inequalities will again show the importance of concrete reproducing kernels.

§1   General principles

We shall consider a RKHS $H_K(E)$ as an input function space of the following nonlinear transform

$$\varphi: \quad f \in H_K(E) \longrightarrow \sum_{n=0}^{\infty} d_n(p)f(p)^n, \tag{1}$$

where $\{d_n(p)\}$ are any complex-valued functions on E.

In this nonlinear transform φ, we shall see that the images $\varphi(f)$, $f \in H_K(E)$, belong to a Hilbert space $\mathbf{H}$ which is naturally determined by the nonlinear transform φ and there exits a natural norm inequality between the two norms $\|\varphi(f)\|_{\mathbf{H}}$ and $\|f\|_{H_K}$, based on [Sa30].

For n-times sums and n-times products, we have, in general, for any $f_j \in H_{K_j}$ $(j = 1, 2, ..., N)$

$$\left\| \sum_{j=1}^{N} f_j \right\|^2_{H_{(\sum_{j=1}^{N} K_j)}} \leq \sum_{j=1}^{N} \|f_j\|^2_{H_{K_j}} \tag{2}$$

and

$$\|f^n\|^2_{H_{K^n}} \leq \|f\|^{2n}_{H_K}, \tag{3}$$

in Theorem 2.3.2 and Theorem 2.3.3, respectively. Recall also Corollary 2.3.3. Then, we obtain, in general

Theorem A   *If*

$$\sum_{n=0}^{\infty} |d_n(p)|^2 K(p,p)^n < \infty \quad on \quad E$$

and if, for $f \in H_K$

$$\sum_{n=0}^{\infty} (\|f\|_{H_K})^{2n} < \infty,$$

243

then, for the nonlinear transform $\varphi(f)$ in (1),

$$\varphi(f) = \sum_{n=0}^{\infty} d_n(p) f(p)^n$$

converges absolutely on E, and

$$\varphi(f) \in H_{\mathbf{K_d}} \tag{4}$$

and

$$\|\varphi(f)\|_{H_{\mathbf{K_d}}}^2 \leq \sum_{n=0}^{\infty} (\|f\|_{H_K})^{2n}, \tag{5}$$

where $H_{\mathbf{K_d}}$ is the RKHS admitting the reproducing kernel

$$\mathbf{K_d}(p,q) = \sum_{n=0}^{\infty} d_n(p)\overline{d_n(q)} K(p,q)^n \quad on \quad E,$$

which converges absolutely on $E \times E$.

In particular, $\{d_n\}$ are constants, and if

$$\sum_{n=0}^{\infty} K(p,p)^n < \infty \quad on \quad E$$

and if, for $f \in H_K$

$$\sum_{n=0}^{\infty} |d_n|^2 (\|f\|_{H_K})^{2n} < \infty,$$

then

$$\varphi(f) = \sum_{n=0}^{\infty} d_n f(p)^n$$

converges absolutely on E, and

$$\varphi(f) \in H_{\mathbf{K}}$$

and

$$\|\varphi(f)\|_{H_{\mathbf{K}}}^2 \leq \sum_{n=0}^{\infty} |d_n|^2 (\|f\|_{H_K})^{2n}, \tag{6}$$

where $H_{\mathbf{K}}$ is the RKHS admitting the reproducing kernel

$$\mathbf{K}(p,q) = \sum_{n=0}^{\infty} K(p,q)^n \quad on \quad E.$$

In Theorem A, the concrete realization of the norm in the RKHS H_{K_d} is, in general, involved. See, [[He]] and [Sa1] for a profound result for $\hat{K}(z, \bar{u})^2$ in the case of the classical Szegö reproducing kernel $\hat{K}(z, \bar{u})$ and for a prototype result of Theorem A, respectively.

At this moment, recall Corollary 2.3.1 that if

$$\mathbf{K_d}(p, q) \ll \tilde{K}(p, q) \quad \text{on} \quad E; \tag{7}$$

that is,

$$\tilde{K}(p, q) - \mathbf{K_d}(p, q)$$

is a positive matrix on E, then we have

$$H_{\mathbf{K_d}} \subset H_{\tilde{K}}$$

(as classes of functions) and

$$\|f\|_{H_{\tilde{K}}} \leq \|f\|_{H_{\mathbf{K_d}}} \quad \text{for all} \quad f \in H_{\mathbf{K_d}}.$$

Hence, for some suitable reproducing kernel $\tilde{K}(p, q)$ satisfying (7) whose norm can be determined, in a reasonable way, we can obtain the inequality

$$\|\varphi(f)\|_{H_{\tilde{K}}}^2 \leq \sum_{n=0}^{\infty} (\|f\|_{H_K})^{2n},$$

in Theorem A.

When all the coefficients $\{d_n\}$ are constants, as a typical reproducing kernel $\tilde{K}(p, q)$ satisfying (7) and a large reproducing kernel for $K(p, q)$ in the sense (7), we can consider the exponential of $K(p, q)$

$$\exp K(p, q) = 1 + K(p, q) + \frac{K(p, q)^2}{2!} + \cdots$$

which is a positive matrix on E. Note here that the constants $(n!)^{-1}$ are not essential in our arguments.

Indeed, we have, in general, the isometrical relation, for any positive constant α,

$$\|f\|_{H_{\alpha K}}^2 = \frac{1}{\alpha} \|f\|_{H_K}^2,$$

245

for the reproducing kernels $\alpha K(p,q)$ and $K(p,q)$.

Then, from the expression

$$\varphi(f) = \sum_{n=0}^{\infty} (d_n n!) \frac{f(p)^n}{n!},$$

we have, in particular

Corollary 1   *If $\{d_n(p)\}$ are all constants, we have, in Theorem A*

$$\|\varphi(f)\|^2_{H_{\exp K}} \leq \sum_{n=0}^{\infty} |d_n|^2 n! (\|f\|_{H_K})^{2n},$$

if the right hand side converges.
In particular, we have

$$\|f^n\|^2_{H_{\exp K}} \leq n! \|f\|^{2n}_{H_K} \quad (n \geq 1).$$

Corollary 2   *Let*

$$N(z) = \sum_{n=0}^{\infty} a_n z^n$$

be analytic around $z = 0$ which converges on the disc $\{|z| < R\}$. We define the analytic function $N^+(z)$ on $\{|z| < R\}$ by

$$N^+(z) = \sum_{n=0}^{\infty} |a_n| z^n.$$

We assume that for a reproducing kernel $K(p,q)$ on E,

$$K(p,p) < R.$$

Then, $N^+(K(p,q))$ converges absolutely on $E \times E$ and is a positive matrix on E. For the RKHS $H_{N^+(K)}$ admitting the reproducing kernel $N^+(K(p,q))$ and for a function f in H_K satisfying

$$N^+(\|f\|^2_{H_K}) < \infty,$$

we have the norm inequality

$$\|N(f)\|^2_{H_{N^+(K)}} \leq N^+(\|f\|^2_{H_K}).$$

In Corollary 2, for example, we have

$$\| \sin f \|_{H_{\exp K}}^2 \leq \sin^+ (\| f \|_{H_K}^2)$$

and

$$\| \cos f \|_{H_{\exp K}}^2 \leq \cos^+ (\| f \|_{H_K}^2),$$

if the right hand sides converge, respectively.

In the theory of nonlinear partial differential equations, we meet nonlinear transforms, for example, for $u(x,t)$

$$u \longrightarrow u_t + 6uu_x + u_{xxx}$$

and

$$u \longrightarrow u_{tt} - u_{xx} + m^2 \sin u \quad (m > 0; \text{constant}).$$

For such nonlinear transforms we shall show that similar results are valid as in Theorem A.

In order simply to state the result, we shall assume that E is an open interval on $\mathbf{R}$. Then, for the smoothness of a RKHS $H_K(E)$, note that if

$$\frac{\partial^{(j+j')} K(x,y)}{\partial x^j \partial y^{j'}} \quad (j, j' \leq n)$$

are continuously differentiable on $E \times E$, then for any member f of $H_K(E)$, $f^{(j)}(j \leq n)$ are also continuously differentiable on E (Theorem 2.2.9), and we have

$$f^{(n)} \in H_{K^{n,n}}$$

and

$$\| f^{(n)} \|_{K^{n,n}} \leq \| f \|_{H_K},$$

for the RKHS $H_{K^{n,n}}$ admitting the reproducing kernel

$$K^{n,n}(x,y) = \frac{\partial^{2n} K(x,y)}{\partial x^n \partial y^n} \quad \text{on} \quad E$$

by Theorem 2.3.4. Hence, for example, in the nonlinear transform

$$\psi : f \in H_K(E) \longrightarrow h_1(x) f''(x) + h_2(x) f'(x)^2 + h_3(x) |f(x)|^2 \tag{8}$$

for any complex-valued functions $\{h_j(x)\}$ on E, the images $\psi(f)$ belong to the RKHS $H_{\psi^+(K)}$ admitting the reproducing kernel

$$\psi^+(K(x,y)) = h_1(x)\overline{h_1(y)} K^{2,2}(x,y) + h_2(x)\overline{h_2(y)} K^{1,1}(x,y)^2$$
$$+ h_3(x)\overline{h_3(y)} K(x,y)\overline{K(x,y)}, \tag{9}$$

247

and we obtain the inequality

$$\|\psi(f)\|^2_{H_{\psi+(K)}} \leq \|f\|^2_{H_K} (1 + 2\|f\|^2_{H_K}). \tag{10}$$

In some general linear transform of Hilbert spaces we could get essentially isometrical identities between the input and the output function spaces, but in our nonlinear transforms we get norm inequalities, essentially, and to determine the cases making the equalities hold in the inequalities is, in general, involved, in even the case of a finite dimensional RKHS H_K and we need case by case arguments to determine the cases. See, for example, [Sa1–3, 15, 18, 24, 27]. However, for many cases (not always), for the reproducing kernels $f(p) = K(p,q)$ $(q \in E)$ equalities hold in our inequalities. See [Sa5], for example.

§2   Concrete examples

Subsequently we derive typical norm inequalities in nonlinear transforms by applying Theorem A to several typical reproducing kernels.

1 ℓ^2 space

The Kronecker delta $\delta_{jj'}$ $(j, j' \geq 1;$ integers) is the reproducing kernel for the usual ℓ^2 space and, for

$$\mathbf{c} = \{c_j\}_{j=1}^{\infty} \in \ell^2$$

and

$$\delta_j = (0, 0, ..., 0, 1, 0, ...),$$

$$(\mathbf{c}, \delta_j)_{\ell^2} = c_j.$$

We shall consider the nonlinear transform of ℓ^2

$$\varphi(\mathbf{c}) = \sum_{n=0}^{\infty} d_n \mathbf{c}^n \quad (d_n : \text{constants}) \tag{11}$$

in the sense of Hadamard's product

$$\mathbf{c}^n = \{c_j^n\}_{j=1}^{\infty}.$$

We consider the positive matrix

$$\Delta_{jj'} = \sum_{n=0}^{\infty} p_n (\delta_{jj'})^n \quad (p_n > 0; \text{constants}),$$

248

with

$$p = \sum_{n=1}^{\infty} p_n < \infty.$$

Then, by Theorem A we obtain the inequality

$$\frac{1}{p_0}|d_0|^2 + \frac{1}{p}\sum_{j=1}^{\infty}|\sum_{n=1}^{\infty} d_n c_j^n|^2 \le \sum_{n=0}^{\infty} \frac{|d_n|^2}{p_n}(\sum_{j=1}^{\infty}|c_j|^2)^n, \tag{12}$$

if the right hand side converges.

2 The Hilbert space consisting of almost periodic functions

The delta function on **R**

$$\delta(x - y) = \begin{cases} 1 & x = y \\ 0 & x \ne y \end{cases}$$

is the reproducing kernel for the Hilbert space **AP** consisting of all almost periodic functions $f(x)$ on **R** with finite norms

$$\{\lim_{L\to\infty} \frac{1}{2L}\int_{-L}^{L} |f(x)|^2 dx\}^{\frac{1}{2}} < \infty.$$

We shall consider the nonlinear transform for **AP** functions

$$\varphi(f) = \sum_{n=0}^{\infty} d_n f(x)^n \quad (d_n : \text{constants})$$

and the corresponding reproducing kernel

$$\Delta(x - y) = \sum_{n=0}^{\infty} p_n \delta(x - y)^n \quad (p_n > 0; \text{constants})$$

with

$$p = \sum_{n=1}^{\infty} p_n < \infty.$$

Then, by Theorem A we have the inequality

$$\frac{1}{p_0}|d_0|^2 + \frac{1}{p}\{\lim_{L\to\infty} \frac{1}{2L}\int_{-L}^{L} |\sum_{n=1}^{\infty} d_n f(x)^n|^2 dx\}$$

$$\le \sum_{n=0}^{\infty} \frac{|d_n|^2}{p_n}\{\lim_{L\to\infty} \frac{1}{2L}\int_{-L}^{L} |f(x)|^2 dx\}^n, \tag{13}$$

249

if the right hand side converges.

3   Positive-definite Hermitian matrices

Recall first that to every positive-definite Hermitian matrix, there corresponds a uniquely determined inner product space admitting a reproducing kernel.

Let E be the set $\{1, 2, ..., N\}$. For any positive-definite Hermitian matrix $A = \|a_{\nu\mu}\|_{N \times N}$ we set $\tilde{A} = \overline{A^{-1}} = \|\tilde{a}_{\nu\mu}\|$. We think of $\mathbf{C}^N$ as the vectors composed of all functions on E which is the space of $N \times 1$ matrices and denote the space by $H[A]$ when it is equipped with the inner product

$$(\mathbf{x}, \mathbf{y})_{H[A]} = \mathbf{y}^* A \mathbf{x}. \tag{14}$$

We consider the function $K(\nu, \mu)$ on $E \times E$ defined by

$$K(\nu, \mu) = \tilde{a}_{\nu\mu}. \tag{15}$$

This function is the reproducing kernel for the space $H[A] = H_K$; that is, if we consider (15) as a row of column vectors

$$\mathbf{K}_\mu = [\tilde{a}_{1\mu}, ..., \tilde{a}_{N\mu}]^T,$$

then for any vector $\mathbf{x} = (x_1, x_2, ..., x_N)^T$, we have

$$(\mathbf{x}, \mathbf{K}_\mu)_{H[A]} = x_\mu; \mu = 1, 2, ..., N. \tag{16}$$

For a nonlinear transform of $\mathbf{C}^N$

$$\varphi(\mathbf{x}) = \sum_{n=0}^{\infty} d_n \mathbf{x}^n \quad (d_n : \text{constants}) \tag{17}$$

in the sense of Hadamard's product such that

$$\mathbf{x}^n = (x_1^n, x_2^n, ..., x_N^n)^T,$$

and for the corresponding positive matrix on E

$$\mathbf{K}(\nu, \mu) = \sum_{n=0}^{\infty} p_n (\tilde{a}_{\nu\mu})^n \quad (p_n > 0; \text{constants}) \tag{18}$$

with

$$\sum_{n=0}^{\infty} p_n |(\tilde{a}_{\nu\nu})^n| < \infty \quad \text{on} \quad E,$$

we have the inequality, by Theorem A

$$\left(\sum_{n=0}^{\infty} d_n \mathbf{x}^n\right)^* \left(\sum_{n=0}^{\infty} p_n \tilde{A}^n\right)^\sim \left(\sum_{n=0}^{\infty} d_n \mathbf{x}^n\right) \leq \sum_{n=0}^{\infty} \frac{|d_n|^2}{p_n}(\mathbf{x}^* A \mathbf{x})^n, \tag{19}$$

if the right hand side converges. Here, $\tilde{A}^n = \|\tilde{a}^n_{\nu\mu}\|$. For some special cases, see [Sa3, 15].

4 The simplest Sobolev Hilbert space

Note that

$$G(x,y) = \frac{1}{2}e^{-|x-y|} = \frac{1}{2\pi}\int_{-\infty}^{\infty}\frac{1}{\xi^2+1}e^{i\xi(x-y)}d\xi \tag{20}$$

is the reproducing kernel for the Sobolev Hilbert space S on $\mathbf{R}$ consisting of all real-valued and absolutely continuous functions $f(x)$ on $\mathbf{R}$ with finite norms

$$\left\{\int_{-\infty}^{\infty}(f'(x)^2 + f(x)^2)dx\right\}^{\frac{1}{2}} < \infty. \tag{21}$$

Then, we have

$$\begin{aligned}
K(x,y) &= \sum_{n=1}^{\infty} G(x,y)^n \\
&= \sum_{n=1}^{\infty}\frac{1}{2^n}e^{-n|x-y|} \\
&= \sum_{n=1}^{\infty}\frac{2n}{2^n}\frac{1}{2\pi}\int_{-\infty}^{\infty}\frac{1}{\xi^2+n^2}e^{i\xi(x-y)}d\xi \\
&\ll \frac{1}{2\pi}\int_{-\infty}^{\infty}\left(\sum_{n=1}^{\infty}\frac{2n}{2^n}\right)\frac{1}{\xi^2+1}e^{i\xi(x-y)}d\xi \\
&= D\cdot G(x,y); D = 2\sum_{n=1}^{\infty}\frac{n}{2^n} > 0.
\end{aligned} \tag{22}$$

Hence, for the nonlinear transform of $f \in S$

$$\varphi(f) = \sum_{n=1}^{\infty} d_n f(x)^n \quad (d_n : \text{constants}),$$

we have the inequality

$$\frac{1}{D}\int_{-\infty}^{\infty}\left\{\left(\sum_{n=1}^{\infty}d_n f(x)^n\right)'^2 + \left(\sum_{n=1}^{\infty}d_n f(x)^n\right)^2\right\}dx \tag{23}$$

251

$$\leq \sum_{n=1}^{\infty} |d_n|^2 \left\{ \int_{-\infty}^{\infty} (f'(x)^2 + f(x)^2) dx \right\}^n,$$

if the right hand side converges.

5   The reproducing kernel $\log \dfrac{\min(x,y)}{a}$

As we can see directly

$$K(x,y) = \log \frac{\min(x,y)}{a} \quad (0 < a \leq x, y \leq b \leq \infty) \tag{24}$$

is the reproducing kernel for the Hilbert space $H(a,b)$ consisting of all real-valued and absolutely continuous functions $f(x)$ on $[a,b]$ such that $f(a) = 0$ and f have finite norms

$$\left\{ \int_a^b f'(x)^2 x \, dx \right\}^{\frac{1}{2}} < \infty. \tag{25}$$

Meanwhile,

$$\mathbf{K}(x,y) = \frac{\min(x,y)}{a} - 1 \quad (0 < a \leq x, y \leq b \leq \infty) \tag{26}$$

is the reproducing kernel for the Hilbert space $H_{\mathbf{K}}$ consisting of all real-valued and absolutely continuous functions $f(x)$ on $[a,b]$ such that $f(a) = 0$ and f have finite norms

$$\left\{ a \int_a^b f'(x)^2 dx \right\}^{\frac{1}{2}} < \infty.$$

Then we have the identity

$$\exp K(x,y) = \mathbf{K}(x,y) + 1. \tag{27}$$

Since

$$H_{\mathbf{K}} \cap \mathbf{C} = \{0\},$$

the RKHS admitting the reproducing kernel $\mathbf{K}(x,y) + 1$ is the direct sum of $H_{\mathbf{K}}$ and $\mathbf{C}$. Hence, for the nonlinear transform of $f \in H(a,b)$

$$\varphi(f) = \sum_{n=0}^{\infty} d_n f(x)^n \quad (d_n : \text{constants}),$$

we have the inequality

$$|d_0|^2 + a \int_a^b \left(\sum_{n=1}^{\infty} d_n f(x)^n \right)'^{\,2} dx$$

252

$$\le \sum_{n=0}^{\infty} |d_n|^2 n! \left(\int_a^b f'(x)^2 x dx \right)^n, \tag{28}$$

if the right hand side converges.

In particular, we have

$$a \int_a^b (f(x)^n)'^2 dx \le n! \left(\int_a^b f'(x)^2 x dx \right)^n \tag{29}$$

and

$$1 + a \int_a^b (\exp f(x))'^2 dx \le \exp\left\{ \int_a^b f'(x)^2 x dx \right\}. \tag{30}$$

For some special cases, see [Sa27].

6   The reproducing kernel $\min(x, y)$

Note that

$$K(x, y) = \min(x, y) \quad (0 \le x, y)$$

is the reproducing kernel for the Hilbert space H_0 consisting of all real-valued and absolutely continuous functions $f(x)$ on $[0, \infty)$ such that $f(0) = 0$ and f have finite norms

$$\left\{ \int_0^\infty f'(x)^2 dx \right\}^{\frac{1}{2}} < \infty.$$

Then,

$$\mathbf{K}_N(x, y) = \sum_{n=1}^N K(x, y)^n = \min\left\{ \frac{x(1 - x^N)}{1 - x}, \frac{y(1 - y^N)}{1 - y} \right\} \tag{31}$$

is the reproducing kernel for the Hilbert space $H_{0,N}$ consisting of all real-valued and absolutely continuous functions $f(x)$ on $[0, \infty)$ such that $f(0) = 0$ and f have finite norms

$$\left\{ \int_0^\infty f'(x)^2 \left(\frac{x(1 - x^N)}{1 - x} \right)'^{-1} dx \right\}^{\frac{1}{2}} < \infty. \tag{32}$$

Hence, for the nonlinear transform of $f \in H_0$

$$\varphi(f) = \sum_{n=1}^N f(x)^n$$

$$= \frac{f(x)(1 - f(x)^N)}{1 - f(x)},$$

we have the inequality

$$\int_0^\infty \left(\frac{f(x)(1 - f(x)^N)}{1 - f(x)} \right)'^2 \left(\frac{x(1 - x^N)}{1 - x} \right)'^{-1} dx$$

$$\leq \frac{\int_0^\infty f'(x)^2 dx \left\{ 1 - \left(\int_0^\infty f'(x)^2 dx \right)^N \right\}}{1 - \int_0^\infty f'(x)^2 dx}. \tag{33}$$

In particular, for $f \in H_0$ satisfying

$$0 < \int_0^1 f'(x)^2 dx < 1,$$

we have the inequality, by letting $N \to \infty$

$$\int_0^1 \left(\frac{f(x)}{1 - f(x)} \right)'^2 (1 - x)^2 dx \leq \frac{\int_0^1 f'(x)^2 dx}{1 - \int_0^1 f'(x)^2 dx}. \tag{34}$$

7 The reproducing kernel $\exp\{\min(x, y)\}$

From the identity

$$\exp\{\min(x, y)\} = \{\min(e^x, e^y) - 1\} + 1 \tag{35}$$

and from the arguments in (5), for the nonlinear transform of $f \in H_0$ (the RKHS in (6))

$$\varphi(f) = \sum_{n=0}^\infty d_n f(x)^n \quad (d_n : \text{constants}),$$

we have the inequality

$$|d_0|^2 + \int_0^\infty \left(\sum_{n=1}^\infty d_n f(x)^n \right)'^2 e^{-x} dx$$

$$\leq \sum_{n=0}^\infty |d_n|^2 n! \left\{ \int_0^\infty f'(x)^2 dx \right\}^n. \tag{36}$$

In particular, we have

$$\int_0^\infty (f(x)^n)'^2 e^{-x} dx \leq n! \left\{ \int_0^\infty f'(x)^2 dx \right\}^n \tag{37}$$

254

and

$$1 + \int_0^\infty \left(e^{f(x)}\right)'^2 e^{-x}\, dx \le e^{\int_0^\infty f'(x)^2\, dx}.\tag{38}$$

8   The reproducing kernel $\log \dfrac{1}{(1-\bar{u}z)^2}$

Note that

$$K(z,\bar{u}) = \log \frac{1}{(1-\bar{u}z)^2} \quad \text{(principal value)}\tag{39}$$

is the reproducing kernel for the Hilbert space H_D consisting of all analytic functions $f(z)$ on the unit disc $\Delta = \{|z| < 1\}$ such that $f(0) = 0$ and $f(z)$ have finite norms

$$\left\{\frac{1}{2\pi} \iint_\Delta |f'(z)|^2 dx dy\right\}^{\frac{1}{2}} < \infty.$$

Meanwhile,

$$\frac{1}{(1-\bar{u}z)^2} = \exp K(z,\bar{u})$$

is the reproducing (Bergman) kernel for the Hilbert space consisting of all analytic functions $f(z)$ on Δ having finite norms

$$\left\{\frac{1}{\pi} \iint_\Delta |f(z)|^2 dx dy\right\}^{\frac{1}{2}} < \infty.$$

Hence, for the nonlinear transform for $f \in H_D$

$$\varphi(f) = \sum_{n=0}^\infty d_n f(x)^n \quad (d_n : \text{constants}),$$

we have the inequality

$$|d_0|^2 + \frac{1}{\pi} \iint_\Delta \left| \sum_{n=1}^\infty d_n f(z)^n \right|^2 dx dy$$

$$\le \sum_{n=0}^\infty |d_n|^2 n! \left\{ \frac{1}{2\pi} \iint_\Delta |f'(z)|^2 dx dy \right\}^n.\tag{40}$$

In particular, we have

$$\frac{1}{\pi} \iint_\Delta |f(z)^n|^2 dx dy \le n! \left(\frac{1}{2\pi} \iint_\Delta |f'(z)|^2 dx dy \right)^n\tag{41}$$

255

and

$$\frac{1}{\pi} \iint_\Delta |e^{f(z)}|^2 \, dxdy \le \exp\left\{ \frac{1}{2\pi} \iint_\Delta |f'(z)|^2 \, dxdy \right\} \qquad (42)$$

([Sa2]).

For more general reproducing kernels of exponential type, see [[Sa]] and [Sa18]. See also [Br] for some connections with the theory of univalent functions on Δ.

9 The Bergman–Selberg kernels

For $q > \frac{1}{2}$,

$$K_q(z, \overline{u}) = \frac{\Gamma(2q)}{(z + \overline{u})^{2q}}$$

is the Bergman–Selberg reproducing kernel on the half plane $R^+ = \{\text{Re}\, z > 0\}$ consisting of all analytic functions $f(z)$ on R^+ with finite norms

$$\|f\|_{HK_q}^2 = \frac{1}{\pi\Gamma(2q-1)} \iint_{R^+} |f(z)|^2 [2\text{Re}\, z]^{2q-2} \, dxdy.$$

For $q = \frac{1}{2}$, $K_{1/2}(z, \overline{u})$ is the Szegö reproducing kernel on R^+ consisting of all analytic functions $f(z)$ on R^+ with finite norms

$$\|f\|_{HK_{1/2}}^2 = \frac{1}{2\pi} \sup_{x>0} \int_{-\infty}^{\infty} |f(x+iy)|^2 \, dy.$$

Then, a member $f(z)$ of $H_{K_{1/2}}$ has nontangential boundary values on the imaginary axis belonging to L_2 and we have

$$\|f\|_{HK_{1/2}}^2 = \frac{1}{2\pi} \int_{-\infty}^{\infty} |f(iy)|^2 \, dy.$$

Note that

$$\frac{\partial^2 K_1(z, \overline{u})}{\partial z \partial \overline{u}} = \frac{6}{(z+\overline{u})^4}, \quad \frac{\partial^4 K_1(z, \overline{u})}{\partial z^2 \partial \overline{u}^2} = \frac{120}{(z+\overline{u})^6},$$

$$\frac{\partial^2 K_{1/2}(z, \overline{u})}{\partial z \partial \overline{u}} = \frac{2}{(z+\overline{u})^3}, \quad \text{and} \quad \frac{\partial^4 K_{1/2}(z, \overline{u})}{\partial z^2 \partial \overline{u}^2} = \frac{24}{(z+\overline{u})^5}.$$

Hence, for the nonlinear transforms of $f \in H_{K_1}$

$$d_1 f' + d_2 f^2,$$
$$d_1 f'' + d_2 f' f + d_3 f^3,$$
$$d_1 f f'' + d_2 (f')^2 + d_3 f^4,$$
$$d_1 f' f'' + d_2 f(f')^2 + d_3 f^5,$$

256

and
$$d_1(f'')^2 + d_2(f')^3 + d_3(f')^2 f^2 + d_4 f^6,$$

we have the specially simple norm inequalities. For example, we have the inequalities

$$\frac{6}{7}\|d_1 f' + d_2 f^2\|^2_{HK_2} \leq \|f\|^2_{HK_1}(\frac{1}{6}|d_1|^2 + |d_2|^2\|f\|^2_{HK_1}),$$

and

$$\frac{120}{127}\|d_1 f'' + d_2 f' f + d_3 f^3\|^2_{HK_3}$$

$$\leq \|f\|^2_{HK_1}(\frac{1}{120}|d_1|^2 + \frac{1}{6}|d_2|^2\|f\|^2_{HK_1} + |d_3|^2\|f\|^4_{HK_1}).$$

For the nonlinear transforms of $f \in H_{K_{1/2}}$

$$d_1 f' + d_2 f^3, d_1 f f' + d_2 f^4, d_1(f')^2 + d_2 f' f^3 + d_3 f^6, d_1 f'' + d_2 f' f^2 + d_3 f^5,$$

and

$$d_1 f f'' + d_2 (f')^2 + d_3 f^6,$$

we have the corresponding and specially simple norm inequalities.

10 A nonanalytic transform of analytic functions

Note that

$$\frac{1}{\pi}\log\frac{1}{|1 - \overline{u}z|} = \frac{1}{4\pi}\left[\log\frac{1}{(1 - \overline{u}z)^2} + \overline{\log\frac{1}{(1 - \overline{u}z)^2}}\right]$$

is the reproducing kernel on the unit disc Δ consisting of all harmonic functions $h(z)$ satisfying $h(0) = 0$ and equipped with finite norms

$$\left\{\iint_{\Delta}(u_x^2 + h_y^2)dxdy\right\}^{\frac{1}{2}} < \infty.$$

Hence, for the nonanalytic mapping of H_D functions f in (39)

$$f \longrightarrow \frac{1}{2}\left[f + \overline{f}\right] = \operatorname{Re}f = u,$$

we have the isometrical identity (not inequality)

$$\iint_{\Delta}(u_x^2 + u_y^2)dxdy = \iint_{\Delta}|f'(z)|^2 dxdy.$$

257

11 A transform with nonconstant coefficients

As we see directly by using the Taylor expansion,

$$\frac{\pi(1+\overline{u}z)}{1-\overline{u}z} = \frac{\pi}{1-\overline{u}z} + \frac{\pi\overline{u}z}{1-\overline{u}z}$$

is the reproducing kernel consisting of all analytic functions $f(z)$ on Δ with finite norms

$$\frac{1}{2\pi}\left\{\int_{\partial\Delta} |f(z)|^2|dz| + 2\pi|f(0)|^2\right\}^{\frac{1}{2}} < \infty$$

(2.4.73). Hence, for the transform of the Szegö space on Δ

$$f \longrightarrow f + zf = (1+z)f,$$

we have the inequality

$$\int_{\partial\Delta} |(1+z)f(z)|^2|dz| + 2\pi|f(0)|^2 \leq 4\int_{\partial\Delta} |f(z)|^2|dz|. \tag{43}$$

Meanwhile,

$$K(z,\overline{u}) = \frac{1}{\overline{u}z} \log \frac{1}{1-\overline{u}z}$$

is the reproducing kernel for the Hilbert space H_K consisting of all analytic functions $f(z)$ on Δ with finite norms

$$\left\{\frac{1}{\pi}\iint_\Delta |f'(z)|^2\,dx\,dy + \frac{1}{2\pi}\int_{\partial\Delta} |f(z)|^2|dz|\right\}^{\frac{1}{2}} < \infty$$

(2.4.78). Hence, from the identity

$$\frac{1}{1-\overline{u}z} = e^{z\overline{u}K(z,\overline{u})},$$

in the transform of H_K functions

$$f \longrightarrow e^{zf},$$

we have the inequality

$$\frac{1}{2\pi}\int_{\partial\Delta} |e^{zf(z)}|^2|dz| \leq \exp\left[\frac{1}{\pi}\iint_\Delta |f'(z)|^2\,dx\,dy + \frac{1}{2\pi}\int_{\partial\Delta} |f(z)|^2|dz|\right]. \tag{44}$$

Appendix 3
Stability of Lipschitz type in determination of initial heat distribution

For the solution $u(x,t) = u(f)(x,t)$ of the equations

$$\begin{cases} u'(x,t) = \Delta u(x,t), & x \in \Omega, \ t > 0 \\ u(x,0) = f(x), & x \in \Omega \\ u(x,t) = 0, & x \in \partial\Omega, \ t > 0, \end{cases}$$

where $\Omega \subset \mathbb{R}^r$, $2 \leq r \leq 3$, is a bounded domain with C^2-boundary and for an appropriate subboundary Γ of Ω we derive a Lipschitz estimate of $\|f\|_{L^2(\Omega)}$ as an application of analytic extension formulas and real inversion formulas for the Laplace transform in Chapter 5. For $\mu \in (1, \frac{5}{4})$ and for a positive constant C

$$C^{-1}\|f\|_{L^2(\Omega)} \leq \left\| \frac{\partial u(f)}{\partial \nu} \right\|_{B_\mu(\Gamma \times (0,\infty))} \equiv \int_\Gamma \left\{ \sum_{n=0}^{\infty} \frac{1}{n! \Gamma(n + 2\mu + 1)} \right.$$

$$\times \left. \int_0^{\infty} \left| (p\partial_p^{n+1} + n\partial_p^n) p^{-\frac{3}{2}} \frac{\partial u(f)}{\partial \nu} \left(x, \frac{1}{4p}\right) \right|^2 p^{2n + 2\mu - 1} dp \right\} dS.$$

The norm $\| \cdot \|_{B_\mu(\Gamma \times (0,\infty))}$ is involved and strong, but it is a natural one in our situation relating to the Bergman–Selberg norm for analytic functions. Furthermore, it is acceptable in the sense that $\left\| \frac{\partial u(f)}{\partial \nu} \right\|_{B_\mu(\Gamma \times (0,\infty))} \leq C\|f\|_{H^2(\Omega)}$ holds. The materials are taken from [SY1].

§1   Introduction and theorem

We consider an initial value problem for the heat equation:

$$\begin{cases} u'(x,t) = \Delta u(x,t), & x \in \Omega, \ t > 0 \\ u(x,0) = f(x), & x \in \Omega \\ u(x,t) = 0, & x \in \partial\Omega, \ t > 0, \end{cases} \tag{1}$$

where $\Omega \subset \mathbb{R}^r$, $2 \leq r \leq 3$, is a bounded domain with C^2-boundary $\partial\Omega$, $u' = \frac{\partial u}{\partial t}$, Δ is the Laplacian, and $\frac{\partial}{\partial \nu}$ is a trace operator (e.g. [[A]]), that is, if $u \in C^1(\overline{\Omega})$, then

$$\frac{\partial u}{\partial \nu}(x) = \sum_{i=1}^{r} \nu_i(x) \frac{\partial u}{\partial x_i}(x), \quad x \in \partial\Omega,$$

259

$\nu(x) = (\nu_1(x), ..., \nu_r(x))$ being the outward unit normal to $\partial\Omega$ at x.

For $f \in L^2(\Omega)$, there exists a unique (strong) solution

$$u = u(f) \in C^0([0,\infty); L^2(\Omega)) \cap C^1((0,\infty); L^2(\Omega))$$

such that $\Delta u \in C^0((0,\infty); L^2(\Omega))$ and $u(\cdot, t)_{|\partial\Omega} = 0$, $t > 0$ (e.g. [[Pa]]).

In this situation, we have the following problems when we consider the heat flux $\frac{\partial u(f)}{\partial\nu}(x,t)$ on a subboundary Γ of Ω as measurements for $t > 0$.

(I) (Uniqueness) For what kind of set $\Gamma \subset \partial\Omega$ does

$$\frac{\partial u(f)}{\partial\nu}(x,t), \quad x \in \Gamma, t > 0$$

determine $f(x)$, $x \in \Omega$ uniquely?

(II) (Construction) We wish to represent the initial heat distribution $f(x)$ on Ω in terms of $\frac{\partial u(f)}{\partial\nu}(x,t)$, $x \in \Gamma$, $t > 0$.

(III) (Stability) Can we estimate $\|f\|_{L^2(\Omega)}$ by $\frac{\partial u(f)}{\partial\nu}(x,t)$, $x \in \Gamma$, $t > 0$? Moreover what norm of $\frac{\partial u(f)}{\partial\nu}(x,t)$, $x \in \Gamma$, $t > 0$ should we choose for the estimation of $\|f\|_{L^2(\Omega)}$?

The determination of the initial heat distribution is called an *observation problem*. The observation problem is important also in the theory of control, because it is a dual problem to the controllability problem (e.g. [DR]). For the observation problem in the heat equation, we can refer to [[Ca]]. For similar types of problems, the reader can consult [Do], [MS], and [Sak].

For the uniqueness, the answer is known under a comfortable assumption:

Proposition 1 *Let $\Gamma \subset \partial\Omega$ satisfy $\Gamma = \partial\Omega \cap U \neq \{\phi\}$ for an open set $U \subset \mathbb{R}^r$. If*

$$\frac{\partial u(f)}{\partial\nu}(x,t) = 0, \quad x \in \Gamma, t > 0,$$

then $f(x) = 0$, $x \in \Omega$.

This is proved by using the eigenfunction expansion of the solution and the unique continuation theorem for the elliptic operator (e.g. [[Mi]]), and we can further refer to [SW] for the proof.

Now we proceed to the stability. It is easily expected that the stability is very delicate, because the map $f \longmapsto \frac{\partial u(f)}{\partial\nu}$ advances the regularity by the 'smoothing' property of the parabolic equation (e.g. [[Fr]]). We have to take a strong norm $\|\cdot\|_*$

for $\frac{\partial u(f)}{\partial \nu}(x, t)$, $x \in \Gamma$, $t > 0$, in order to get an upper estimate of $\|f\|_{L^2(\Omega)}$. More precisely, there are two ways.

(a) We search for a norm $\| \cdot \|_*$ of functions on $\Gamma \times (0, \infty)$ such that

$$\|f\|_{L^2(\Omega)} \le C \left\| \frac{\partial u(f)}{\partial \nu} \right\|_* .$$

(b) Taking the usual norm for $\frac{\partial u(f)}{\partial \nu}$ such as

$$\left\| \frac{\partial u(f)}{\partial \nu} \right\|_{L^2(\Gamma \times (0, \infty))} ,$$

we search for a stability modulus $\omega \in C^0[0, \infty)$ which is monotone increasing and $\omega(0) = 0$ such that

$$\|f\|_{L^2(\Omega)} \le \omega \left(\left\| \frac{\partial u(f)}{\partial \nu} \right\|_{L^2(\Gamma \times (0, \infty))} \right) .$$

In (a) we insist on stability of Lipschitz type, while we have to admit the choice of a strong norm $\| \cdot \|_*$. In (b), we insist on the usual norm for measurements $\frac{\partial u(f)}{\partial \nu}$, at the cost of a worse stability modulus ω. The latter way (b) seems to have been more popular in existing papers (e.g. Exercise 11.4 (pp.144–145) in [[Ca]]), and the estimate of the type

$$\|f\|_{L^2(\Omega)} = O \left([\log \|\mathrm{Data}\|^{-1}]^\alpha \right) ,$$

with some constant $\alpha > 0$, is typical (e.g. see p.147 in [[Ca]]).

The purpose of this appendix is to follow (a). Our choice of the norm $\| \cdot \|_*$ is, of course, stronger than the $L^2(\Gamma \times (0, \infty))$-norm, but is not extreme in the sense that

$$\left\| \frac{\partial u(f)}{\partial \nu} \right\|_* = O(\|f\|_{L^2(\Omega)}).$$

We shall define

$$\|g(x, p)\|^2_{B_\mu(\Gamma \times (0, \infty))} = \int_\Gamma \left\| p^{-\frac{3}{2}} g \left(x, \frac{1}{4p} \right) \right\|^2_{H_{K_\mu}(R^+)} dS,$$

the right hand side being convergent. See Theorem 5.1.2 for the Bergman–Selberg space $H_{K_\mu}(R^+)$. Then we obtain

261

Theorem A *For an arbitrarily fixed $x_0 \in \mathbb{R}^r$, we set*

$$\Gamma = \{x \in \partial\Omega; (x - x_0) \cdot \nu(x) > 0\} \tag{2}$$

and take

$$\mu \in \left(1, \frac{5}{4}\right). \tag{3}$$

We assume

$$r \quad (= the\ spatial\ dimension) \leq 3 \tag{4}$$

and

$$f \in H^2(\Omega) \cap H_0^1(\Omega). \tag{5}$$

Then, there exists a constant $C = C(\Omega, \Gamma, \mu) > 0$ such that

$$C^{-1}\|f\|_{L^2(\Omega)} \leq \left\|\frac{\partial u(f)}{\partial\nu}\right\|_{B_\mu(\Gamma\times(0,\infty))} \leq C\|f\|_{H^2(\Omega)}. \tag{6}$$

In Theorem A, if $r = 1$, then Γ is taken as a boundary point of the interval Ω. Here for simplicity, we assume (4), that is, the spatial dimension is less than or equal to 3. This is not essential and the condition (5) should be replaced by $f \in \mathcal{D}(A^\alpha)$ where $\alpha = \left[\frac{r}{4}\right] + 1$, $(Au)(x) = -\Delta u(x)$ with $\mathcal{D}(A) = \{u \in H^2(\Omega); u_{|\partial\Omega} = 0\}$; $[\beta]$ denotes the greatest integer not exceeding $\beta \in \mathbb{R}$.

§2   Proof of theorem

The proof will be divided into three steps.

1   First step

We shall discuss the regularity of solutions to the wave equation corresponding to (1). First by (5) we see that $u(f) \in C^1([0, \infty) \times \overline{\Omega})$ and $\Delta u(f) \in C^0([0, \infty) \times \overline{\Omega})$ (e.g. Theorems 4.3.5 and 4.3.6 in [[Pa]] and the Sobolev embedding theorem (e.g. [[A]])). We shall consider the corresponding wave equation:

$$\begin{cases} w''(x, t) = \Delta w(x, t), & x \in \Omega, t > 0 \\ w(x, 0) = 0, \quad w'(x, 0) = f(x), & x \in \Omega \\ w(x, t) = 0, & x \in \partial\Omega, t > 0. \end{cases} \tag{7}$$

Again applying the regularity assumption (5), by the eigenfunction expansion of the solution w and the Sobolev embedding theorem, we see that there exists a unique solution

$$w = w(f) \in C^0([0, \infty); H^3(\Omega) \cap H_0^1(\Omega)) \cap C^1([0, \infty); H^2(\Omega) \cap H_0^1(\Omega))$$
$$\cap C^2([0, \infty); H_0^1(\Omega)), \tag{8}$$

262

and

$$\|w(f)(\cdot,t)\|_{H^2(\Omega)}, \quad \|w(f)'(\cdot,t)\|_{H^2(\Omega)} \le \|f\|_{H^2(\Omega)}, \quad t > 0 \qquad (9)$$

(e.g. Theorem 1.1.1 in [[Kom]]). The inequalities in (9) follow from conservation of energy. We set

$$W(t) = \int_\Gamma \left(\frac{\partial w(f)}{\partial \nu}(x,t)\right)^2 dS \equiv \left\|\frac{\partial w(f)}{\partial \nu}(\cdot,t)\right\|_{L^2(\Gamma)}^2, \quad t > 0.$$

By (8) and the trace theorem (e.g. [[A]]), we have

$$W \in C^1[0,\infty), \quad |W'(t)|, W(t) \le C_1\|f\|_{H^2(\Omega)}^2, \qquad t > 0, \qquad (10)$$

where $C_1 = C_1(\Omega) > 0$ is a constant independent of $t > 0$.

In fact, $W \in C^0[0,\infty)$ and $W(t) \le C_1\|f\|_{H^2(\Omega)}^2$ is straightforward from (9) and the trace theorem. Next, by (9),

$$\left(\frac{\partial w(f)}{\partial \nu}\right)' \in C^0([0,\infty); L^2(\Gamma))$$

and

$$\int_\Gamma \left(\left(\frac{\partial w(f)}{\partial \nu}\right)'(x,t)\right)^2 dS \le C_1'\|f\|_{H^2(\Omega)}^2, \quad t > 0,$$

so that

$$W'(t) = 2\int_\Gamma \frac{\partial w(f)}{\partial \nu}(x,t)\left(\frac{\partial w(f)}{\partial \nu}\right)'(x,t) dS, \quad t > 0.$$

Hence $W' \in C^0[0,\infty)$ and by Schwarz's inequality

$$|W'(t)| \le 2\left\|\frac{\partial w(f)}{\partial \nu}(\cdot,t)\right\|_{L^2(\Gamma)}\left\|\left(\frac{\partial w(f)}{\partial \nu}\right)'(\cdot,t)\right\|_{L^2(\Gamma)}$$

$$\le 2\sqrt{C_1'}\|f\|_{H^2(\Omega)}\sqrt{C_1'}\|f\|_{H^2(\Omega)}, \quad t > 0,$$

which proves (10).

It follows from (10) and $w(f)(\cdot,0) = 0$ that $W(0) = 0$. Furthermore, we have, using (10) and the mean value theorem

$$W(t) = W(t) - W(0) \le t \sup_{0 \le s \le t} |W'(s)|$$

$$\le C_1 t\|f\|_{H^2(\Omega)}^2, \quad t > 0. \qquad (11)$$

By (11) and (3) we have

$$\int_0^\infty W(t)t^{3-4\mu}dt$$

$$= \int_0^1 W(t)t^{3-4\mu}dt + \int_1^\infty W(t)t^{3-4\mu}dt$$

$$\leq C_1\|f\|_{H^2(\Omega)}^2 \left(\int_0^1 t^{4-4\mu}dt + \int_1^\infty t^{3-4\mu}dt \right)$$

$$= C_1\|f\|_{H^2(\Omega)}^2 \left(\frac{1}{5-4\mu} + \frac{1}{4\mu-4} \right)$$

$$\equiv C_2\|f\|_{H^2(\Omega)}^2.$$

That is,

$$\int_0^\infty \int_\Gamma \left(\frac{\partial w(f)}{\partial \nu}(x,t) \right)^2 t^{3-4\mu}dSdt \leq C_2\|f\|_{H^2(\Omega)}^2. \tag{12}$$

2 Second step

By the transform formula of Reznitskaya, we get

$$u(f)(x,t) = \frac{1}{2\sqrt{\pi t^3}} \int_0^\infty \eta \exp\left(-\frac{\eta^2}{4t}\right) w(f)(x,\eta)d\eta, \quad x \in \Omega, \, t > 0 \tag{13}$$

(e.g. §5 in Chapter VII in [[LRS]]). By (8), (9) and the Sobolev embedding theorem, we have

$$\left| \frac{\partial w(f)}{\partial x_i}(x,t) \right| \leq C_1\|f\|_{H^2(\Omega)}, \quad x \in \Omega, \, t > 0,$$

that is,

$$\left| \frac{\partial w(f)}{\partial \nu}(x,t) \right| \leq C_1\|f\|_{H^2(\Omega)}, \quad x \in \Gamma, \, t > 0.$$

Therefore in (13), we can exchange $\int_0^\infty ...d\eta$ and $\frac{\partial}{\partial \nu}$, so that

$$2\sqrt{\pi t^3}\frac{\partial u(f)}{\partial \nu}(x,t) = \int_0^\infty \eta \exp\left(-\frac{\eta^2}{4t}\right) \frac{\partial w(f)}{\partial \nu}(x,\eta)d\eta,$$

$$x \in \Gamma, \, t > 0.$$

Setting $t = \frac{1}{4p}$ and changing independent variables by $s = \eta^2$, we have

$$\frac{\sqrt{\pi}}{2} \frac{1}{p^{\frac{3}{2}}} \frac{\partial u(f)}{\partial \nu}(x,\frac{1}{4p}) = \int_0^\infty \exp(-sp) \frac{\partial w(f)}{\partial \nu}(x,\sqrt{s})ds,$$

$$x \in \Gamma, \, p > 0. \tag{14}$$

We set

$$\tilde{u}(x, p) = \frac{\sqrt{\pi}}{2} \frac{1}{p^{\frac{3}{2}}} \frac{\partial u(f)}{\partial \nu}(x, \frac{1}{4p})$$

and

$$\tilde{w}(x, s) = \frac{\partial w(f)}{\partial \nu}(x, \sqrt{s}).$$

Then, by Theorem 3.1.1 and Theorem 5.1.2, we obtain

$$\int_0^\infty (\tilde{w}(x, t))^2 t^{1-2\mu} dt = \|\tilde{u}(x, \cdot)\|_{H_\mu(R+)}^2, \quad x \in \Gamma, \tag{15}$$

provided that either side is convergent.

Since

$$\int_0^\infty (\tilde{w}(x, t))^2 t^{1-2\mu} dt = 2 \int_0^\infty \left(\frac{\partial w(f)}{\partial \nu}(x, s) \right)^2 s^{3-4\mu} ds, \quad x \in \Gamma,$$

it follows from (12) and Fubini's theorem that

$$\int_0^\infty (\tilde{w}(x, t))^2 t^{1-2\mu} dt < \infty$$

for almost all $x \in \Gamma$. Consequently application of (15) yields

$$\int_0^\infty \left(\frac{\partial w(f)}{\partial \nu}(x, s) \right)^2 s^{3-4\mu} ds$$

$$= \frac{1}{2} \|\tilde{u}(x, \cdot)\|_{H_\mu(R+)}^2, \quad a.e. \quad x \in \Gamma,$$

and hence

$$\int_\Gamma \int_0^\infty \left(\frac{\partial w(f)}{\partial \nu}(x, t) \right)^2 t^{3-4\mu} dt dS$$

$$= \frac{\pi}{8} \int_\Gamma \left\| p^{-\frac{3}{2}} \frac{\partial u(f)}{\partial \nu}(x, \frac{1}{4p}) \right\|_{H_\mu(R+)}^2 dS$$

$$= \frac{\pi}{8} \left\| \frac{\partial u(f)}{\partial \nu} \right\|_{B_\mu(\Gamma \times (0,\infty))}^2. \tag{16}$$

265

3   Third step

In this step, we apply the stability estimate for the wave equation (1):

Proposition 2 (Observability inequality)   *Let* $\Gamma \subset \partial\Omega$ *be defined by (2) and let*

$$T > 2\sup_{x\in\Omega}|x - x_0| \tag{17}$$

where $x_0 \in \mathbb{R}^r$ *is a point which is arbitrarily chosen for specifying the observation subboundary* Γ. *Then there exists a constant* $C_3 = C_3(\Omega, T, \Gamma) > 0$ *such that*

$$\|f\|_{L^2(\Omega)} \le C_3 \left\| \frac{\partial w(f)}{\partial \nu} \right\|_{L^2(\Gamma\times(0,T))}. \tag{18}$$

The estimate (18) is proved in [Ho] and [[Li]]. See also [[Kom]].

Now, by combining (12) with (16), we have the second inequality in (6). Next, we fix $T > 0$ satisfying (17). Then, by Proposition 2, since

$$\int_\Gamma \int_0^T \left(\frac{\partial w(f)}{\partial \nu}(x, t) \right)^2 dt\,dS$$

$$\le T^{4\mu-3} \int_\Gamma \int_0^T \left(\frac{\partial w(f)}{\partial \nu}(x, t) \right)^2 t^{3-4\mu} dt\,dS$$

$$\le T^{4\mu-3} \int_\Gamma \int_0^\infty \left(\frac{\partial w(f)}{\partial \nu}(x, t) \right)^2 t^{3-4\mu} dt\,dS,$$

we obtain

$$\|f\|_{L^2(\Omega)}^2 \le C_3^2 T^{4\mu-3} \int_\Gamma \int_0^\infty \left(\frac{\partial w(f)}{\partial \nu}(x, t) \right)^2 t^{3-4\mu} dt\,dS$$

$$= \frac{\pi}{8} C_3^2 T^{4\mu-3} \left\| \frac{\partial u(f)}{\partial \nu} \right\|_{B_\mu(\Gamma\times(0,\infty))}^2$$

by (16), which is the first inequality in (6). Thus we complete the proof of Theorem A.

For the inversion of the Laplace transform, the complex case is well known. The complex form is, however, not adequate in our problem, because the observation data is real-valued and we have to extend the data analytically, which makes the stability unclear.

One of our keys is the transform formula between the heat equation and the wave equation, through which we reduce the stability in the heat problem to that in the wave problem. A similar technique is used also in [Ya3]. Thus we do not use the eigenfunction expansion of the solution to the heat equation, which is used in Exercise 11.4 in [[Ca]], [Do], [MS] and [Sak].

References

Books

[[A]]. Adams, R. A., *Sobolev Spaces*, Academic Press, New York (1975).

[[AG]]. Akhiezer, N. I., Glazman, I. M., *Theory of linear operators in Hilbert space*, volume 1, Translated from the Russian by M. Nestell, Frederick Ungar Publishing Co., New York (1961).

[[An]]. Ando, T., *Reproducing kernel spaces and quadratic inequalities*, Hokkaido University, Sapporo, Japan (1987).

[[AS]]. Abramowitz, M., Stegun, I. A., *Handbook of Mathematical Functions With Formulas, Graphs and Mathematical Tables*, U. S. Government Printing Office, Washington, D. C. (1972).

[[AY]]. Aĭzenberg, I. A., Yuzhakov, A. P., *Integral Representations and Residues in Multidimensional Complex Analysis*, Translations of Mathematical Monographs 58, American Mathematical Society, Providence, R. I. (1983).

[[B]]. Berezanskiĭ, Ju. M., *Expansions in eigenfunctions of selfadjoint operators*, Transl. Math. Monographs 17, Amer. Math. Soc., Providence, R. I. (1968).

[[BaG]]. Bakushinsky, A., Goncharsky, A., *Ill-Posed Problems: Theory and Applications*, Kluwer Academic Publishers, Dordrecht (1994).

[[Ber]]. Bergman, S., *The kernel function and conformal mapping*, Amer. Math. Soc., Providence, R. I. (1950, 1970).

[[Bert]]. Bertero, M., *Linear inverse and ill-posed problems*, INFNITC-8812, 19, Gennaio (1988).

[[BG]]. Begehr, H., Gilbert, R. P.
    1. *Transformations, transmutations and kernel functions*, Vol. 1, Pitman Monographs and Surveys in Pure and Applied Mathematics, **58**, Longman Scientific & Technical, UK (1992).
    2. *Transformations, transmutations and kernel functions*, Vol. 2, Pitman Monographs and Surveys in Pure and Applied Mathematics, **59**, Longman Scientific & Technical, UK (1993).

[[Bi]]. Bitsadze, A. V., *Integral equations of first kind*, World Scientific, Singapore (1995).

[[BM]]. Bochner, S., Martin, W. T., *Several complex variables*, Princeton University Press, Princeton, N. J. (1948).

[[BuM]]. Burbea, J., Masani, P., *Banach and Hilbert spaces of vector-valued functions, their general theory and applications to holomorphy*, Pitman Research Notes in Math., Pitman, London, **90** (1984).

[[BN]]. Butzer, P. L., Nessel, R. J., *Fourier analysis and approximation*, Academic Press, New York (1971).

[[Br]]. Branges, L. de., *Hilbert spaces of entire functions*, Prentice-Hall, Englewood Cliffs, N. J. (1968).

[[Bru]]. Bruinsma, P., *Interpolation problems for Schur and Nevanlinna pairs*, Thesis, Rijksuniversiteit Groningen (1964).

[[BS]]. Bergman, S., Schiffer, M., *Kernel functions and elliptic differential equations*, Academic Press. New York (1953).

[[By]]. Byun, D.-W., *Integral transforms and approximation of functions*, Thesis, Gunma University, Japan (1993).

[[Ca]]. Cannon, J. R., *The One-Dimensional Heat Equation*, Addison-Wesley Publishing, Reading, Mass. (1984).

[[Che]]. Cherednichenko, V. G., *Inverse Logarithmic Potential Problem*, VSP, Utrecht, The Netherlands (1996).

[[Chu]]. Chui, C. K., *An introduction to wavelets*, Academic Press, New York (1992).

[[CJ]]. Carslaw, H. S., Jaeger, J. C., *Conduction of heat in solids*, Clarendon Press, Oxford (1959).

[[CK]]. Colton, D., Kress, R., *Inverse acoustic and electromagnetic scattering theory*, Springer-Verlag, Berlin (1992).

[[Co]]. Corduneanu, C., *Integral equations and applications*, Cambridge University Press. (1991).

[[D]]. Daubechies, I., *Ten lectures on wavelets*, Society for Industrial and Applied Mathematics, Philadelphia (1992).

[[Da]]. Davis, P. J., *Interpolation & approximation*, Dover Books on Advanced Mathematics, New York (1973).

[[Do]]. Doetsch, G., *Introduction to the Theory and Applications of the Laplace Transformation*, (English translation), Springer-Verlag, Berlin (1974).

[[Don]]. Donoghue, W. F.
  1. *Distributions and Fourier transforms*, Academic Press, New York and London (1969).
  2. *Monotone matrix functions and analytic continuation*, Springer-Verlag, Berlin (1974).

[[DS]]. Djrbashian, A. E., Shamoian, F. A., *Topics in the theory of A^p_α spaces*, TEUBNER -TEXTE zur Mathematik, Leipzig **105** (1988).

[[Du]]. Duren, P. L., *Theory of H^p spaces*, Academic Press, New York and London (1970).

[[Dy]]. Dym, H., *J contractive matrix functions, reproducing kernel Hilbert spaces and interpolation*, AMS, Providence, R. I. (1989).

268

[[EMOT]]. Erdélyi, A., Magnus, W., Oberhettinger, F., Tricomi, F. G., *Tables of integral transforms*, McGraw-Hill Book Company Inc., New York, **1** (1954).

[[Fa]]. Fay, J. D.
1. *Theta functions on Riemann surfaces*, Lecture Notes in Math., 352, Springer-Verlag, Berlin (1973).
2. *Kernel Functions, Analytic Torsion, and Moduli Spaces*, Memoirs of the American Mathematical Society, Providence, R. I. **464** (1992).

[[Fo]]. Folland, G. B., *Introduction to partial differential equations*, Math. Notes, Princeton University Press., N. J. (1976).

[[Fr]]. Friedman, A., *Partial Differential Equations of Parabolic Type*, (Reprint Edition), Krieger Publishing, Malabar, Florida (1983).

[[Fu]]. Fuks, B. A.
1. *Introduction to the theory of analytic functions of several complex variables*, Transl. Math. Mono. vol. 8, Amer. Math. Soc., Providence, R.I. (1963).
2. *Special chapters in the theory of analytic functions of several complex variables*, Transl. Math. Mono. vol. 14, Amer. Math. Soc., Providence, R.I. (1965).

[[G]]. Garabedian, P. R., *Partial differential equations*, John Wiley & Sons, Inc., New York (1964).

[[Gr]]. Groetsch, C. W., *Inverse problems in the mathematical sciences*, Vieweg Mathematics for Scientists and Engineers, Vieweg (1993).

[[GR]]. Gradshteyn, I. S., Ryzhik, I. M., *Table of Integrals*, Series, and Products, Academic Press, New York (1980).

[[He]]. Hejhal, D. A., *Theta functions, kernel functions and Abel integrals*, Memoirs of Amer. Math. Soc., Providence, R. I. **129** (1972).

[[Hi]]. Higgins, J. R., *Sampling theory in Fourier and signal analysis foundations*, Oxford Science Publications (1996).

[[Hö]]. Hörmander, L., *An introduction to complex analysis in several variables*, D. VAN NOSTRAND COMPANY, INC. Princeton, N. J. (1967).

[[Hu]]. Hua, L. K., *Harmonic analysis of functions of several complex variables in the classical domains*, Transl. Math. Mono., vol. 6, Amer. Math. Soc., Providence, R. I. (1963).

[[I]]. Isakov, V., *Inverse source problems*, Mathematical Surveys and Monographs, Number 34, American Mathematical Society, Providence, R. I. (1990).

[[Kom]]. Komornik, V., *Exact Controllability and Stabilization–the multiplier method*, Masson, Paris (1994).

[[Kon]]. Kondo, J., *Integral equations*, Kodanshia, Tokyo (1991) and Clarendon Press, Oxford (1991).

[[Kr]]. Kra, I., *Automorphic Forms and Kleinian Groups*, W. A. Benjamin, Inc., Mass. (1972).

269

[[Kö]]. Körner, T. W., *Fourier Analysis*, Cambridge University Press (1988).

[[L]]. Lions, J.-L., *Contrôlabilité Exacte Perturbations et Stabilisation de Systèmes Distribués*, Masson, Paris **1** (1988).

[[Lo]]. Loève, M., *Probability theory I, II*, 4th Edition, Springer-Verlag, Berlin (1977).

[[LRS]]. Lavrentiev, M. M., Romanov, V. G., Shishat·skiǐ., *Ill-posed Problems of Mathematical Physics and Analysis*, (English translation), American Mathematical Society, Providence, R. I. (1986).

[[M]]. Meschkowski, H., *Hilbertsche Räume mit Kernfunktion*, Springer-Verlag, Berlin (1962).

[[MF]]. Morse, P. M., Feshbach, H., *Methods of theoretical physics*, McGraw-Hill Book Company Inc., New York (1953).

[[Mi]]. Mizohata, S., *The Theory of Partial Differential Equations*, Cambridge University Press, London (1973).

[[Ne]]. Nehari, Z., *Conformal mapping*, McGraw-Hill Book Company Inc., New York (1952).

[[Pa]]. Papoulis, A., *Probability, random variables and stochastic processes*, Mc-Graw-Hill Book Company, Inc., New York (1965).

[[Paz]]. Pazy, A., *Semigroups of Linear Operators and Applications to Partial Differential Equations*, Springer-Verlag, Berlin (1983).

[[PBM]]. Prudnikov, A. P., Brychkov, Yu. A., Marichev, O. I., *Integrals and series*, 1. Translated from the Russian by N. M. Queen, Gordon and Breach Sci. Publishers, New York (1986).

[[PS]]. Parthasarathy, K. R., Schmidt, K., *Positive Definite Kernels, Continuous Tensor Products, and Central Limit Theorems of Probability Theory*, Springer Lecture Notes in Math., Berlin, **107** (1972).

[[PW]]. Paley, R.E.A.C., Wiener, N., *Fourier transforms in the complex plane*, Amer. Math. Soc. Colloq. Publ., Vol. 19, Amer. Math. Soc., Providence, R. I. (1934).

[[Ra]]. Ramm, A.G.
   1. *Random fields estimation theory*, Pitman Monographs and Surveys in Pure and Applied Mathematics, **48**, Longman Scientific & Technical, UK (1990).
   2. *Multidimensional inverse scattering problems*, Pitman Monographs and Surveys in Pure and Applied Mathematics, **51**, Longman Scientific & Technical, UK (1992).

[[Re]]. Reimer, M., *Constructive theory of multivariate functions with an application to tomography*, Wissenschaftsverlag, Mannheim/Wien/Zürich (1990).

[[Ro]]. Roach, G. F., *Green's functions*, Cambridge University Press (1982).

[[Ron]]. Ronkin, L. I., *Introduction to the theory of entire functions of several variables*, Transl. Math. Mono., vol. 44, Amer. Math. Soc., Providence, R. I. (1974).

[[Sa]]. Saitoh, S., *Theory of reproducing kernels and its applications*, Pitman Research Notes in Mathematics Series, **189**, Longman Scientific & Technical, UK (1988).

[[Sh]]. Shapiro, H. S., *Topics in approximation theory*, VIII, Springer Lecture Notes in Mathematics, Berlin, **275** (1971).

[[Sn]]. Sneddon, I. N., *The use of integral transforms*, McGraw-Hill Book Company Inc., New York (1972).

[[SS]]. Schiffer, M., Spencer, D. C., *Functionals of finite Riemann surfaces*, Princeton University Press, N. J. (1954).

[[SW]]. Stein, E. M., Weiss, G., *Introduction to Fourier analysis on Euclidean spaces*, Princeton University Press, Princeton, N. J. (1971).

[[Sz]]. Szmydt, Z., *Fourier transformation and linear differential equations*, D. Reidel Publishing Company, Boston (1977).

[[TA]]. Tikhonov, A. N., Arsenin, V. Y., *Solution of ill-posed problems*, John Wiley & Sons, Inc., New York (1977).

[[Ti]]. Titchmarsh, E. C., *An introduction to the theory of Fourier integrals*, 2nd ed., Oxford University Press (1948).

[[Tr]]. Treves, F.
  1. *Linear partial differential equations with constant coefficients*, Gordon and Breach, New York (1966).
  2. *Topological vector spaces, distributions and kernels*, Academic Press, New York (1967).

[[Tri]]. Tricomi, F. G., *Integral equations*, Pure and Applied Mathematics Vol. 5, New York (1957).

[[VTC]]. Vakhania, N. N., Tarieladze, V. I., Chobanyan, S. A., *Probability Distributions on Banach Spaces*, D. Reidel Publishing Company, Dordrecht (1987).

[[W]]. Watson, G. N., *Theory of Bessel functions*, Cambridge University Press (1980).

[[WG]]. Wang, Z. X., Guo, D. R., *Special functions*, World Scientific, Singapore (1989).

[[Wi]]. Widder, D. V., *The Laplace transform*, Princeton University Press, Princeton, N. J. (1972).

[[Yo]]. Yoshida, K., *Lectures on differential and integral equations*, Pure and Applied Mathematics, Interscience Publishers, New York (1960).

Articles

[A].   Adachi, K., *Continuation of holomorphic functions from subvarieties, Geometric Complex Analysis*, edited by Junjiro Noguchi et al., World Scientific Publishing Co. (1996), 9-17.

[ABDS]. Alpay, D., Bruinsma, P., Dijksma, A., Snoo, H.d.

   1. *Interpolation problems, extensions of symmetric operators and reproducing kernel spaces I*, Operator Theory: Adv. Appl. **50** (1991), 35-82.

   2. *Interpolation problems, extensions of symmetric operators and reproducing kernel spaces II*, Integral Equations and Operator Theory **14** (1991), 465-500.

[AD]. Alpay, D., Dym, H., *On applications of reproducing kernel spaces to the Schur algorithm and rotation J unitary factorization*, Operator Theory: Adv. Appl. **18** (1986), 89-159.

[AFP]. Arazy, J., Fisher, S. D., Peetre, J., *Hankel operators on weighted Bergman spaces*, Amer. J. Math. **110** (1988), 989-1054.

[AHOS]. Aikawa, H., Hayashi, N., Onda, I., Saitoh, S., *Analytical extensions of the members of the Bergman and Szegő spaces on some tube domains*, Arch. Math. **56** (1991), 362-369.

[AHS]. Aikawa, H., Hayashi, N., Saitoh, S.

   1. *The Bergman space on a sector and the heat equation*, Complex Variables **15** (1990), 27-36.

   2. *Isometrical identities for the Bergman and the Szegő spaces on a sector*, J. Math. Soc. Japan **43** (1991), 196-201.

[Ai]. Aikawa, H., *Infinite order Sobolev spaces, analytic continuation and polynomial expansions*, Complex Variables **18** (1992), 253-266.

[Ag]. Agler, J., *Nevanlinna-Pick interpolation on Sobolev space*, Proc. Amer. Math. Soc. **108** (1990), 341-351.

[Ar]. Aronszajn, N.

   1. *Theory of reproducing kernels*, Trans. Amer. Math. Soc. **68** (1950), 337-404.

   2. *Green's functions and reproducing kernels*, Proc. of the Symposium on Spectoral Theory and Differential Problems, Oklahoma A. and M. College, Oklahoma (1951), 355-411.

[ArSm]. Aronszajn, N., Smith, K. T., *Theory of Bessel potentials. Part I*, Ann. Inst. Fourier, Grenoble **11** (1961), 385-475.

[AS]. Ando, T., Saitoh, S., *Restrictions of reproducing kernel Hilbert spaces to subsets*, Preliminary reports, Suri Kaiseki Kenkyu Jo, Koukyu Roku **743** (1991), 164-187.

[B].   Bargmann, V.

   1. *On a Hilbert space of analytic functions and an associated integral transform, part I*, Commun. Pure and Appl. Math. **14** (1961), 187-214.

   2. *On a Hilbert space of analytic functions and an associated integral trans-form, part II*, Commun. Pure and Appl. Math. **20** (1967), 1-101.

[Ba]. Barker II, W. H., *Kernel functions on domains with hyperelliptic double,* Trans. Amer. Math. Soc. **231** (1977), 339-347.

[BHK]. Bouard, A. de, Hayashi, N., Kato, K., *Gevrey regularizing effect for the (generalized) Korteweg-de Vries equation and nonlinear Schrödinger equations,* Ann. Henri Inst. Poincare Analyse nonlinear **12** (1995), 673-725.

[BL]. Boor, C. De., Lynch, R. E., *On splines and their minimum properties,* J. Math. Mech. **15** (1966), 953-969.

[Br]. Branges, L. de, *Underlying concepts in the proof of the Bieberbach conjecture,* Proc. of International Congress of Math., Berkeley, Cal. (1986), 25-42.

[BS]. Byun, D.-W., Saitoh, S.
   1. *A real inversion formula for the Laplace transform,* Zeitschrift für Analysis und ihre Anwendungen **12** (1993), 597-603.
   2. *Approximation by the solutions of the heat equation,* J. Approximation Theory **78** (1994), 226-238.
   3. *Best approximation in reproducing kernel Hilbert spaces,* Proc. of the 2th International Colloquium on Numerical Analysis, VSP-Holland (1994), 55-61.
   4. *Analytic extensions of functions on the real line to entire functions,* Complex Variables **26** (1995), 277-281.

[Bu]. Burbea, J., *Total positivity of certain reproducing kernels,* Pacific J. Math. **67** (1976), 101-130.

[BW]. Boas, R. P., Widder, D. V., *An inverse formula for the Laplace integral,* Duke Math. J. **6** (1940), 1-26.

[By]. Byun, D.-W.
   1. *Isometrical mappings between the Szegö and the Bergman-Selberg spaces,* Complex Variables **20** (1992), 13-17.
   2. *Relationship between the analytic solutions of the heat equation,* Math. Japonica **38** (1993), 477-481.
   3. *Inversions of Hermite semigroup,* Proc. Amer. Math. Soc. **118** (1993), 437-445.

[C]. Chalmers, B., *Subspace kernels and minimum problems in Hilbert spaces with kernel functions,* Pacific J. Math. **31** (1969), 619-628.

[CG]. Christ, M., Geller, D., *Counterexamples to analytic hypoellipticity for domains of finite type,* Ann. of Math. **235** (1992), 551-566.

[D]. Devinatz, A.
   1. *Integral representations of positive definite functions,* Trans. Amer. Math. Soc. **74** (1953), 56-76.
   2. *On the extensions of positive definite functions,* Acta Math. **102** (1959), 109-134.

[DD]. Dzhrbashyan, M. M., Dzhrbashyan, A. È., *Integral representation for some classes of analytic functions in a half-plane*, Soviet Math. Dokl. **32** (1985), 727-730.

[Do]. Dolecki, S., *Observability for the one-dimensional heat equation*, Studia Mathematica **48** (1973), 291-305.

[DR]. Dolecki, S., Russell, D. L., *A general theory of observation and control*, SIAM J. Control and Optimization **15** (1977), 185-220.

[E]. Earle, C. J., *A reproducing formula for integrable automorphic forms*, Amer. J. Math. **88** (1966), 867-870.

[G]. Garabedian, P. R.
  1. *Schwarz's lemma and the Szegö kernel function*, Trans. Amer. Math. Soc. **67** (1949), 1-35.
  2. *A partial differential equation arising in conformal mapping*, Pacific J. Math. **1** (1951), 485-524.

[GH]. Gilbert, R. P., Hile, G. N., *Hilbert function modules with reproducing kernels*, Nonlinear Anal. **1(2)** (1977), 135-150.

[Gi]. Gindikin, S. G.
  1. *Integral formulas for Siegel domains of the second kind*, Soviet Math. Dokl. **2** (1961), 1480-1483.
  2. *Analytic functions in tubelar regions*, Soviet Math. Dokl. **3** (1962), 1178 -1183.

[GL]. Gilbert, R. P., Lin, W., *Wavelet solutions for time harmonic acoustic waves in a finite ocean*, J. of Computational Acoustics **1** (1993), 31-60.

[GW]. Gilbert, R. P., Weinacht, R. J., *Reproducing kernels for elliptic systems*, J. of Approximation Theory **15** (1975), 243-255.

[H]. Hadamard, J.
  1. *Sur les problème aux dérivées partielles et leur signification physique*, Bull. Univ. Princeton **13** (1902).
  2. *Lectures on Cauchy's Problems in Linear Partial Differential Equations*, New Haven, Yale University Press. **348** (1923).

[Ha]. Haslinger, F.
  1. *Szegö kernels of certain unbounded domains in C^2*, Rev. Roumain Math. Pures Appl. **39** (1994), 939-950.
  2. *Singularities of the Szegö kernels for certain weakly pseudoconvex domains in C^2*, J. Funct. Analysis **129** (1995), 406-427.

[Hay]. Hayashi, N.
  1. *Global existence of small analytic solutions to nonlinear Schrödinger equations*, Duke Math. J. **60** (1990), 717-727.
  2. *Solutions of the (generalized) Korteweg-de Vries equation in the Bergman and the Szegö spaces on a sector*, Duke Math. J. **62** (1991), 575-591.

[Hig]. Higgins, J. R., *Five short stories about the cardinal series*, Bull. of Amer. Math. Soc. **12** (1985), 45-89.

[Hil]. Hilgers, J. W., *On the equivalence of regularization and certain reproducing kernel Hilbert space approaches for solving first kind problems*, SIAM J. Numer. Anal. **13** (1976), 172-184.

[HK]. Hayashi, N., Kato, K., *Regularity of solutions in time to nonlinear Schrödinger equations*, J. Funct. Anal. **128** (1995), 255-277.

[Ho]. Ho, L. F., *Observabilité frontière de l'équation des ondes*, C. R. Acad. Sci. Paris, Sér. I Math. **302** (1986), 443-446.

[Hö]. Hörmander, L., L^2 *estimates and existence theorems for the $\bar{\partial}$-operator*, Acta Math. **113** (1965), 89-152.

[HS]. Hayashi, N., Saitoh, S.
   1. *Analyticity and smoothing effect for the Schrödinger equation*, Ann. Inst. Henri Poincaré **52** (1990), 163-173.
   2. *Analyticity and global existence of small solutions to some nonlinear Schrödinger equation*, Commun. Math. Phys. **139** (1990), 27-41.

[I]. Innis, G. S. Jr., *Some reproducing kernels for the unit disc*, Pacific J. Math. **14** (1964), 177-187.

[J]. Jerri, A. J., *The Shannon sampling theorem – its various extensions and applications*, a tutorial review, Proceedings of the IEEE **65** (1977), 1565-1596.

[Jo]. Jorgensen, P. E. T., *Integral representations for locally defined positive definite functions on Lie group*, International J. of Math. **2** (1991), 257-286.

[JPR]. Janson, S., Peetre, J., Rochberg, R., *Hankel forms and the Fock space*, Revista Mat. Iberoamericana, **3** (1987), 61-138.

[K]. Kamimoto, J.
   1. *On the non-analytic examples of Christ and Geller*, Proc. Japan Acad. Ser. A Math. Sci. **72** (1996), 51-52.
   2. *Breakdown of analyticity for $\bar{\partial}_b$ and the Szegö kernels*, Preprint Series **96-4** (1996), Graduate School of Mathematical Sciences, The University of Tokyo.
   3. *Asymptotic expansion of the Bergman kernel for weakly pseudoconvex tube domains in C^2*, Preprint Series **96-42** (1996), Graduate School of Mathematical Sciences, The University of Tokyo.

[Klo]. Klopfenstein, K. F., *A note on Hilbert spaces of factorial series*, Indiana University Math. J. **25** (1976), 1073-1081.

[Klu]. Klusch, D., *The sampling theorem, Dirichlet series and Hankel transforms*, J. of Computational and Applied Math. **44** (1992), 261-273.

[Ko]. Korányi, A., *The Bergman kernel function for tubes over convex cones*, Pacific J. Math. **12** (1962), 1355-1359.

[Kos]. Košelev, A. D., *On the kernel function of the Hilbert space of functions polyanalytic in a disc*, Soviet Math. Dokl. **18** (1977), 56-62.

[Kö]. Körezlioğlu, H., *Reproducing kernels in separable Hilbert spaces*, Pacific J. Math. **25** (1968), 305-314.

[Kr]. Kramer, H., *A generalized sampling theorem*, J. Math. Phys. **38** (1959), 68-72.

[Kre]. Kreĭn, M. G., *Hermitian positive kernels on homogeneous spaces*, I. Amer. Math. Soc. Transl.(2) **34** (1963), 69-108.

[KS]. Korányi, A., Stein, E. M., H^2 *spaces of generalized half-planes*, Studia Math. T. XLIV (1972), 381-388.

[KT]. Kajiwara, J., Tsuji, M.

    1. *Program for the numerical analysis of inverse formula for the Laplace transform*, Proceedings of the Second Korean-Japanese Colloquium on Finite or Infinite Dimensional Complex Analysis, (1994), 93-107.

    2. *Inverse formula for Laplace transform*, Proceedings of the 5th International Colloquium on Differential Equations, VSP-Holland (1995), 163-172.

[KW]. Kimeldorf, G., Wahba, G., *Some results on Tchebycheffian spline functions*, J. Math. and Appl. **33** (1971), 82-95.

[L]. Linde, A. Van Der.

    1. *Interpolation of regression functions in reproducing kernel Hilbert spaces*, Statistics **16** (1985), 351-361.

    2. *Rethinking factor analysis as an interpolation problem*, Statistics **19** (1988), 359-367.

    3. *A note on smoothing splines as Baysian estimates*, Statistics & Decisions **11** (1993), 61-67.

[LLT]. Lasiecka, I., Lions, J.-L., Triggiani, R., *Non homogeous boundary value problems for second order hyperbolic operators*, J. Math. Pure Appl. **65** (1986), 149-192.

[Lue]. Luecking, D. H., *Representation and duality in weighted spaces of analytic functions*, Indian Univ. Math. J., **34** (1985), 319-336.

[M]. Martin, W. T., *Analytic functions and multiple Fourier integrals*, Amer. J. Math. **62** (1940), 673-679.

[MAFG]. Morlet, J., Arens, G., Fourgeau, I., Giard, D., *Wave propagation and sampling theory*, Geophysics **47** (1982), 203-236.

[Mo]. Morlet, J., *Sampling theory and wave propagation*, C. H. Chen, ed., Issues in Acoustic Signal Image Processing and Recognition, NATO ASI Series, Vol. 1, Springer-Verlag, Berlin (1983), 233-261.

[MRR]. Markett, C., Rosenblum, M., Rovnyak, J., *A Plancherel theory for Newton spaces*, Integral Equations and Operator Theory **9** (1986), 831-862.

[MS]. Mizel, V. J., Seidman, T. I., *Observation and prediction for the heat equation II*, J. Math. Anal. Appl. **38** (1972), 149-166.

[N]. Nagel, A., *Vector fields and nonisotropic metrics*, Beijing Lectures in Harmonic Analysis , (E. M. Stein, ed), Princeton University Press, Princeton, NXJ (1986), 241-306.

[NS]. Nakai, M., Sario, L., *Square integrable harmonic functions on plane regions*, Ann. Acad. Sci. Fenn. **4** (1978/1979), 193-201.

276

[NW]. Nashed, M. Z., Wahba, G.

1. *Convergence rates of approximate least squares solutions of linear integral and operator equations of the first kind*, Math. Computation **28** (1974), 69-80.
2. *Regularization and approximation of linear operator equations and reproducing kernel spaces*, Bull. of Amer. Math. Soc. **80** (1974), 1213-1218.
3. *Generalized inverses in reproducing kernel spaces: an approach to regularization of linear operator equations*, SIAM J. Math. Anal. **5** (1974), 974-987.

[Oh]. Ohsawa, T.

1. *On the Extension of L^2 Holomorphic Functions II*, Publ. RIMS, Kyoto Univ. **24** (1988), 265-275.
2. *On the extension of L^2 holomorphic functions III*, negligible weights, Math. Z. **219** (1995), 215-225.
3. *Applications of L^2 estimates and some complex Geometry, Geometric Complex Analysis*, edited by Junjiro Noguchi et al., World Scientific Publishing Co. (1996), 505-523.

[OT]. Ohsawa, T., Takegoshi, K., *On the extension of L^2 holomorphic functions*, Math. Zeit. **195** (1987), 197-204.

[Oz]. Ozawa, M., *Fredholm eigen value problem for general domains*, Kōdai Math. Sem. Rep., **12** (1960), 38-43.

[P]. Parzen, E., *Statical inference on time series by RKHS methods*, Proc. of 12th Biennial Seminar of the Canadian Mathematical Congress, Amer. Math. Soc., Providence, R. I. (1970).

[Pe]. Peetre, J., *Some calculations related to Fock space and the Shale-Weil representation*, Integral Equations and Operator Theory **12** (1989), 67-81.

[PP]. Plancherel, M., Pólya, G., *Founctions entières et integrals de Fourier multiple*, Comment. Math. Helv. **9** (1936-37), 224-248;, **10** (1937-38), 110-163.

[PY]. Puel, J.-P., Yamamoto, M., *Applications of Exact Controllability to Some Inverse Problems for the Wave Equations*, Control of partial differential equations and applications, Proceedings of the IFIP TC7/WG-7.2, International Conference, Laredo, Spain (1995).

[R]. Rovnyak, J., *Euler spaces of analytic functions*, Trans. Amer. Math. Soc. **308** (1988), 197-208.

[Sa]. Saitoh, S.

1. *The Bergman norm and the Szegö norm*, Trans. Amer. Math. Soc. **249** (1979), 261-279.
2. *Some inequalities for analytic functions with a finite Dirichlet integral on the unit disc*, Math. Ann. **246** (1979), 69-77.
3. *Positive definite Hermitian matrices and reproducing kernels*, Linear Algebra Appl. **48** (1982).

4. *Integral transforms in Hilbert spaces*, Proc. Japan Acad. **58** (1982), 361 -364.

5. *Reproducing kernels of the direct product of two Hilbert spaces*, Riazi J. Kar. Math. Assoc. **4** (1982), 1-20.

6. *Hilbert spaces induced by Hilbert space valued functions*, Proc. Amer. Math. Soc. **89** (1983), 74-78.

7. *The Weierstrass transform and an isometry in the heat equation*, Applicable Analysis **16** (1983), 1-6.

8. *Fourier transforms with weighted functions and the Green's functions*, Applicable Analysis **16** (1983), 123-130.

9. *Some fundamental interpolation problems for analytic and harmonic functions of class L_2*, Applicable Analysis **17** (1984), 87-106.

10. *Some isometrical identities in the wave equation*, International J. Math. and Math. Sci. **7** (1984), 117-130.

11. *Integral transforms by Green's function on R^n*, Applicable Analysis **17** (1984), 157-167.

12. *Integral transforms in linear equations of parabolic type with constant coefficients*, Applicable Analysis **18** (1984), 13-27.

13. *The Laplace transform of L^p functions with weights*, Applicable Analysis **22** (1986), 103-109.

14. *Generalizations of Paley-Wiener's theorem for entire functions of exponential type*, Proc. Amer. Math. Soc. **99** (1987), 465-471.

15. *Quadratic norm inequalities deduced from the theory of reproducing kernels*, Linear Algebra Appl. **93** (1987), 171-178.

16. *Cauchy integrals for L_2 functions*, Arch. Math. **51** (1988), 451-454.

17. *Fourier-Laplace transforms and the Bergman spaces*, Proc. Amer. Math. Soc. **102** (1988), 985-992.

18. *Inequalities of exponential type for analytic functions with finite Dirichlet integrals*, Complex Variables **10** (1988), 123-139.

19. *Isometrical identities and inverse formulas in the one-dimensional Schrödinger equation*, Complex Variables **15** (1990), 135-148.

20. *Isometrical identities and inverse formulas in the one-dimensional heat equation*, Applicable Analysis **40** (1991), 139-149.

21. *Inequalities for the solutions of the heat equation*, General Inequalities 6, Birkhäuser Verlag, Basel Boston (1992), 351-359.

22. *Representations of the norms in Bergman-Selberg spaces on strips and half planes*, Complex Variables **19** (1992), 231-241.

23. *The Hilbert spaces of Szegö type and Fourier-Laplace transforms on $\mathbb{R}^n$*, Generalized Functions and Their Applications, Plenum Publishing Corporation, New York (1993), 197-212.

24. *Inequalities in the most simple Sobolev space and convolutions of L_2 functions with weights*, Proc. Amer. Math. Soc. **118** (1993), 515-520.

278

25. *Decreasing principles in transforms of reproducing kernel spaces*, Mathematica Montisnigri **1** (1993), 99-109.

26. *Analyticity of the solutions of the heat equation on the half space* $\mathbb{R}^n_+$, Proc. of the 4th International Colloquium on Differential Equations, VSP-Holland (1994), 265-275.

27. *An integral inequality of exponential type for real-valued functions*, World Scientific Series in Applicable Analysis 3, Inequalities and Applications, World Scientific (1994), 537-541.

28. *Inverse source problems in Poisson's equation*, Suri Kaiseki Kenkyu Jo, Koukyu Roku **890** (1994), 19-30; J. of Inverse and Ill-posed Problems. (to appear).

29. *One approach to some general integral transforms and its applications*, Integral Transforms and Special Functions **3** (1995), 49-84.

30. *Natural norm inequalities in nonlinear transforms*, General Inequalities 7, Birkhäuser Verlag, Basel, Boston. (to appear)

31. *Representations of inverse functions*, Proc. Amer. Math. Soc. (to appear).

32. *Linear integro-differential equations and the theory of reproducing kernels*, (to appear).

[Sak]. Sakawa, Y., *Observability and related problems for partial differential equations of parabolic type*, SIAM J. Control **13** (1975), 14-27.

[Sch]. Schiffer, M.
1. *The kernel function of an orthonormal system*, Duke Math. J. **13** (1946), 529-540.
2. *On various types of orthonormalization*, Duke Math. J. **17** (1950), 329-366.

[Schw]. Schwartz, L., *Sous-espaces hilbertiens d'espaces vectoriels topologiques et noyaux associés (noyaux reproduisants)*, J. Analyse Math. **13** (1964), 115-256.

[Se]. Seip, K.
1. *Density theorems for sampling and interpolation in the Bargmann-Fock space I*, J. reine angew. Math. **429** (1992), 91-96.
2. *Density theorems for sampling and interpolation in the Bargmann-Fock space II*, J. reine angew. Math. **429** (1992), 107-113.

[Sel]. Selberg, A., *Harmonic analysis and discontinuous groups*, J. India. Math. Soc. **20** (1956), 47-87.

[Sh]. Shannon, C. E., *A mathematical theory of communication*, Bell System Tech. J. **27** (1948), 379-423, 623-656.

[SS]. Shapiro, H. S., Shields, A. L., *On the zeros of functions with finite Dirichlet integral and some related function spaces*, Math. Z. **80** (1962), 217-229.

[Stg]. Stenger, F.
1. *Numerical methods based on Whittaker cardinal, or sinc functions*, SIAM Review **23** (1981), 165-225.

2. *Numerical Methods Based on Sinc and Analytic Functions*, Springer-Verlag, New York (1993).

[Stw]. Stewart, J., *Positive definite functions and generations and historical survey*, Rocky Mountain J. of Math. **6** (1976), 409-434.

[SW]. George Schmidt, E. J. P., Weck, N., *On the boundary behavior of solutions to elliptic and parabolic equations–with applications to boundary control for parabolic equations*, SIAM J. Control and Optimization **16** (1978), 593-598.

[SY]. Saitoh, S., Yamamoto, M.
   1. *Stability of Lipschitz type in determination of initial heat distribution*, J. of Inequalities and Applications **1** (1997), 73-83.
   2. *Integral transforms by smooth functions*, (to appear).

[T]. Tsuji, K., *An algorithm for sum of floating-point numbers without rounding error*, In Abstracts of the Third International Colloquium on Numerical Analysis, Bulgaria (1994).

[W]. Wahba, G.
   1. *On the approximate solution of Fredholm integral equations of the first kind*, Tech. Summ. Rep. 990, Mathematics Research Center, Univ. of Wisconsin-Madison (1969).
   2. *Convergence rates of certain approximate solutions to Fredholm integral equations of the first kind*, J. of Approx. Theory **7** (1973), 167-185.
   3. *A class of approximate solutions to linear operator equations*, J. of Approx. Theory **9** (1973), 61-77.
   4. *Practical approximate solutions to linear operators equations when the data are noisy*, SIAM, J. Numer. Anal. **14** (1977), 651-667.

[Y]. Yao, K., *Application of reproducing kernel Hilbert spaces – Bandlimited signal models*, Information and Control **11** (1967), 429-444.

[Ya]. Yamamoto, M.
   1. *Conditional Stability in Determination of Force Terms of Heat Equations in a Rectangle*, Mathl. Comput. Modelling **18** (1993), 79-88.
   2. *Stability, reconstruction formula and regularization for an inverse source hyperbolic problem by a control method*, Inverse Problems **11** (1995), 481-496.
   3. *Multidimensional Inverse Problems for Partial Differential Equations: Controllability Method and Perturbation Method.* (in preparation).

[Z]. Zhu, K., *A Forelli-Rudin type theorem with applications*, Complex Variables **16** (1991), 107-113.